NEUROMETHODS

Series Editor
Wolfgang Walz
University of Saskatchewan
Saskatoon, SK, Canada

For further volumes:
http://www.springer.com/series/7657

Neuromethods publishes cutting-edge methods and protocols in all areas of neuroscience as well as translational neurological and mental research. Each volume in the series offers tested laboratory protocols, *step by step* methods for reproducible lab experiments and addresses methodological controversies and pitfalls in order to aid neuroscientists in experimentation. Neuromethods focuses on traditional and emerging topics with *wide ranging* implications to brain function, such as electrophysiology, neuroimaging, behavioral analysis, genomics, neurodegeneration, translational research and clinical trials. *Neuromethods* provides investigators and trainees with highly useful compendiums of key strategies and approaches for successful research in animal and human brain function including translational "bench to bedside" approaches to mental and neurological diseases.

Basic Neurobiology Techniques

Edited by

Nicholas J. D. Wright

Levine College of Health Sciences, Wingate University, Wingate, NC, USA

Humana Press

Editor
Nicholas J. D. Wright
Levine College of Health Sciences
Wingate University
Wingate, NC, USA

ISSN 0893-2336 ISSN 1940-6045 (electronic)
Neuromethods
ISBN 978-1-4939-9946-0 ISBN 978-1-4939-9944-6 (eBook)
https://doi.org/10.1007/978-1-4939-9944-6

This Humana imprint is published by the registered company Springer Science+Business Media, LLC part of Springer Nature.
The registered company address is: 233 Spring Street, New York, NY 10013, U.S.A.

Preface to the Series

Experimental life sciences have two basic foundations: concepts and tools. The *Neuromethods* series focuses on the tools and techniques unique to the investigation of the nervous system and excitable cells. It will not, however, shortchange the concept side of things as care has been taken to integrate these tools within the context of the concepts and questions under investigation. In this way, the series is unique in that it not only collects protocols but also includes theoretical background information and critiques which led to the methods and their development. Thus it gives the reader a better understanding of the origin of the techniques and their potential future development. The *Neuromethods* publishing program strikes a balance between recent and exciting developments like those concerning new animal models of disease, imaging, *in vivo* methods, and more established techniques, including, for example, immunocytochemistry and electrophysiological technologies. New trainees in neurosciences still need a sound footing in these older methods in order to apply a critical approach to their results.

Under the guidance of its founders, Alan Boulton and Glen Baker, the *Neuromethods* series has been a success since its first volume published through Humana Press in 1985. The series continues to flourish through many changes over the years. It is now published under the umbrella of Springer Protocols. While methods involving brain research have changed a lot since the series started, the publishing environment and technology have changed even more radically. Neuromethods has the distinct layout and style of the Springer Protocols program, designed specifically for readability and ease of reference in a laboratory setting.

The careful application of methods is potentially the most important step in the process of scientific inquiry. In the past, new methodologies led the way in developing new disciplines in the biological and medical sciences. For example, Physiology emerged out of Anatomy in the nineteenth century by harnessing new methods based on the newly discovered phenomenon of electricity. Nowadays, the relationships between disciplines and methods are more complex. Methods are now widely shared between disciplines and research areas. New developments in electronic publishing make it possible for scientists that encounter new methods to quickly find sources of information electronically. The design of individual volumes and chapters in this series takes this new access technology into account. Springer Protocols makes it possible to download single protocols separately. In addition, Springer makes its print-on-demand technology available globally. A print copy can therefore be acquired quickly and for a competitive price anywhere in the world.

Saskatoon, SK, Canada ***Wolfgang Walz***

Preface

As I prepared to put pen to paper (figuratively speaking of course) to complete the final part of this epic project, I decided to check when I had started and was amazed to see that I was actually recruited back in August 2014. I realized I had spent nearly 5 years on this project which, at times, seemed much harder than herding my 4 cats. All manner of problems both for myself and the authors frequently intervened, and there were times when I wondered if I would ever get this book published. Despite that, I am overjoyed that I managed to convince the 31 contributors that produced the 13 superb chapters to stay the course. I have always considered the *Neuromethods* series to be an invaluable tool especially in an area that is so dependent on technical applications and innovations. The idea of a complete, single-source, technical "cookbook" is an incredible resource. I think I have managed to assemble a fine balance of new, novel techniques as well as significant improvements to existing ones and also some really exciting methodologies to explore some important pathological conditions. In Chapter 1, Field et al. discuss the novel application of semiconductor quantum dots as photostable fluorophores used for labeling, tracking, and voltage sensing, which is illustrated with some exceptional figures. Daniel Llano and his coworkers (Chapters 2 and 3) use optogenetic and chemogenetic probes and GCaMP6 transcranial imaging to monitor neural activity, while the next two chapters (4 and 5; Alkadhi and Umans, respectively) introduce very interesting models to study crucial topics such as Alzheimer's disease and blood-brain barrier challenges. The following two chapters by Ohmori, and Suarez-Delgardo and Islas (Chapters 6 and 7) utilize significant evolutions of the patch-clamping technique allowing simultaneous monitoring of membrane currents and optical signals. Chapter 8 by Zhang et al. contains some exciting work on a stroke model, while Chapter 9 (Salim) provides an excellent model for studying psychological trauma. Buskila et al. (Chapter 10) have contributed a very fine chapter of particular interest as it discusses extending the life span of the ubiquitous neuronal slice that has served the neurophysiological community wonderfully now for years. One of the slice's obvious limitations was its limited life span of typically only a few hours. Chapter 11 (Korkotian and Segal) discusses and develops the powerful technique of the near instantaneous release of caged compounds by flash photolysis while Chapter 12 (Du et al.) looks at the unsung heroes of the nervous system, namely the glia, and their function as supportive networks. The final chapter by Kabadi and Byrnes (Chapter 13) is of great social significance as it attempts to quantify behavioral dysfunction caused by traumatic brain injury, something plaguing both our military and our athletes. This volume contains some truly excellent research that I really hope is not only useful but inspiring. Yes, this project was grueling at times but now I think it was totally worthwhile. Enjoy!

Wingate, NC, USA *Nicholas J. D. Wright*

Contents

Preface to the Series . . . . *v*
Preface . . . . *vii*
Contributors . . . . *xi*

1 Semiconductor Quantum Dots for Visualization and Sensing in Neuronal Cell Systems 1
Lauren D. Field, Yung Chia Chen, and James B. Delehanty

2 Injection of Adeno-Associated Virus Containing Optogenetic and Chemogenetic Probes into the Neonatal Mouse Brain 19
Nhan C. Huynh, Baher A. Ibrahim, Christopher M. Lee, Mickeal N. Key, and Daniel A. Llano

3 Guide to Transcranial Imaging of Sound-Evoked Activity in the Auditory Cortex of GCaMP6s Mice In Vivo 45
Georgiy Yudintsev, Christopher M. Lee, Alexander R. Asilador, and Daniel A. Llano

4 A Novel Preclinical Rat Model of Alzheimer's Disease 69
Karim A. Alkadhi

5 Using Zebrafish to Study and Visualize Barriers of the CNS 99
Robyn Anne Umans

6 Photometric Patch Electrode to Simultaneously Measure Neural Electrical Activity and Optical Signal in the Brain Tissue 131
Harunori Ohmori

7 Patch-Clamp Fluorometry and Its Applications to the Study of Ion Channels 155
Esteban Suárez-Delgado and León D. Islas

8 Modeling Early-Onset Postischemic Seizures in Adult and Aging Mice 185
Liang Zhang, Chiping Wu, and Justin Wang

9 Modeling Psychological Trauma in Rats 213
Samina Salim

10 Extending the Life Span of Acute Neuronal Tissue for Imaging and Electrophysiological Studies 235
Yossi Buskila, Alba Bellot-Saez, Orsolya Kékesi, Morven Cameron, and John Morley

11 Confocal Imaging of Fast Flash Photolysis of Caged Compounds in Cultured Neurons 261
Eduard Korkotian and Menahem Segal

12 Analysis of the Functional States of an Astrocyte Syncytium 285
Yixing Du, Conrad M. Kiyoshi, David Terman, and Min Zhou

13 Assessment of Behavioral Dysfunction Following Experimental Traumatic Brain Injury (TBI) 315
Shruti V. Kabadi and Kimberly R. Byrnes

Index *351*

Contributors

KARIM A. ALKADHI • *Department of Pharmacological and Pharmaceutical Sciences, College of Pharmacy, University of Houston, Houston, TX, USA*

ALEXANDER R. ASILADOR • *Neuroscience Program, University of Illinois at Urbana-Champaign, Urbana, IL, USA; Beckman Institute for Advanced Science and Technology, Urbana, IL, USA*

ALBA BELLOT-SAEZ • *The MARCS Institute, Western Sydney University, Sydney, NSW, Australia; School of Medicine, Western Sydney University, Sydney, NSW, Australia*

YOSSI BUSKILA • *The MARCS Institute, Western Sydney University, Sydney, NSW, Australia; School of Medicine, Western Sydney University, Sydney, NSW, Australia*

KIMBERLY R. BYRNES • *Department of Anatomy, Physiology and Genetics, Uniformed Services University, Bethesda, MD, USA*

MORVEN CAMERON • *School of Medicine, Western Sydney University, Sydney, NSW, Australia*

YUNG CHIA CHEN • *US Naval Research Laboratory, Center for Bio/Molecular Science and Engineering, Washington, DC, USA*

JAMES B. DELEHANTY • *US Naval Research Laboratory, Center for Bio/Molecular Science and Engineering, Washington, DC, USA*

YIXING DU • *Department of Neuroscience, Ohio State University, Columbus, OH, USA*

LAUREN D. FIELD • *US Naval Research Laboratory, Center for Bio/Molecular Science and Engineering, Washington, DC, USA*

NHAN C. HUYNH • *Department of Molecular and Integrative Physiology, University of Illinois at Urbana-Champaign, Urbana, IL, USA; Beckman Institute for Advanced Science and Technology, Urbana, IL, USA; Massachusetts Institute of Technology, Cambridge, MA, USA*

BAHER A. IBRAHIM • *Department of Molecular and Integrative Physiology, University of Illinois at Urbana-Champaign, Urbana, IL, USA; Beckman Institute for Advanced Science and Technology, Urbana, IL, USA*

LEÓN D. ISLAS • *Facultad de Medicina, Departamento de Fisiología, Universidad Nacional Autónoma de México, Ciudad de México, Mexico*

SHRUTI V. KABADI • *Center for Food Safety and Applied Nutrition (CFSAN), United States Food and Drug Administration (US FDA), College Park, MD, USA*

ORSOLYA KÉKESI • *The MARCS Institute, Western Sydney University, Sydney, NSW, Australia; School of Medicine, Western Sydney University, Sydney, NSW, Australia*

MICKEAL N. KEY • *Beckman Institute for Advanced Science and Technology, Urbana, IL, USA; Neuroscience Program, University of Illinois at Urbana-Champaign, Urbana, IL, USA*

CONRAD M. KIYOSHI • *Department of Neuroscience, Ohio State University, Columbus, OH, USA*

EDUARD KORKOTIAN • *Department of Neurobiology, The Weizmann Institute, Rehovot, Israel*

CHRISTOPHER M. LEE • *Beckman Institute for Advanced Science and Technology, Urbana, IL, USA; Department of Molecular and Integrative Physiology, University of Illinois at Urbana-Champaign, Urbana, IL, USA*

DANIEL A. LLANO • *Department of Molecular and Integrative Physiology, University of Illinois at Urbana-Champaign, Urbana, IL, USA; Beckman Institute for Advanced*

Science and Technology, Urbana, IL, USA; Neuroscience Program, University of Illinois at Urbana-Champaign, Urbana, IL, USA

JOHN MORLEY • *School of Medicine, Western Sydney University, Sydney, NSW, Australia*

HARUNORI OHMORI • *Faculty of Medicine, Department of Neurobiology and Physiology, Kyoto University, Kyoto, Japan*

SAMINA SALIM • *Department of Pharmacological and Pharmaceutical Sciences, College of Pharmacy, University of Houston, Houston, TX, USA*

MENAHEM SEGAL • *Department of Neurobiology, The Weizmann Institute, Rehovot, Israel*

ESTEBAN SUÁREZ-DELGADO • *Facultad de Medicina, Departamento de Fisiología, Universidad Nacional Autónoma de México, Ciudad de México, Mexico*

DAVID TERMAN • *Department of Mathematics, Ohio State University, Columbus, OH, USA*

ROBYN ANNE UMANS • *The Fralin Biomedical Research Institute at VTC, 2 Riverside Circle, Roanoke, VA, USA*

JUSTIN WANG • *Krembil Research Institute, University Health Network, Toronto Western Hospital, Toronto, ON, Canada*

CHIPING WU • *Krembil Research Institute, University Health Network, Toronto Western Hospital, Toronto, ON, Canada*

GEORGIY YUDINTSEV • *Neuroscience Program, University of Illinois at Urbana-Champaign, Urbana, IL, USA; Beckman Institute for Advanced Science and Technology, Urbana, IL, USA*

LIANG ZHANG • *Krembil Research Institute, University Health Network, Toronto Western Hospital, Toronto, ON, Canada; Division of Neurology, Department of Medicine, University of Toronto, Toronto, ON, Canada*

MIN ZHOU • *Department of Neuroscience, Ohio State University, Columbus, OH, USA*

Chapter 1

Semiconductor Quantum Dots for Visualization and Sensing in Neuronal Cell Systems

Lauren D. Field, Yung Chia Chen, and James B. Delehanty

Abstract

Fluorescence imaging continues to play an increasingly vital role in neurobiology from the use of organic fluorophore dyes to genetically encoded proteins. Semiconductor nanocrystals or quantum dots have emerged as a new class of photostable fluorophores for use in a wide array of biological applications ranging from labeling and imaging to sensing and drug delivery. Here, we highlight several applications of quantum dots for imaging and sensing across a variety of neuronal cell platforms. These include the specific labeling of neurons tissue slices, the tracking of neuron movement in brain development, enhanced voltage sensing, and the guided patch clamp of neurons in vivo during electrophysiology. Our goal is to provide the reader with a survey of the use of quantum dots in these applications along with experimental notes and guidelines for their successful use in these applications.

Key words Fluorescence, Nanoparticle, Neuron, Quantum dot, Voltage, Brain, Patch clamp

1 Introduction: The Role of Fluorescence in Neurobiology

1.1 Fluorescence in Neurobiology

The use of fluorescence-based techniques has long played an important role in cellular biology. The use of fluorescent tags and fusion proteins can be very sensitive and quantitative and it is compatible with a myriad optical techniques ranging from fluorescence spectroscopy and cell sorting to the various forms of fluorescence microscopy [1, 2]. From the probing and interrogation of in situ protein expression using fluorophore-conjugated antibodies in fixed cells to the use of fluorescent protein fusions (e.g., green fluorescent protein (GFP)) for live cell tracking of protein and organelle trafficking and sorting, advancements in fluorescence techniques have led to increasingly sophisticated analyses of both cellular homeostasis and disease. This has indeed been the case in neurobiology where the steady progression of available fluorescence tools has enabled significant advances.

One of the more recently developed fluorescent protein-based tools is the calcium-sensitive gCaMP6 reporter protein that is a

Nicholas J. D. Wright (ed.), *Basic Neurobiology Techniques*, Neuromethods, vol. 152,
https://doi.org/10.1007/978-1-4939-9944-6_1,

designed fusion of the calcium-binding protein, calmodulin, and GFP [3]. This construct reports, in real time, on the generation of action potentials in neurons by sensing intracellular calcium release; binding of calcium by calmodulin results in a significant and reversible increase in fluorescence of the GFP reporter. More recently, advancements in optogenetic tools based on channelrhodopsins and archaerhodopsins have allowed the creation of fluorescence-based genetic tools for the controlled fluorescence-based recording and activation of membrane potential and action potentials [4–6]. Clearly, fluorescence and the emergence of new fluorescence tools have played, and continue to play, a critical role in our understanding and control of neuronal cell systems.

1.2 The Emergence of Quantum Dots as a New Class of Fluorescent Probe

In keeping with the goals of the National Nanotechnology Initiative, instituted in 2001, a fundamental focus has been the ascertaining of the potential utility of a variety of nanoparticle materials to advance such fields as energy harvesting, electronics, and medicine. One research frontier that has emerged is the implementation of nanomaterials with biology; the interfacing of the abiotic with the biotic. One example nanoparticle material that has moved to the fore is luminescent semiconductor nanocrystals or quantum dots (QDs) for use in a variety of biological applications. These materials are small in size (5–30 nm) and are extremely bright, photostable, and can be interfaced with cells for the purposes of labeling, imaging, sensing, and drug delivery [7–12].

The interest in the use of QDs for biological applications has been driven largely by their unique physicochemical properties. These include their high quantum yields, photostability, size-dependent emission, ability to engage in energy and charge transfer, amenability to be decorated with and large two photon absorption cross sections. A number of these attributes are summarized in Fig. 1.

1.3 The Implementation of Quantum Dots in Neurobiology Applications

Our goal in this chapter is to highlight, using specific examples, several disparate yet powerful ways in which QDs have been implemented in neuronal systems at a hierarchy of levels from cellular/tissue slice to whole organism. In each case, the QD materials enable a capability that was not previously available or significantly improve on an experimental methodology.

In the first example, hydrophilic QDs appended with cell-targeting peptides are used to specifically label neurons in a hippocampal slice model. Here, Walters et al. show specific labeling of neurons with the QD conjugates but no delivery to astrocytes, oligodendrocytes, or microglia [13]. The presence of the peptide drives both the uptake and subsequent dispersion of the QDs throughout the cytosol of the neurons. In a follow-on study, this same group confirmed the important role played by the charge nature of the QD capping ligand in mediating the specificity of

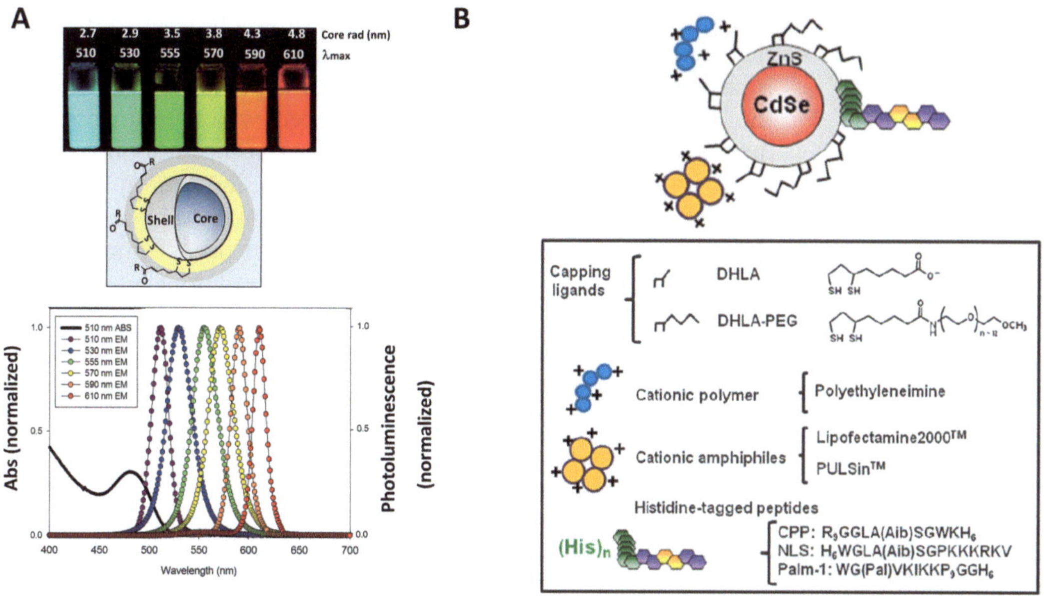

Fig. 1 The unique attributes of quantum dots (QDs) make them enabling tools for use in neuronal systems. (**a**) QDs are characterized by their size-dependent emission. Top panel, QD samples shown with their core radius (nm) and emission wavelength. These core–shell nanoparticles (middle panel) exhibit narrow emission profiles that are to the red portion of the spectrum relative to their absorbance (black solid line, bottom panel). (**b**) QDs can be decorated with a variety of biologicals (e.g., polymers, peptides) to facilitate cellular labeling. Panel **b** is reproduced with permission of Royal Society of Chemistry from Ref. [14]

this uptake and cytosolic delivery [14]. The utility of this construct is then extended to the time-resolved tracking of brain development in the chick model [15]. QD–peptide conjugates injected into the spinal cord canal at embryonic day 4 were able to label and track extensions of the spinal cord into the ventricles, migratory neuroblasts, maturing brain cells, and such complex structures as the choroid plexus.

Next, QDs capped with their native hydrophobic ligands are used to coat the glass surface of patch clamp pipettes to aid in the visualization of the pipette during electrophysiology experiments in live mice [6]. Here, the significant two photon absorption cross section of the QD enables deeper tissue imaging (down to ~760 microns) compared to when the pipette tip is visualized using a conventional organic fluorophore dye loaded into the pipette solution. This facilitates the injection of plasmid DNA encoding fluorescent protein for dual label imaging and manipulation. Recording from neurons as deep as ~800 μm is achieved with no deleterious effects and is optically guided through two photon microscopy to specifically target neurons of interest.

Finally, QDs assembled with a peptide–fullerene conjugate are interfaced with the plasma membrane of neurons to enable bright, photostable sensing and imaging of membrane voltage.

The QD–fullerene pair engages in a basal rate of electron transfer that results in partial quenching of the QD PL when the membrane is at resting potential. Upon membrane depolarization, however, the rate of electron transfer is enhanced which augments QD quenching. The large, fast, reversible fluorescence response of the probe is demonstrated in cultured cells, including neurons, and in a cortical electrical stimulation model in live mice. The response is shown to have millisecond responsivity with overall fluorescence change that is 20- to 40-fold greater than traditionally employed voltage sensitive dyes.

Cumulatively, these examples show the breadth of utility of the QD materials as reagents for labeling, tracking, and sensing applications and in the sections that follow we further detail how to implement them in various neuronal systems paying particular attention to key experimental details that are critical for success.

2 Materials

2.1 QDs and Capping Ligands

Hydrophilic QDs used in the hippocampal slice labeling (*see* Subheading 3.1) and in membrane potential imaging (*see* Subheading 3.3) were synthesized according to previous published procedures [16–19]. Briefly, a two-step ligand exchange procedure was used to coat the QDs with DHLA, DHLA-PEG$_{750}$-OMe, or CL ligands in a procedure adopted from previous work [20]. Briefly, the QDs were precipitated with acetone, centrifuged, and the QD pellet was mixed with a mixture of ethanol, chloroform, and 2-(2-aminoethoxy)-ethanol. This was followed by precipitation with a mixture ethyl acetate, 2-propanol, and hexanes. Following centrifugation, the pelted QDs were suspended in ethanol containing the desired ligand at a ~30,000 equivalents of ligand per QD. The capped QDs were then purified according to published protocols [21].

For coating of patch pipettes for electrophysiology, the as-synthesized hydrophobic QDs were precipitated by the addition of an acetone–methanol mixture and pelleted. The QDs were suspended in hexane in preparation for patch pipette coating. After pulling the pipette with a standard commercial puller (Sutter model P-97), positive pressure was applied through the back of the tip while the tip was dipped into methanol. The tip was then air dried prior to dipping into the QD suspension (typically in hexane) for ~2 s, removed and then allowed to air dry. This procedure was repeated until the desired level of PL was achieved on the tip upon illumination with UV light. If the tip appeared clear and expelled free-flowing solution of QDs, it was deemed ready for use.

2.2 Peptides and QD Peptide Assembly

Peptides were synthesized commercially (>80% purity) using standard solid phase synthesis and were provided lyophilized as a trifluoroacetic acid salt. Prior to use, crude peptide was desalted using an oligonucleotide purification cartridge (OPC) as described elsewhere [22]. Peptide JBD1 was conjugated to C_{60} fullerene according to the procedure in Ref. [23]. Briefly, peptide, C_{60} pyrrolidine tris-acid and *N*-hydroxysuccinimide were mixed in 95% DMSO/H_2O. Triethylamine was added and stirred and then *N,N′*-diisopropylcarbodiimide. The conjugate was then purified on Ni^{2+}-nitrilotriacetic acid agarose resin and eluted with 700 mM imidazole. The peptide-C_{60} conjugate was then desalted using an OPC cartridge, eluted with acetonitrile, aliquoted and dried.

QD–peptide (for hippocampal slice labeling) or QD–peptide–C_{60} (for voltage imaging) bioconjugates were generated by adding either the peptide or peptide–C_{60} to delivery media. An aliquot of concentrated QD stock solution was then added to generate QD–peptide conjugates of the desired ratiometric mix of peptide/QD. After allowing conjugate formation (typically 15–20 min incubation time at room temperature), the solutions were briefly spun in a benchtop centrifuge to remove and precipitated conjugates and the clarity of the conjugate solution was checked using a UV light.

2.3 QD–JB577 Peptide Labeling of Hippocampal Tissue Slices

Hippocampal cultures were harvested from Wistar rat pups, isolated and sectioned (350 μm thick). Slices with an intact pyramidal neuron cell layer and dentate gyrus were transferred to uncoated 30 mm Millicell-CM™ tissue culture inserts (Millipore) in six-well culture dishes. Cultures were used between 21 and 35 days in vitro. The formed QD–JB577 bioconjugates (in complete growth medium) were incubated on slice cultures for up to 24 h at 37 °C prior to washing and fixation in preparation for imaging.

2.4 QD–JBD1–C_{60} Peptide Labeling of Cultured Cells for Membrane Depolarization Imaging and Mouse Cortical Stimulation Studies

For cell culture labeling, complete media was removed from HeLa cell, PC12, or neuronal cell culture monolayers and the cells were washed with DMEM–HEPES. Solutions (typically ~150 μL) QD–DHLA–peptide–C60 complexes (20 nM QD appended with 20 peptides/QD; 400 nM peptide concentration) were incubated on the cells for 10–15 min at 37 °C followed by washing with DPBS. The cells were fixed with paraformaldehyde (3.7% in DPBS) for 15 min at room temperature, stored in 0.05% NaN3/PBS at 4 °C, and imaged within 48 h. Live cell imaging during depolarization studies was performed in Live Cell Imaging solution (ThermoFisher). Cells were imaged for fluorescence response to membrane depolarization induced by the addition of isotonic 140 mM KCl which was perfused into the culture dish over 3 min during realtime imaging. The KCl concentration was allowed to stabilize for another 2 min. Fluorescence was quantified at the beginning and end of this regimen. More detailed experimental procedures can be found in the specific references detailed below in each respective Methods description in Subheading 3.

3 Methods for the Use of Quantum Dots in Neuronal Cell Systems

In this section we detail several different uses of QDs in neuronal systems while paying particular attention to highlight the more challenging aspects of their integration into these systems for various applications.

3.1 QDs for Neuronal Cell Labeling and Tracking

The delivery of bioactive peptides to neurons is an attractive avenue for diagnostics and therapeutics. One strategy that has been under consideration to deliver such materials has been via the use of NP scaffolds such as QDs or other suitable core NP material to which the peptide can be attached or assembled. Walters et al. [13] used 625 nm-emitting QDs (~10 nm diameter) as a scaffold for the assembly of multiple copies of the cell delivery peptide Palm-1 (subsequently renamed and referred to herein as peptide JB577). The sequence of the peptide is given in Fig. 1b. This multifunctional, multidomain peptide has a polyhistidine domain that mediates self-assembly of the peptide to the QD surface and a polyproline helix domain that presents two domains distally from the QD surface: 1-a lysine-rich tract for initial association and uptake of the QD–peptide complex and 2-a palmitoyl moiety that drives eventual release of the QD–peptide conjugates from endolysosomal compartments. Using a rat hippocampal tissue slice model, it was shown that when the assembled QD–peptide conjugates were incubated with slices (~350 μm thick) for 24 h, the QD scaffold delivered the peptide specifically to the cell bodies of neurons located in the CA3 layer (Fig. 2a). Further, it was shown that the QD–peptide conjugates were not delivered to astrocytes, microglia, or oligodendrocytes (Fig. 2b–d) and that, within the cell soma, they were distributed diffusely throughout the cytosol (Fig. 2e, f). The ability of the QD to serve as a shuttle for the JB577 peptide delivery to the cytosol is a key facet of this delivery system as the sequestration of materials within endosomes has heretofore been a critical roadblock in the delivery of bioactive therapeutic peptides. Further characterization of the peptide delivery system demonstrated the critical role played by the negative surface charge of the QD scaffold peptide carrier. Here, the net negative charge of the zwitterionic CL4 ligand promoted enhanced uptake and endosomal escape of the QD–peptide complexes compared to other positively-charged or net neutral ligands (Fig. 3).

This QD–peptide system was further implemented in a chick embryo model of the developing brain. The conjugates were injected into the spinal column on embryonic day 4 (E4) and their distribution was followed in a time-resolved manner all through the development process through chick hatching (Fig. 4a, b). Within days of injection, the conjugates had migrated along glial tracks in the midbrain area (Fig. 4c).

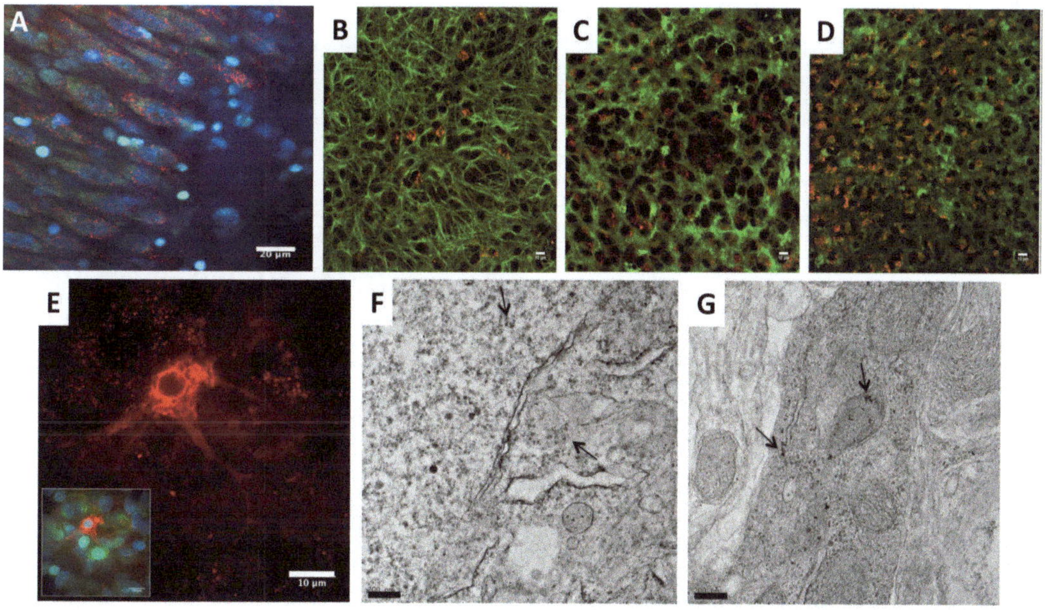

Fig. 2 Neuron-specific labeling in hippocampal tissue slice. (**a**) Cell bodies of neurons located in the CA3 layer of rat hippocampus labeled with 625 nm-emitting QDs (red) capped with CL4 ligands and appended with Palm1/JB577 (referred to herein as JB577) cell uptake peptides. Nissl stain (for neurons) is in green, DAPI (nuclei) is in blue. Scale bar is 20 μm. (**b–d**) QDs (red) do not label or traffic to astrocytes (**b**), microglia (**c**), or oligodendrocytes (**d**). In each panel, cells are labeled with a cell type-specific marker (green). Scale bars in **b–d** are 10 μm. (**e**) Pyramidal neuron of the CA3 layer shows diffuse distribution of QD–JB577 peptide complexes 24 h after delivery. Inset shows the larger image with QDs in red, Nissl stain in green, and DAPI in blue. Scale bar is 10 μm. (**f**, **g**) CL4-capped QDs decorated with JB577 peptides are delivered intracellularly and dispersed throughout the cytosol (black dots indicated by arrows). Scale bars, 200 μm. Figure is adapted from Ref. [13] © the authors

At further time points (E11 through E14), the QDs were found to be localized within the choroid plexus region (Fig. 5a) and transcriptome analysis using mRNA in situ hybridization showed normal transcript expression of critical marker genes in both QD-injected and control chick embryos (Fig. 5b).

In all of the above experiments, there are a number of critical factors that determine successful labeling for visualization and tracking in both the tissue slice and embryo systems. First is the confirmation of the successful assembly of the peptide to the QD surface. This is typically achieved using gel electrophoresis and dynamic light scattering (DLS) [24, 25]. Gel analysis uses an agarose matrix subjected to an applied electric field to confirm the slowed migration of the QDs upon peptide assembly while DLS measures the hydrodynamic diameter of the conjugates, which increases upon peptide assembly to the QD surface. Second, it is also critical to check for evidence of any precipitation of the formed conjugates from the delivery/injection solution. This can be achieved using a combination of centrifugation and UV

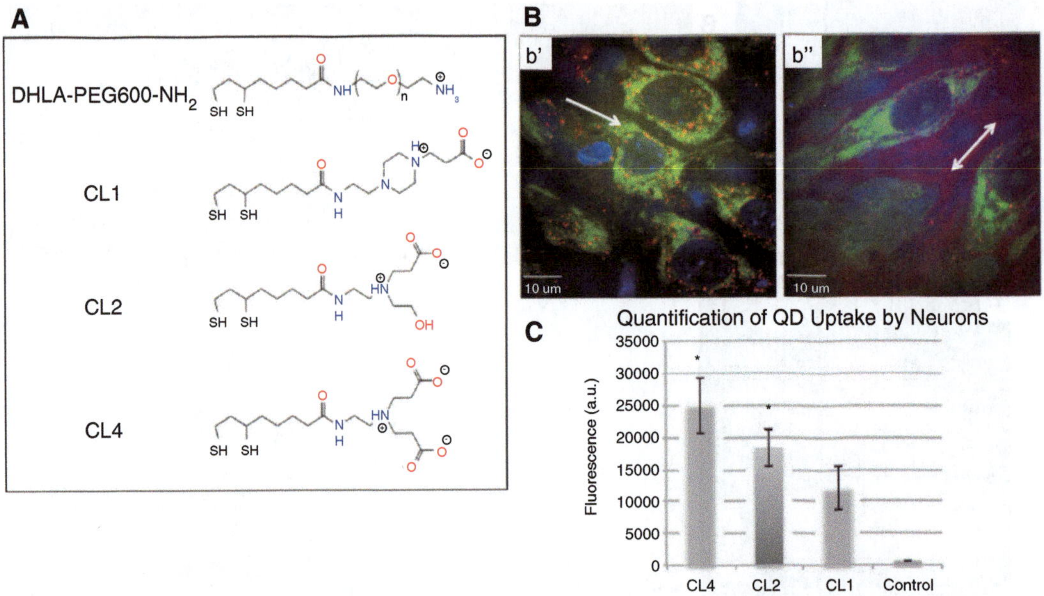

Fig. 3 Role of negative charge in neuronal uptake of QD–peptide complexes. (**a**) The role of QD surface charge in mediating QD uptake was assessed by altering the charge nature of the QD capping ligand. Shown are the structures of four capping ligands: DHLA-PEG$_{600}$-NH$_2$ (net positive), CL1 (zwitterionic, net neutral), CL2 (zwitterionic, net neutral), and CL4 (net positive). (**b**) QDs capped with CL4 (**b'**) or CL1 (**b"**) and appended with JB577 peptides show preferential uptake by hippocampal slice neurons (CA3 region) for the CL4-capped QD as evidenced by arrows indicated red QDs. Scale bars, 20 μm. (**c**) Quantification of neuronal cell uptake of QD-JB577 capped with various ligands. The negatively charged CL4 ligand shows preferential uptake compared to the zwitterionic CL2 and CL1 ligands. Figure is adapted from Ref. [14] © the authors

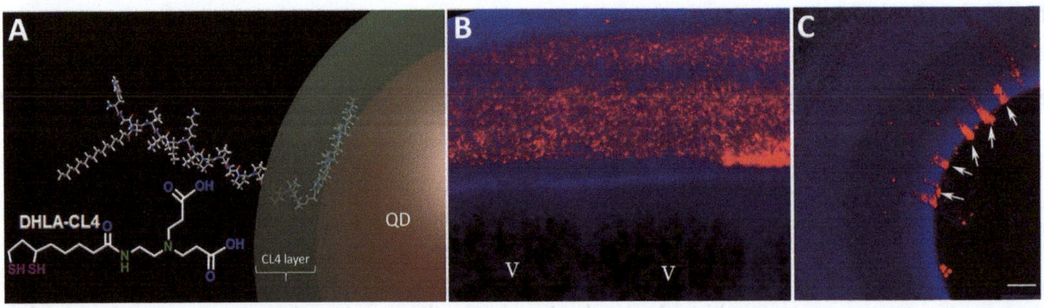

Fig. 4 Distribution of QD–peptide bioconjugates in developing chick embryonic brain. (**a**) 625 nm QD (red sphere) capped with CL4 ligands (green layer) self-assembled with JB577 cell uptake peptide. (**b**) Sagittal (40 μm) section through spinal cord of E6 embryos (injected at E4) showing uniform dispersion of the QDs. Position of the vertebrae is demoted (V). (**c**) Coronal sections through midbrain of chick embryos shows uptake of QDs (red) from the ventricles and long distance migration (shown by arrows) along glial tracks. Scale bar, 100 μm. Image is adapted from Ref. [15] © ACS

illumination as any precipitated conjugates will appear as a fluorescent pellet at the bottom of the tube [26]. In our experience, the best results have been achieved when the peptide is first added to

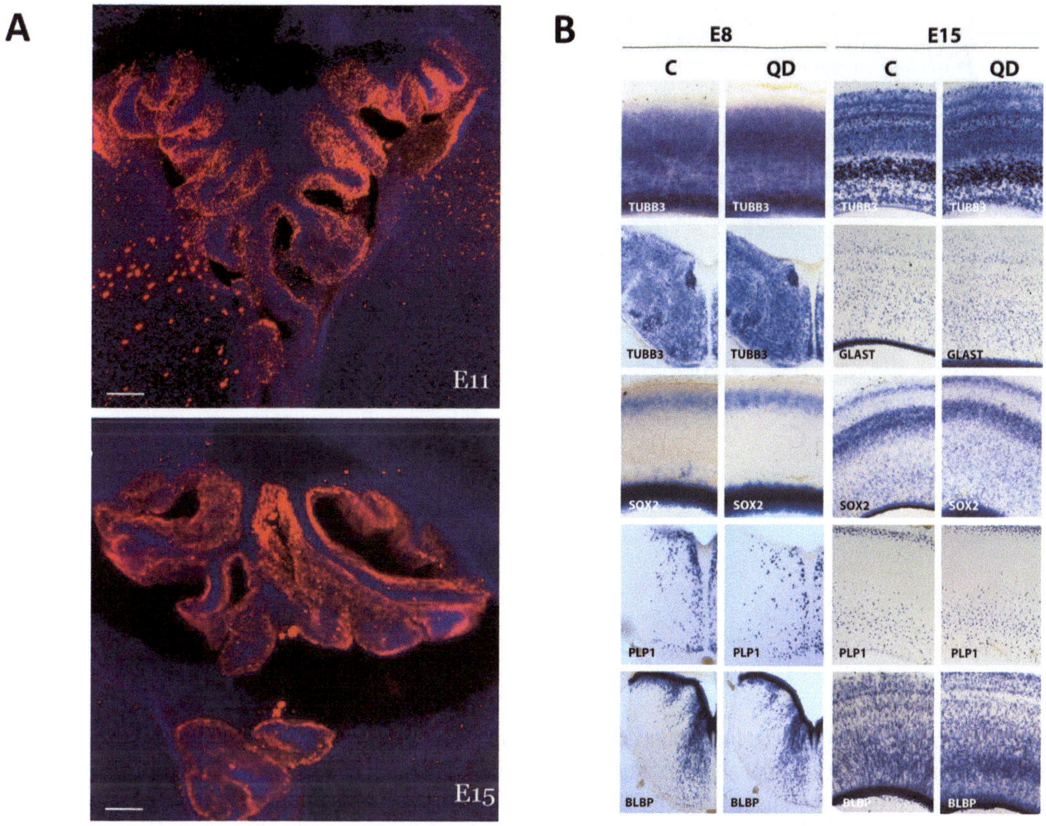

Fig. 5 QD–peptide labeling of the choroid plexus and assessment of neural differentiation after QD incorporation. (**a**) Coronal section through the choroid plexus showing QD labeling at E11 (top) and at E15 (bottom). Scale bars, 200 μm. (**b**) In situ mRNA hybridization to assess expression pattern of SOX2 (transcription factor), proteolipid protein (PLP), class III-tubulin beta (TUBB3), and brain lipid binding protein (BLBP) shows normal expression patterns at E8 and E15 for QD-labeled embryos (QD) compared to control (c). Image adapted from Ref. [15] © ACS

the buffer/medium followed by the QD. Finally, in the hippocampal slice work, the antibody probing of the various cell type markers needs to be optimized empirically to determine the best antibody concentration that produces good signal/staining with minimal background.

3.2 QDs for Combined Two Photon-Assisted Visualization and Electrophysiology

Another area where QDs have been a critically enabling material is in the visualization of patch pipettes during electrophysiological manipulations and recordings in both tissue slices and whole animal settings. One key feature of the QDs that has proven particularly powerful is their significant two photon action absorbance (TPA). QDs possess some of the largest TPA cross sections that are often several orders of magnitude greater than organic dyes or fluorescent proteins [27, 28] and this can significantly enhance deep tissue imaging in both tissue slices and in vivo experimental systems.

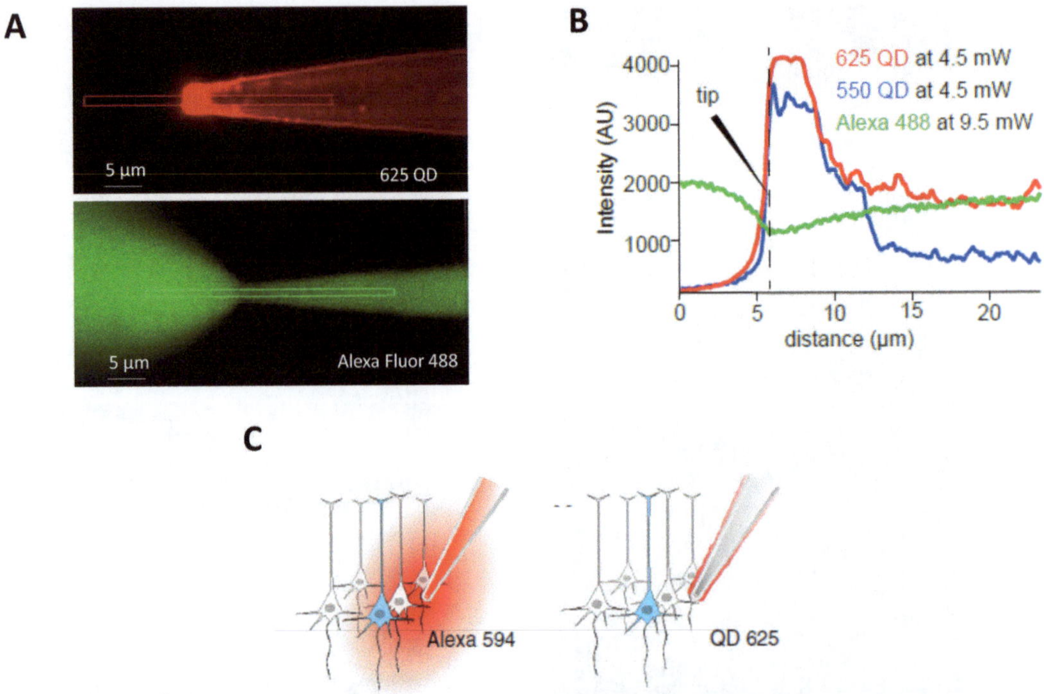

Fig. 6 QD-coated patch pipetts for enhanced visualization during electrophysiology. (**a**) Patch pipette coated with 625 nm-emitting QDs (top) and uncoated pipette ejecting Alexa Fluor 488 dye (bottom). (**b**) Fluorescence intensity measured in the image in (**a**) for 625 QDs and Alexa Fluor 488 in the rectangular area within the pipette tip. Note the significant fluorescence signal outside of the pipette for the Alexa Fluor 488 which contributes to considerable background. (**c**) Schematic representation of the classic technique of expulsion shadowing (with Alexa Fluor 594) versus pipette coating (with QD 625). Image adapted from Ref. [37] © Nature America, Inc.

Patch clamp electrophysiology has long been hampered by the inability to clearly visualize the thin pipette tip as it is advanced through the brain to the targeted neuron. One strategy that has been employed is a technique known as "shadowing" where a solution of fluorescent dye is continuously expelled from the pipette tip, thereby creating a shadow around a neuron. These dyes, however, have limited TPA cross sections which requires bright, extended illumination with high laser powers that leads to photobleaching and phototoxicity and the constant expulsion of the dye leads to dye buildup in the immediate area to be imaged, causing high fluorescence backgrounds.

To address these issue and to enable the integration of QDs into electrophysiology and to demonstrate their utility in two photon imaging, Andrasfalvy and coworkers developed a methodology for coating patch clamp pipettes with QDs capped with their native hydrophobic ligands (e.g., *n*-hexylphosphonic acid, trioctylphosphine oxide, and trioctylphosphine). The as-synthesized QDs were dip-coated onto the pipette tip that was previously primed by

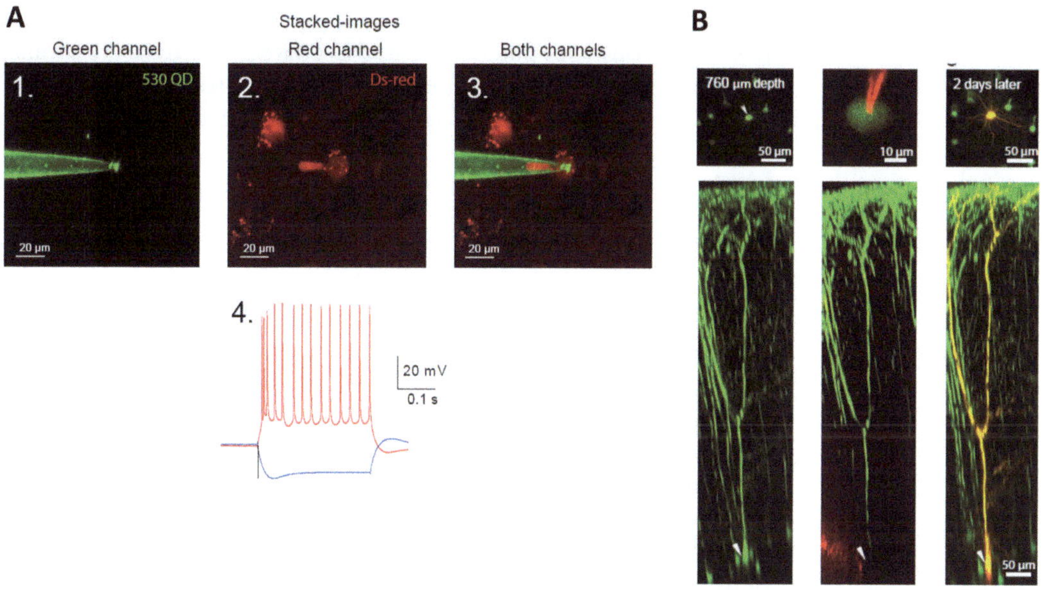

Fig. 7 Two photon monitoring of QD-coated patch pipettes during patching of hippocampal slice and live mouse brain. (**a**) Patching of hippocampal neurons (red) in acute brain slices from a DsRed-expressing transgenic mouse. QD is coated with 530 nm QDs. The recorded activity traces in panel 4 represent 13 cells in 5 animals. (**b**) Top panels, z-projection (80 μm) of targeted neuron soma (arrowhead) at 760-μm depth before electroporation of DsRed-encoding plasmid. Patch pipette is coated with 625 nm QDs (middle panel). Bottom, three-dimensional reconstructed orthogonal view, corresponding to ~800 μm. Green is EGFP-expressing neuron. Arrowhead is the site of pipette contact with the neuron. Right panel shows colocalized DsRed and EGFP expression (yellow). Figure adapted from Ref. [37] © Nature America, Inc.

dipping in a methanol solution. By varying the methanol priming procedure, the pipette could be preferentially coated on the tip or along the length of the pipette (Fig. 6a). Here, the direct labeling of the pipette tip provides continuous, homogenous signal specifically along the tip that improves the visualization of the tip compared to the shadowing technique which is coupled with significant background fluorescence (Fig. 6b, c).

The QD-coated pipettes were first used in electrophysiology experiments on hippocampal brain slices harvested from mice to demonstrate the functionality of the labeled pipettes. The pipette resistance of QD-coated pipettes did not differ from that of uncoated control pipettes [29–31] and they formed gigaseal contacts similar to that of uncoated patch pipettes when the "blow-and-seal" technique was employed. Various cell types in brain slices were patched, including hippocampal DsRed-labeled cholecystokinin-positive and GFP-labeled parvalbumin-positive interneurons (Fig. 7a). The fluorescence intensity of QD-coated pipettes was consistently higher than the intensity of endogenously expressed fluorescent markers. Notably, the sensitivity of the photomultiplier used to detect the QD-coated pipette signal required dialing down to avoid saturation at the laser power required for visualization of the fluorescent proteins. Further, basic electrophysiological properties of neuron types patched with QD-coated pipettes were similar to those recorded using uncoated pipettes (somatic firing and voltage responses to a series of positive and negative current injections) (Fig. 7a).

In vivo, recording using QD-coated patch pipettes was performed on cortical layer 2/3 (L2/3) pyramidal neurons of anesthetized mice expressing the Ca2+ indicator GCaMP6 (Fig. 7b) QD-coated pipettes also successfully electroporated Alexa Fluor 594 dye and a DsRed-encoding plasmid into individually identified L2/3 pyramidal neurons at ~300 μm depth. Sequential electroporation of multiple cells in vivo did not alter the QD coating or produce QD adsorption to the brain parenchyma. Notably, EGFP-expressing neurons were successfully electroporated in vivo at 760-μm depth using QD 625–coated pipettes (Fig. 7b).

There are several critical experimental parameters one needs to consider in this application of QDs. First, the procedure for coating of the patch pipette with the hydrophobic QDs needs to be empirically optimized and iteratively tested to ensure the functionality of the pipette in terms of both electrophysiology and plasmid microinjection/electroporation. The time of dipping the pipette into methanol, air drying, and incubating in the QD solution determine the degree of coating of the patch pipette tip. One critical observation was that in the absence of the methanol pretreatment of the pipette prior to dipping into the QD solution, coating of only the pipette tip (and not the length of the pipette) was observed. Inadequate coating of the pipette length can negatively impact the ability to visualize the pipette at tissue depths greater than 600 μm. The other critical parameter that required optimization was the two laser excitation power and recording gain used for the QDs and fluorophores/fluorescent proteins when the two materials were used in combination. The laser excitation and recording gain had to be adjusted down significantly for the QD due to their large TPA while higher settings were required for the organics dyes and fluorescent proteins.

3.3 QDs for Real-Time Imaging of Membrane Potential

In recent years, much interest has arisen in the development of new voltage imaging/sensing probes with enhanced capabilities relative to currently available materials such as traditional voltage-sensitive organic dyes and fluorescent protein constructs. This has culminated in the inception of the Brain Research through Advancing Innovative Neurotechnologies (BRAIN) Initiative, the goal of which is to create a functional layout or map of the entire "connectome" of the human brain [32]. A large focus of this effort has been on ascertaining the potential of novel nanomaterials and nanomaterial constructs to visualize the activity of thousands of neurons simultaneously with single-cell resolution. Traditionally,

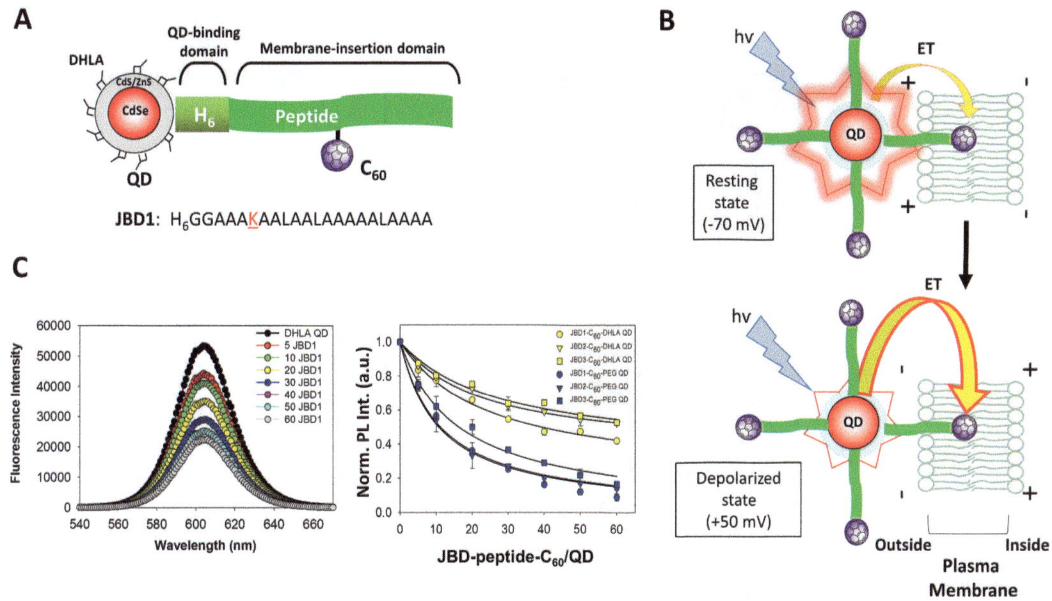

Fig. 8 Schematic and photophysical characterization of voltage-sensitive QD–peptide–C60 conjugates. (**a**) Modular design of QD–peptide–C60 bioconjugates. The QD (electron donor) is a central scaffold around which are arrayed varying numbers of membrane-insertion peptides (JBD1) bearing a C60 fullerene electron acceptor. An N-terminal polyhistidine tract (H6) mediates self-assembly of the peptide to the QD surface. (**b**) Schematic of the response of QD–peptide–C60 bioconjugates to changes in membrane potential. Multiple peptide-C60 conjugates are arrayed around the central QD, all engaging in electron transfer (ET) with the QD. Some of the peptide-C60 are inserted into the membrane bilayer and contribute to depolarization-induced QD quenching. At resting potential, minimal ET from the photoexcited QD to those C60 embedded in the membrane bilayer results in bright QD emission (top). Depolarization of membrane potential augments the rate of ET causing a decrease in QD PL (bottom). (**c**) Photophysical response of photoexcited QD–peptide–C60 conjugates. Left, 625 nm-emitting QD displays quenching that tracks inversely with increasing numbers of JBD1 peptides. Right, quenching response of 625 nm QDs capped with either DHLA (yellow) or DHLA-PEG ligands and displaying the various peptides shown. JBD1 peptides were subsequently used for sensing experiments. Figure adapted from Ref. [7] © ACS 2017

VSDs are susceptible to photobleaching, have small change in fluorescence in response to voltage change, or they are toxic.

Here, we have designed a QD-based optical probe that engages in electron transfer and, when interfaced with the plasma membrane of neurons, responds to changes in membrane potential with modulations in QD PL. Figure 8a shows the design of this voltage-responsive QD-bioconjugate system. To the QD surface are assembled multiple copies of a peptide-C_{60} conjugate (JBD1). The peptide sequence is designed to perform two functions: 1-append a C_{60} fullerene electron acceptor near the QD surface and 2-form an alpha helical conformation that mediates the insertion of the C_{60} into hydrophobic portion of the plasma membrane bilayer. When incubated with cells or tissue, the peptide portion is in the bilayer while the hydrophilic QD rests on the exofacial face of plasma

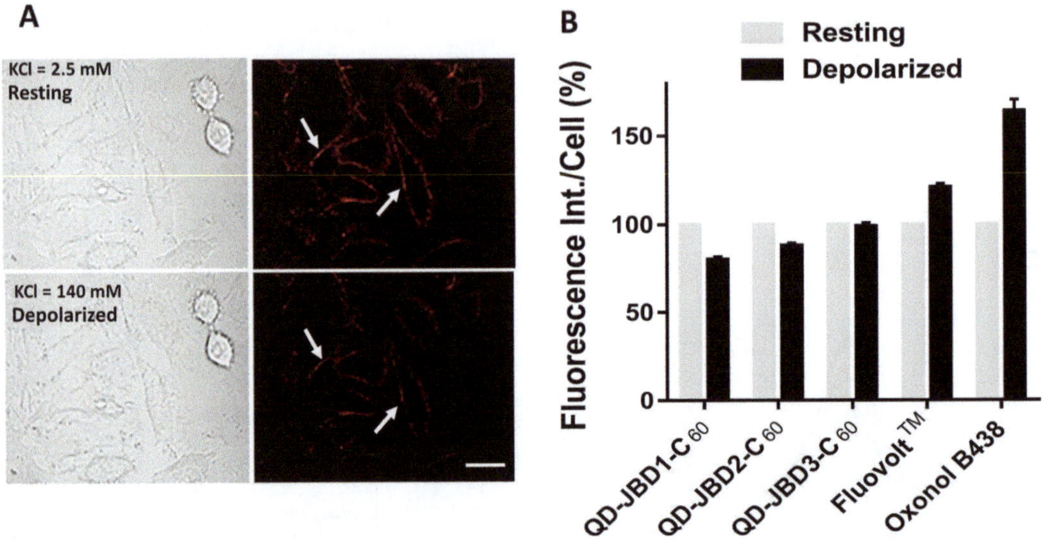

Fig. 9 QD-based imaging of membrane potential in HeLa cells. (**a**) HeLa cell monolayers labeled with QD–peptide–C60 (JBD1 peptide) probes were depolarized by exposure to extracellular KCl solution. The fluorescence images in the top and bottom panels show the fluorescence signal before and after, respectively, the addition of 140 mM KCl. Arrows highlight responsive areas. (**b**) Quantification of the fluorescence response of QD–peptide–C60 probes (various JBD peptides) show a decrease in QD PL in response to KCl-induced depolarization. The response of commercial probes FluoVolt™ and oxonol B438 are shown for comparison. Figure is from Ref. [7] © ACS 2017

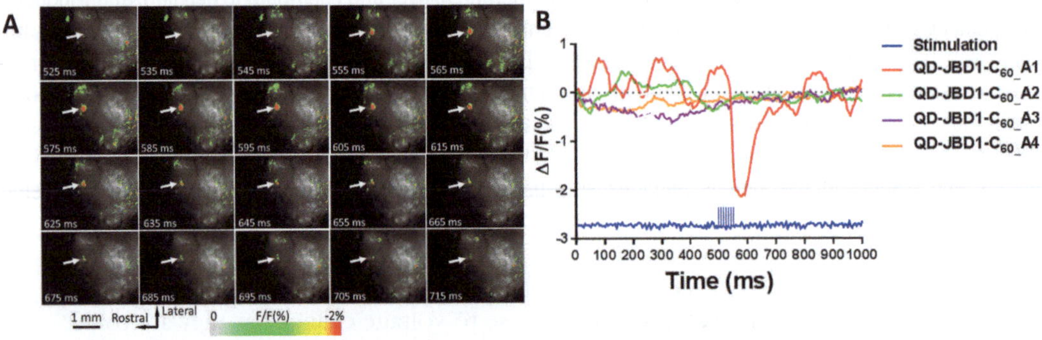

Fig. 10 QD–peptide–C60-based imaging of brain activity in mouse cortex. (**a**) Response of the QD–peptide–C60 probe injected into mouse cortex with applied current stimulation. The arrow indicates the responding area of QD PL decrease (red) and subsequent return to baseline over the time timecourse 525 to 715 ms. (**b**) Quantification of the responsive area from panel **a** (red trace) compared to control areas showing reversible PL response to current stimulation at ~500 ms. Figure is adapted from Ref. [7] © ACS 2017

membrane. At resting membrane potential, there is a basal rate of electron transfer from the photoexcited QD electron donor to the C_{60} electron acceptor that results in an initial level of QD PL. Upon membrane depolarization, the rate of electron transfer from the QD to the C_{60} is augmented, resulting in further, reversible quenching of the QD PL (Fig. 8b). As shown in Fig. 8c, the degree of initial quenching of the QD can be modulated by the number of peptide-C_{60} conjugates assembled onto the QD surface. This degree of "steady state" QD PL quenching tracks inversely with the number of peptide-C_{60} arrayed around the QD.

When delivered to the cell's plasma membrane, the photoexcited construct optically reports on the depolarization of membrane potential as shown in Fig. 9a where HeLa cells were depolarized by the addition of 140 mM KCl which was coupled with a ~20% decrease the PL of the QD (Fig. 9b). Further, when injected into the cortex of live mice, the probe responds nearly immediately and reversibly to applied cortical electrode stimulation with an overall $\Delta F/F_0$ of ~2.1% (Fig. 10). This fluorescence response represents a ~20- to 40-fold larger fluorescence response compared to traditional VSDs used in this same system [33].

There are several critical experimental issues to consider when implementing these QD bioconjugates in neuronal systems. As in the hippocampal slice studies detailed in Subheading 3.1, the assembly of the peptide-C_{60} conjugate to the QD needs to be optimized both in terms of the order of addition of the reagents and the time for which the ensemble complex is allowed to form. Regarding the former, the best success will be achieved when the QD-bioconjugate is formed by first adding the buffer to the tube, then the peptide-C_{60} conjugate, and finally the QD. It has been our experience that the addition of the reagents in this order results in the formation of the most stable QD-bioconjugate with negligible precipitation. It is also preferred that the peptide-C_{60} and QD stock solutions be at high concentration (>300 μM for the peptide-C_{60} conjugate and >5 μM for the QD). This ensures the addition of the smallest volume of the stock solution to the conjugate solution that is to be incubated on the cells. The time of incubation on the cell monolayers also needs to be optimized for efficient labeling; this is typically somewhere in the range of 15–45 min for cell monolayers. Perhaps the most critical experimental parameters that need to be optimized are the final concentration of QD and the number of peptide-C_{60} conjugates that are arrayed around the central QD electron donor. It is imperative to use a QD concentration that affords efficient cellular labeling (for good signal) while using only a modest number of peptide-C_{60} electron acceptors assembled onto the QD. This will impart only a modest amount of QD PL quenching at resting potential (steady state quenching) while ensuring that there is enough dynamic range so that the $\Delta F/F_0$ as a result of membrane depolarization is maximal. In the HeLa cell example described above, the optimal ratio was ~20 peptide-C_{60} per QD (final QD concentration, 20 nM), which resulted in ~30% quenching of QD PL at resting potential ("steady state"). In the mouse cortical stimulation experiment, the QD concentration was increase to ~ 1 μM to account for the diffusion of the QDs in the brain tissue. This QD–peptide–C_{60} probe provides a new imaging platform that offers increased photostability and fluorescence response compared to more traditional voltage-sensitive fluorescent materials while availing deeper tissue sensing using two-photon imaging. One limitation of the probe's current use in vivo is the requirement to inject the probe into the cortical area, rather than delivering it systemically through the circulation or through perfusion from a pad placed on top of the dura. Efforts are underway to develop alternative delivery strategies based on facilitated delivery (e.g., using blood/brain barrier peptides).

4 Conclusions

Over the past two decades, significant advances in the synthesis of biocompatible QDs and the design of more intricate and complex QD-based bioconjugates comprising peptides, proteins, and other biologicals have enabled new advances in cellular labeling, imaging, and sensing [34–36]. These advances have been made across multiple levels of cell and tissue architectures from cultured cell monolayers to tissue slices to whole organisms. This has been driven by both the ability to synthesize QD materials of superior quality, render them stable in aqueous biological media with coating ligands, and deliver them to cells with control.

In this chapter we have reviewed some of the recently demonstrated advanced uses of QDs for imaging and sensing in neuronal cell experimental platforms. These applications have ranged from the specific labeling of neuronal cells in hippocampal tissue slices and the developing chick brain to the QD-assisted imaging of patch pipettes during electrophysiology manipulations to the real-time imaging of plasma membrane potential in cultured cells and in vivo. All of these experimental platforms feature the QD in a central role that completely facilitates the experiment and enables the acquisition of data in ways not previously possible. We envision that these techniques will be extremely valuable and will find utility and wide adoption within the neuroscience community. Specifically, we foresee that the QD-assisted two-photon imaging of pipettes and the QD-based voltage sensing will be particularly popular as they completely enable some of the primary goals set forth in the Human Brain Initiative [32] which seeks to develop the next generation of tools for imaging and recording the activity of all the complex interconnections in the human brain.

Clearly the development of new tools for imaging, tracking, and sensing will enable the acquisition of information from neuronal systems in new ways that can shed light on neuronal function and communication. We anticipate that the QD-based materials and methods highlighted in this chapter will facilitate new discoveries and the development of new drugs and therapies for the treatment of diseases and disorders of the nervous system.

Acknowledgments

This research was supported by funding from the NRL Nanoscience Institute and Base Funding Program. LDF is a PhD candidate in the Fischell Department of Bioengineering, University of Maryland, MD, USA. YC was supported by a postdoctoral research associateship through the American Association for Engineering Education.

References

1. Combs CA (2010) Fluorescence microscopy: a concise guide to current imaging methods. Curr Protoc Neurosci 50(1):2–1
2. Taraska JW, Zagotta WN (2010) Fluorescence applications in molecular neurobiology. Neuron 66(2):170–189
3. Chen T-W et al (2013) Ultrasensitive fluorescent proteins for imaging neuronal activity. Nature 499(7458):295–300
4. Venkatachalam V, Cohen AE (2014) Imaging GFP-based reporters in neurons with multiwavelength optogenetic control. Biophys J 107(7):1554–1563
5. Nakajima R et al (2016) Optogenetic monitoring of synaptic activity with genetically encoded voltage indicators. Front Synaptic Neurosci 8:22
6. Hochbaum DR et al (2014) All-optical electrophysiology in mammalian neurons using engineered microbial rhodopsins. Nat Methods 11 (8):825–833
7. Boeneman K et al (2009) Sensing caspase 3 activity with quantum dot–fluorescent protein assemblies. J Am Chem Soc 131 (11):3828–3829
8. Delehanty JB et al (2013) Site-specific cellular delivery of quantum dots with chemoselectively-assembled modular peptides. Chem Commun (Camb) 49(72):7878–7880
9. Delehanty JB et al (2010) Delivering quantum dot-peptide bioconjugates to the cellular cytosol: escaping from the endolysosomal system. Integr Biol 2(5–6):265–277
10. Delehanty JB et al (2011) Spatiotemporal multicolor labeling of individual cells using peptide-functionalized quantum dots and mixed delivery techniques. J Am Chem Soc 133(27):10482–10489
11. Field L et al (2015) Modulation of intracellular quantum dot to fluorescent protein Förster resonance energy transfer via customized ligands and spatial control of donor–acceptor assembly. Sensors 15(12):29810
12. Medintz IL et al (2008) Intracellular delivery of quantum dot-protein cargos mediated by cell penetrating peptides. Bioconjug Chem 19:1785–1795
13. Walters R et al (2012) Nanoparticle targeting to neurons in a rat hippocampal slice culture model. ASN Neuro 4(6):383–392
14. Walters R et al (2015) The role of negative charge in the delivery of quantum dots to neurons. ASN Neuro 7(4):1–12
15. Agarwal R et al (2015) Delivery and tracking of quantum dot peptide bioconjugates in an intact developing avian brain. ACS Chem Nerosci 6(3):494–504
16. Clapp AR, Goldman ER, Mattoussi H (2006) Capping of CdSe-ZnS quantum dots with DHLA and subsequent conjugation with proteins. Nat Protoc 1(3):1258–1266
17. Snee PT et al (2005) Whispering-gallery-mode lasing from a semiconductor nanocrystal/microsphere resonator composite. Adv Mater 17:1131
18. Li JJ et al (2003) Large-scale synthesis of nearly monodisperse CdSe/CdS Core/Shell nanocrystals using air-stable reagents via successive ion layer adsorption and reaction. J Am Chem Soc 125:12567
19. Blackman B, Battaglia D, Peng XG (2008) Bright and water-soluble near Ir-emitting Cdse/Cdte/ZnSe type-II/type-I nanocrystals, tuning the efficiency and stability by growth. Chem Mater 20:4847
20. Susumu K et al (2014) A new family of pyridine-appended multidentate polymers as hydrophilic surface ligands for preparing stable biocompatible quantum dots. Chem Mater 26 (18):5327–5344
21. Mei BC et al (2009) Polyethylene glycol-based bidentate ligands to enhance quantum dot and gold nanoparticle stability in biological media. Nat Protoc 4:412–423
22. Nag OK et al (2017) Quantum dot–peptide–fullerene bioconjugates for visualization of

in vitro and in vivo cellular membrane potential. ACS Nano 11(6):5598–5613

23. Delehanty JB et al (2013) Controlling the actuation of therapeutic nanomaterials: enabling nanoparticle-mediated drug delivery. Ther Deliv 4(11):1411–1429
24. Prasuhn DE et al (2010) Polyvalent display and packing of peptides and proteins on semiconductor quantum dots: predicted versus experimental results. Small 6(4):555–564
25. Field LD et al (2015) Modulation of intracellular quantum dot to fluorescent protein Forster resonance energy transfer via customized ligands and spatial control of donor-acceptor assembly. Sensors 15(12):30457–30468
26. Stewart MH et al (2010) Multidentate poly (ethylene glycol) ligands provide colloidal stability to semiconductor and metallic nanocrystals in extreme conditions. J Am Chem Soc 132:9804–9813
27. Clapp AR et al (2007) Two-photon excitation of quantum dot-based fluorescence resonance energy transfer and its applications. Adv Mater 19:1921
28. Resch-Genger U et al (2008) Quantum dots versus organic dyes as fluorescent labels. Nat Methods 5(9):763–775
29. Margrie TW et al (2003) Targeted whole-cell recordings in the mammalian brain in vivo. Neuron 39(6):911–918
30. Komai S et al (2006) Two-photon targeted patching (TPTP) in vivo. Nat Protoc 1 (2):647–652
31. Kitamura K et al (2008) Targeted patch-clamp recordings and single-cell electroporation of unlabeled neurons in vivo. Nat Methods 5 (1):61–67
32. Weiss PS (2013) President Obama announces the BRAIN initiative. ACS Nano 7 (4):2873–2874
33. Tsytsarev V et al (2008) Imaging cortical electrical stimulation in vivo: fast intrinsic optical signal versus voltage-sensitive dyes. Opt Lett 33(9):1032–1034
34. Delehanty JB, Mattoussi H, Medintz IL (2009) Delivering quantum dots into cells: strategies, progress and remaining issues. Anal Bioanal Chem 393(4):1091–1105
35. Breger J, Delehanty JB, Medintz IL (2015) Continuing progress toward controlled intracellular delivery of semiconductor quantum dots. Wiley Interdiscip Rev Nanomed Nanobiotechnol 7(2):131–151
36. Delehanty JB et al (2009) Quantum dots: a powerful tool for understanding the intricacies of nanoparticle-mediated drug delivery. Expert Opin Drug Deliv 6:1091–1112
37. Andrasfalvy BK et al (2014) Quantum dot-based multiphoton fluorescent pipettes for targeted neuronal electrophysiology. Nat Methods 11(12):1237–1241

Chapter 2

Injection of Adeno-Associated Virus Containing Optogenetic and Chemogenetic Probes into the Neonatal Mouse Brain

Nhan C. Huynh, Baher A. Ibrahim, Christopher M. Lee, Mickeal N. Key, and Daniel A. Llano

Abstract

Optogenetics and chemogenetics are neuromodulation techniques used to study neuronal pathways. Viral vectors containing optogenetic or chemogenetic probes require at least a week after injection to produce effective expression levels in neurons. Therefore, injections need to be done in neonatal mice to enable experiments that require mice younger than 30 days of age. Here, we describe a protocol for performing surgery on neonatal mice and brain injection of a viral vector. The procedure utilizes cryoanesthesia and a pressure injector with a micropipette that can be directly injected through the skull of neonatal mice. Compared to other approaches, this protocol is relatively easy to implement and takes only minutes to perform, additionally allowing for increased numbers of injections. In the examples shown in this chapter, viral vectors were successfully delivered into the auditory cortex and the hippocampus as indicated by labeled expression of the optogenetic/chemogenetic probes. This technique provides a method for investigators to perform surgical injections and optogenetic/chemogenetic experiments in neonatal mice.

Key words Optogenetics, Chemogenetics, Surgery, Rodent, Neonatal mouse, Cryoanesthesia, Adeno-associated virus

1 Introduction

The aim of this method is to produce stable expression of optogenetic or chemogenetic probes in the brain of the neonatal mouse. Optogenetics is a neuromodulation technique that utilizes light to control neurons that have been genetically modified to express light-sensitive ion channels. Many diverse optogenetic tools have enabled temporally precise control over neuronal activity in spatially defined cell types. This technique has greatly promoted the study of the central nervous system through a variety of applications in the field [1]. The basics of optogenetics are development of light-sensitive proteins and strategies for delivering the genes to specific neurons. The first demonstrations involved using channelrhodopsin-2, a light-activated cation channel from algae. Illumination of

Nicholas J. D. Wright (ed.), *Basic Neurobiology Techniques*, Neuromethods, vol. 152,
https://doi.org/10.1007/978-1-4939-9944-6_2,

channelrhodopsin-2-positive neurons with blue light evokes inward currents that drive excitation of neural circuitry [2]. Since the inception of optogenetics, the toolbox has expanded to include halorhodopsins and other molecules that inhibit neural activity [3] as well as opsins with faster kinetics and different excitatory spectra [4]. Additionally, several methods for delivering the genes have been developed. Genes for the light-sensitive molecules can be transferred to target cells by transfection, viral transduction, or creation of transgenic animal lines. With viral transduction, mice are typically injected with adeno-associated viruses (AAV) that contain the opsin genes and the virus infects all areas where it is injected and expresses the light-sensitive protein constructs. To accomplish targeted cell-type specific expression of the opsins, viral transduction can be used in conjunction with a Cre-lox system. Mouse lines have been created to express Cre recombinase in specific cell-types. The AAVs that are injected contain the opsin genes in an anti-sense orientation and flanked by loxP sites. The Cre recombinase then inverts the double-floxed site into the sense orientation for transcription. The virus infects all areas where it is injected, but the construct is only expressed in the specific cell types that contain Cre recombinase [5, 6]. These tools have enabled precise control of neuronal electrophysiological activity to study the functions of neuronal networks.

Chemogenetics is another neuromodulation approach that utilizes genetically modified proteins that can only be activated by small endogenously inert chemical molecules (actuators) [7]. The most widely used chemogenetic probes are the designer receptors exclusively activated by designer drugs (DREADDs) which are the third generation of modified G-protein coupled receptors (GPCR) [8, 9]. Enhancement or inhibition of neuronal firing depends on the type of G-protein subunit. Three modified human muscarinic receptors, $hM1D_q$, $hM3D_q$ (the most frequently used), and $hM5D_q$, are used to enhance the neuronal firing through G_q signaling [8, 10]. In contrast, three Gi-coupled receptors, hM2Di, hM4Di (the most frequently used) and the κ-opioid-derived DREADD (KORD), are used to inhibit the neuronal activity via hyperpolarization by Gβ/γ-mediated activation of G-protein inwardly rectifying potassium channels (GIRKs) and inhibition of the presynaptic release of neurotransmitters [8, 11, 12]. In addition, Gs and β-arrestin coupled DREADDs are used to study the behavior linked to Gs and β-arrestin downstream signaling [13, 14]. Clozapine-N-oxide (CNO) is an example of endogenously inert compound in rodents that solely activates all DREADDs except KORD which can only be activated by salvinorin B [8, 11, 15, 16]. In addition, Compound 21, a non-CNO compound, was introduced to activate DREADDs with an equal potency as CNO against hM3Dq to avoid the potential back metabolism of CNO to clozapine, an atypical antipsychotic agent, in nonrodent species [7, 17, 18]. Like opsins and optogenetics,

DREADDs probes are delivered through many types of AAVs and can use FLEX switch technology that allows for Cre-mediated cell-type-specific expression in any cell type for which there is a Cre-driver line available [19, 20]. Unlike opsins, which use light for activation in a milliseconds time frame, DREADDs provide a longer time frame (seconds and hours) to modulate neuronal activity mediating a certain animal behavior in a freely moving animal [11, 12, 21]. Furthermore, use of DREADDs is a neuromodulation method that does not interfere with in vitro and in vivo imaging. Different types of DREADDs activated by different actuators offer a wide platform for multiplexed and bidirectional chemogenetic modulation of neural activity and behavior neuromodulation [11]. However, as any GPCR, DREADDs could have multiple states of activation such as inactive, partially active, or fully active, which is dependent on the ability of the receptor to recruit certain signaling complex [22, 23]. Further, they may not tolerate the change of ligand concentration due to possible receptor down-regulation or desensitization [10, 24]. These concerns motivated the investigators to examine DREADDs expression level as well as conducting control experiments when using chemogenetic approach.

AAVs require at least a week after injection to produce effective expression levels of optogenetic/chemogenetic probes in neurons. Longer wait times of up to 2–5 weeks may even be necessary for desired expression levels [5]. For experiments that require expression in young mice, such as patch clamp electrophysiology experiments in highly metabolic or myelinated regions that make cell visualization difficult in adult animals, the injections have to be done in neonatal mice to allow adequate time for the viral genes to express. In this chapter, we demonstrate expression in the auditory cortex (AC) and the hippocampus. This method can be applied for other cortical and subcortical injections into neonatal mice.

The neonatal mouse injection protocol is different from the common technique of adult stereotaxic viral transduction. Adult injections involve anesthetizing the animal, performing surgery to cut open the skin and drill holes for the injection sites, injecting the solution containing the viral vector, and finally suturing the skin above the injection site [5]. Neonatal injections require different anesthetic and injection approaches. Neonatal mice (less than 5 days old) are cryoanesthetized by placing them on a small aluminum plate over ice to induce hypothermia anesthesia after 5–10 min. Their small body size makes rapid cooling feasible; they are more resistant to the arrest of blood supply to the brain and can tolerate an extended period of low body temperature without known negative effects [25]. A cocktail of ketamine/xylazine could also be used in neonatal mice, but may not be as safe as cryoanesthesia [26]. After being anesthetized, the neonatal mice are kept on a cold surface and a pressure injector with a micropipette containing the AAV can be directly injected through the skull and into the brain. This injection can be done freehand or with a stereotaxic injection device. After the injection of virus is completed, the micropipette is withdrawn and the neonatal mouse is placed on a warming pad to recover. Compared to other approaches, neonatal injections are relatively easy to implement and take only minutes to perform, additionally allowing for increased numbers of mice to be injected during a surgical session [27].

Proper viral preparation is crucial for successful transduction as improper preparations will result in significantly diminished neuronal infectivity. One factor is serotype. AAV constructs are available in a variety of serotypes that vary based on their cell surface antigens. The effectiveness of the AAV as gene vectors differs depending on the serotype, target cell type, and age of the animal. AAVs with serotype 5 (AAV5) are effective at gene transfer to many structures in an adult mouse brain, but may be ineffective in neonatal mice [28]. In contrast, injections with AAV1 or AAV2 on neonatal mice within 2 days after birth have resulted in widespread expression [28, 29]. For our neonatal mice injections, Deisseroth optical excitation channelrhodopsin-2 AAV2 (AAV-CaMKIIa-hChR2(H134R)-mCherry) and optical inhibition halorhodopsin "eNpHR3.0" AAV1 (AAV-EF1a-DIO-eNpHR3.0-mCherry) (Ha-AAV) constructs were obtained from the University of North Carolina (UNC) Vector Core (Chapel Hill, NC) and the Roth Gi-coupled hM4Di DREADDs AAV8 (AAV8-DIO-hSyn-hM4Di-mCherry) (Di-AAV) construct was obtained from Addgene (Cambridge, MA).

2 Materials (*See* Table 1)

2.1 Animal Preparations

For auditory cortical injections, we used transgenic mice that were 4 days old and obtained after cross breeding between either a heterozygous Ntsr1-Cre (Cre only expressed in cortical layer 6) or Rbp4-Cre (Cre only expressed in cortical layer 5) mouse and a C57BL/6 wild type mouse. For hippocampal injection, we used BALB/c neonatal mice (2.5–5 g, Envigo, Indianapolis, IN) of both sexes that were 3–5 days old. Mice were bred and housed at the local animal facility approved by the American Association for Accreditation of Laboratory Animal Care (AAALAC), and all surgical procedures were approved by the university Institutional Animal Care and Use Committee (IACUC, protocol #16164). Every attempt was made to minimize the number of animals used and to reduce suffering at all stages of this study.

2.2 Neonatal Mice Cryoanesthesia

Neonatal mice were removed from the cage and placed on a warming pad to await injection (Fig. 1; *see* **Note 1**). An insulated ice

Table 1
List of materials

Name	Company	Catalog number	Comments
Ntsr1-Cre mice (C57BL/6 background)	Generous gift from Dr Gordon Shepherd, Northwestern University	N/A	Litters delivered at P4
Rbp4-Cre mice (C57BL/6 background)	From cryopreserved stock from the Mutant Mouse Resource and Research Center	Stock number 031125-UCD	
BALB/c mice	Envigo	Order code #047	Litters delivered at P3–P5
Deltaphase isothermal pads, 6″ × 7.5″	Braintree Scientific	DPIP	
Magic touch 2 ice bucket with lid, 2.5 L capacity	Sigma-Aldrich	BAM168072002	
Aluminum plate	McMaster Carr	8975K439	Shaped to 2.5″ × 5″
Kimberly-Clark Kimwipes delicate task wipes	Fisher Scientific	06-666A	
Optical excitation: Channelrhodopsin-2 AAV-CaMKIIa-hChR2(H134R)-mCherry, serotype 2	UNC Vector Core	—	100 uL aliquot
Third-generation Optogenetic inhibition: eNpHR3.0 AAV-EF1a-DIO-eNpHR3.0-mCherry, serotype 1	UNC Vector Core	—	100 uL aliquot
pAAV-hSyn-DIO-hM4D(Gi)-mCherry, serotype 8	Addgene	44362-AAV8	100 uL aliquot
P-97 flaming/brown micropipette puller	Sutter Instruments	P-97	
3.5-in. glass capillaries, 1.14 mm OD	World Precision Instruments	504949	
Mineral oil	Thermo Fisher Scientific	E10159	

(continued)

Table 1
(continued)

Name	Company	Catalog number	Comments
Syringe, 1 mL, Luer slip, plastic	Grainger	45UC66	Pkg. of 100; used to fill glass micropipettes with mineral oil. Any syringe with a needle that fits micropipettes would work
Needle, gauge 19, 1.5 in.	Grainger	19G376	Pkg. of 100; used with 1 mL plastic syringe
Nanoliter 2010 injector	World Precision Instruments	NANOLITER2010	
Micro4 controller	World Precision Instruments	SYS-MICRO4	
Small animal stereotaxic instrument with digital display console	David Kopf Instruments	Model 940	
Universal syringe holder	David Kopf Instruments	Model 1772	
Mouse and neonatal rat adaptor	Stoelting	51625	
Standard ear bars and rubber tips for mouse stereotaxic	Stoelting	51648	
Reversible temperature labels, multi-temp crystal strips	Omega	RLC-60-26/56-10	16 temperature levels label, 26–56 °F/−3 °C to 13 °C (2 °F increments) in package of 10
6/0 thread size, nylon sutures	CP Medical	1273B	
10 μL syringe, Neuros Model 1701 RN, point style 4, SYR	Hamilton	65460-06	Other method of injecting neonatal mice. Recommended to be used with UMP3 UltraMicroPump and Micro4 controller
UMP3 UltraMicroPump	World Precision Instruments	UMP3	

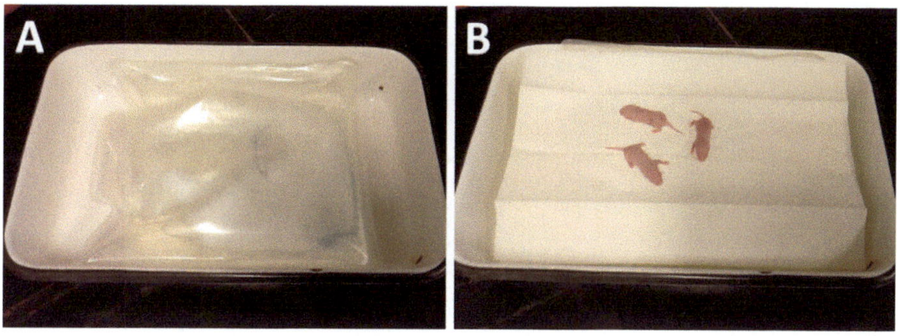

Fig. 1 Neonatal mice preparation (**a**) Warming pad was heated in the microwave and placed in a dish (**b**) Neonatal mice transferred onto paper towel placed on warming pad in preparation for surgery

Fig. 2 Cryoanesthesia preparation (**a**) Cryoanesthesia setup with ice bucket filled with ice (**b**) A small aluminum plate was placed on ice and (**c**) covered with a task wipe

bucket (2.5 L) was filled with ice and a small aluminum plate (2.5″ × 5″) was placed over the ice to cool. A dry task wipe was placed on top of the plate to protect the neonatal mouse's skin from the cold metal (Fig. 2). Neonatal mice were transferred from the warming pad to the cold metal plate and laid on their backs (Fig. 3; *see* **Note 2**). Cryoanesthesia was induced after 5–10 min. A toe pinch was done to confirm that the mice were fully anesthetized. It has been observed by our laboratory that neonatal mice can withstand up to 25 min of cryoanesthesia, though generally shorter periods are used for surgery (10–15 min).

2.3 Adeno-Associated Virus (AAV) Aliquots

The optical excitation channelrhodopsin-2 AAV2 (AAV-CaMKIIa-hChR2(H134R)-mCherry) (Ch-AAV) and the optical inhibition halorhodopsin "eNpHR3.0" AAV1 (AAV-EF1a-DIO-eNpHR3.0-mCherry) (Ha-AAV) constructs from UNC Vector Core (Chapel Hill, NC) as well as Gi-coupled hM4Di DREADDs AAV8 (AAV8-DIO-hSyn-hM4Di-mCherry) (Di-AAV) construct from Addgene (Cambridge, MA) were used for the injections. These vectors

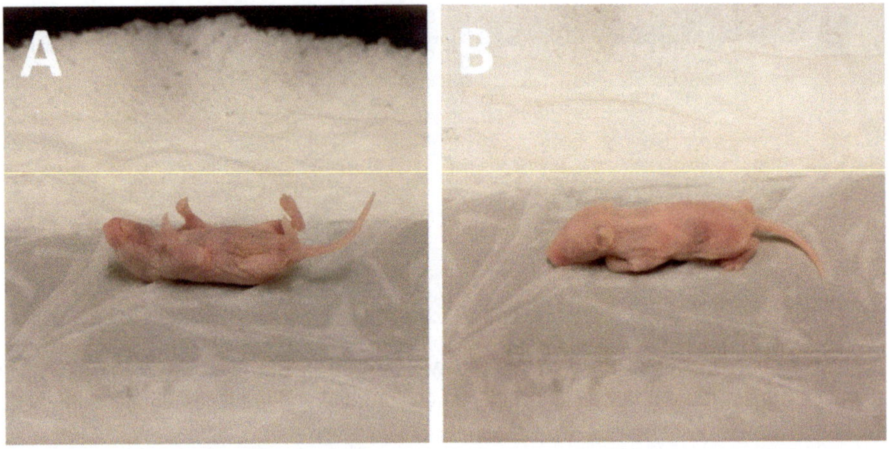

Fig. 3 Neonatal mice cryoanesthesia (**a**) Initial placement of the neonatal mouse on its back on the task wipe covered, cooled aluminum plate until insensate and motionless (**b**) Neonatal mouse rotated onto its ventral surface for surgery

express mCherry red fluorescent protein as a reporter protein along with the main optically or chemically modified probes. The initial 100 μL aliquot was thawed over ice for 20 min and then divided into 10 μL aliquots and refrozen at −80 °C (*see* **Note 3**). We have found that the use of polybrene for dilution of the viral vector does not affect the expression levels (*see* **Note 4**). The addition of Fluoro-Emerald tracer and Trypan Blue dye to the injectate for immediate visualization of the injection reduces expression of the optogenetic probe (*see* **Note 5**).

2.4 Pressure Injector Preparations

Glass micropipettes (3.5-inches, World Precision Instruments, Sarasota, FL) were pulled using a micropipette puller (Fig. 4a; P-97, Sutter Instruments, Novato, CA) and broken back to a tip diameter between 50 and 150 μm (*see* **Note 6**). A black permanent marker was used to mark the micropipette up to 1.5 mm from the tip leaving only the tip unmarked (Fig. 4b; *see* **Note 7**). The micropipette was filled with mineral oil (Thermo Fisher Scientific Inc., Waltham, MA) using a plastic syringe with a needle (Fig. 4c) and attached to a pressure injector (Nanoliter 2010, World Precision Instruments, Sarasota, FL) connected to a pump controller (Micro4 Controller, World Precision Instruments, Sarasota, FL). An aliquot of the AAV was thawed over ice for approximately 20 min before being loaded into the glass micropipette. For an auditory cortical injection in one mouse, 1600 nL of AAV needs to be loaded into the micropipette. For a hippocampal injection in one mouse, 500 nL of AAV needs to be loaded. For any amount to be loaded into the micropipette, the same amount of mineral oil first needed to be injected at a rate of 50 nL/s. Then the AAV could be withdrawn also at a rate of 50 nL/s (Fig. 5).

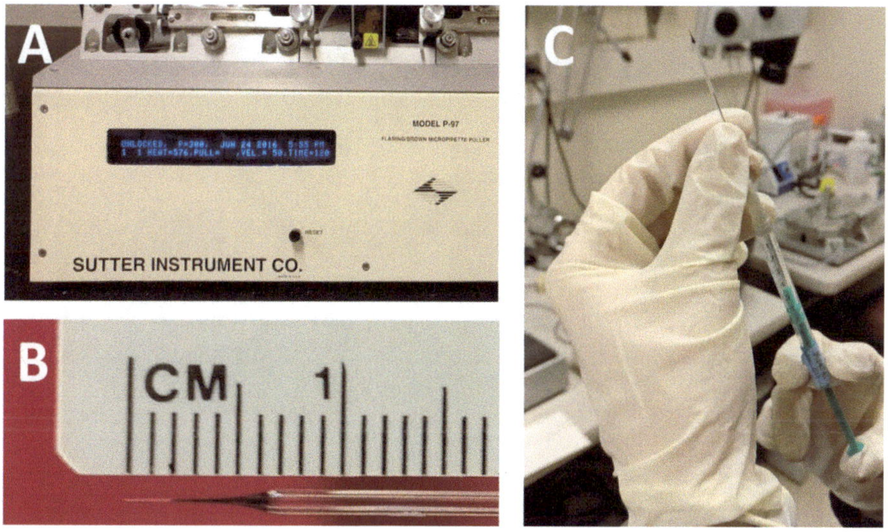

Fig. 4 Glass micropipette pulled, marked and filled with mineral oil (**a**) Micropipette pulled using a P-97 glass micropipette puller with the indicated settings (**b**) Micropipette tip was marked with black permanent marker up to 1.5 mm from the tip leaving only the tip unmarked for easy indication of penetration depth (**c**) Micropipette was filled with mineral oil using a syringe and needle

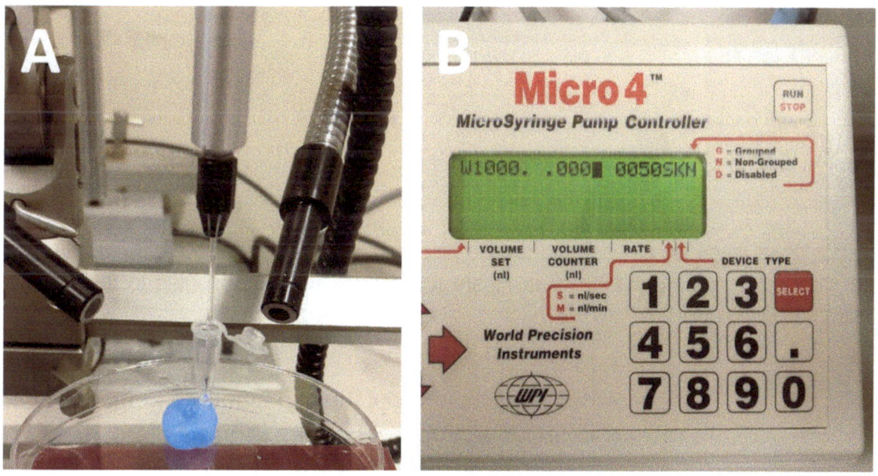

Fig. 5 Glass micropipette filled with AAV (**a**) Glass micropipette was attached to the pressure injector and lowered into an aliquot of AAV (**b**) 1000 nL of AAV was withdrawn using the Micro4 controller at a rate of 50 nL/s

2.5 Stereotaxic Instrument

Neonatal mice injections can be done freehand or with a stereotaxic injection device. However, using a stereotaxic instrument would remove human factors during the procedure and increase the rate of successful injections. A small animal stereotaxic instrument (Fig. 6; David Kopf Instruments, Tujunga, CA) was used with a universal

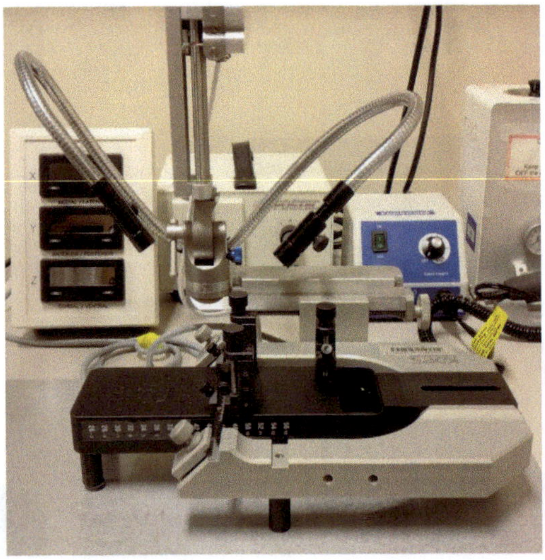

Fig. 6 Stereotaxic instrument

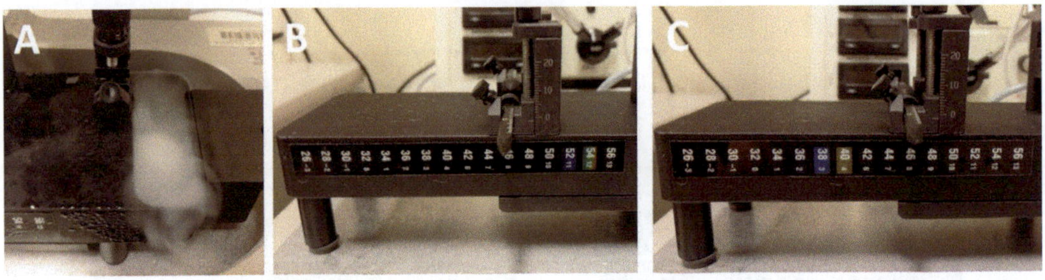

Fig. 7 Cooled adaptor stage (**a**) Ethanol and dry ice pieces were added to the well of the adaptor (**b**) Stage temperature started to cool, around 12 °C in Figure (**c**) Stage temperature cooled to desired temperature range, around 3.5 °C in figure

syringe holder (David Kopf Instruments, Tujunga, CA), a mouse and neonatal rat adaptor (Stoelting, Wood Dale, IL), and standard ear bars with rubber tips (Stoelting, Wood Dale, IL). The adaptor stage was cooled by adding ethanol and dry ice to the well. A temperature label (RLC-60-26/56, Omega, Norwalk, CT) was attached to the adaptor to provide the temperature of the stage during cooling (Fig. 7). Green indicates at temperature, brown indicates slightly above temperature, and blue indicates slightly below temperature. The temperature should be maintained above 1 °C to prevent hypothermia or cold-induced skin damage of the neonatal mice and below 8 °C to sustain cryoanesthesia [27].

3 Methods

3.1 Auditory Cortical Injection, Stereotaxic Procedure Using a Pressure Injector and Electrophysiology Procedure

3.1.1 Stereotaxic Procedure Setup

Auditory cortical injections were done with the transgenic neonatal mice. The scalp of the neonatal mouse was cleaned with a 70% ethanol swab (Fig. 8a). An incision (1.5–2 mm) was made at the midline between the ears using dull forceps and clean scissors (Fig. 8b). The skin was moved to expose the skull (Fig. 8c). The neonatal mouse was transferred to the cooled adaptor stage (*see* **Note 8**) and the skull was fixed between the ear bars. The region containing the target brain regions was visually identified. The pressure injector was mounted to the universal syringe holder and moved into position for the injection.

3.1.2 Auditory Cortical Injection

For auditory cortical injections, the micropipette tip was first injected above the auditory cortex of the left hemisphere at 1–1.5 mm anterior to lambda and just at the edge of the skull's flat horizon. The tip was lowered to 1.2 mm from the brain surface, then it was pulled back to 1.0 mm for the first injection where 200 nL of Ha-AAV or Di-AAV was injected at 300 nL/min. After the injection was finished, the micropipette was left in the brain for 1 min before removing to allow the injectate to settle into the brain. Following the first injection, the tip was pulled back stepwise in 0.1 mm increments, and 200 nL of the injectate was injected at every step until the tip reached 0.3 mm from the surface as shown in (Fig. 9). In total, 1600 nL of AAV was injected into the AC.

The incision was sutured using 6/0 thread size, nylon sutures (Fig. 10; CP Medical, Norcross, GA). The neonatal mice were transferred back onto a warming pad to recover. After around 20 min, their skin color should return to normal and there should

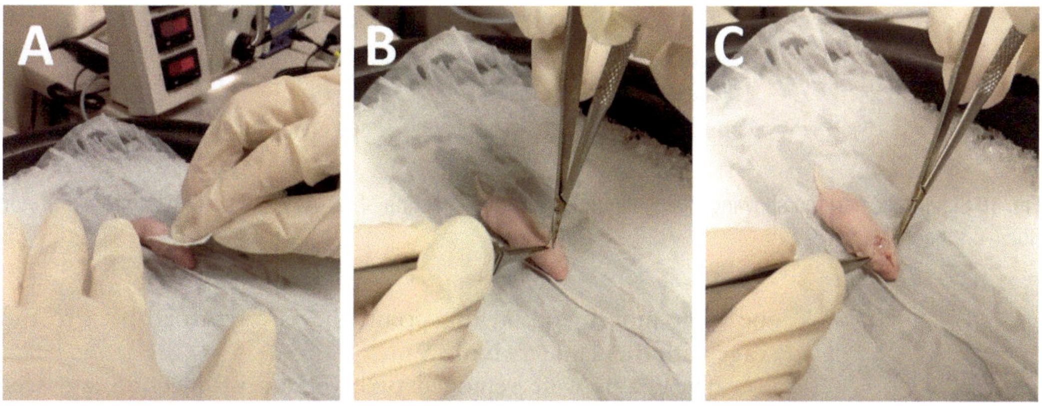

Fig. 8 Surgical procedure to expose the skull (**a**) Scalp of neonatal mouse cleaned with 70% ethanol swab (**b**) Using dull forceps and clean scissors, an incision (1.5–2 mm) was made at the midline between the ears (**c**) Skin was moved away for easy visualization of physical landmarks and the injection site

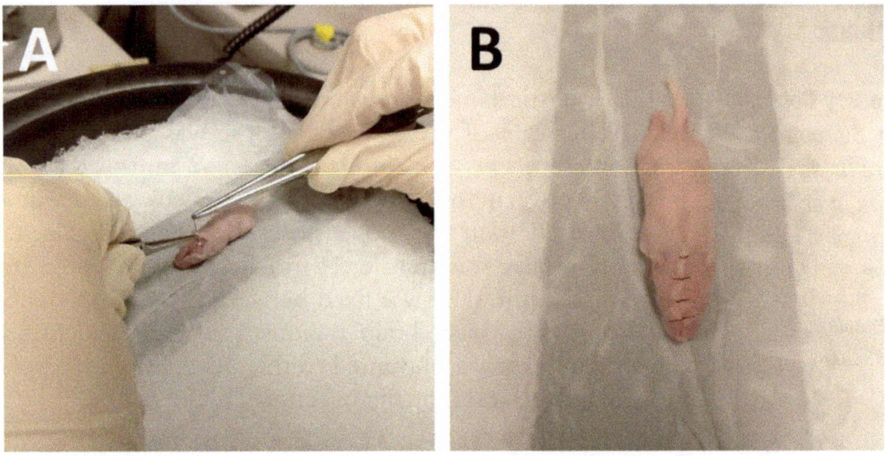

Fig. 9 Sutures (**a**) Neonatal mouse transferred to aluminum plate on ice bucket for easier maneuvering and two forceps were used to suture the incision site (**b**) The incision site should be tightly sutured to decrease the chance of cannibalization by the dam

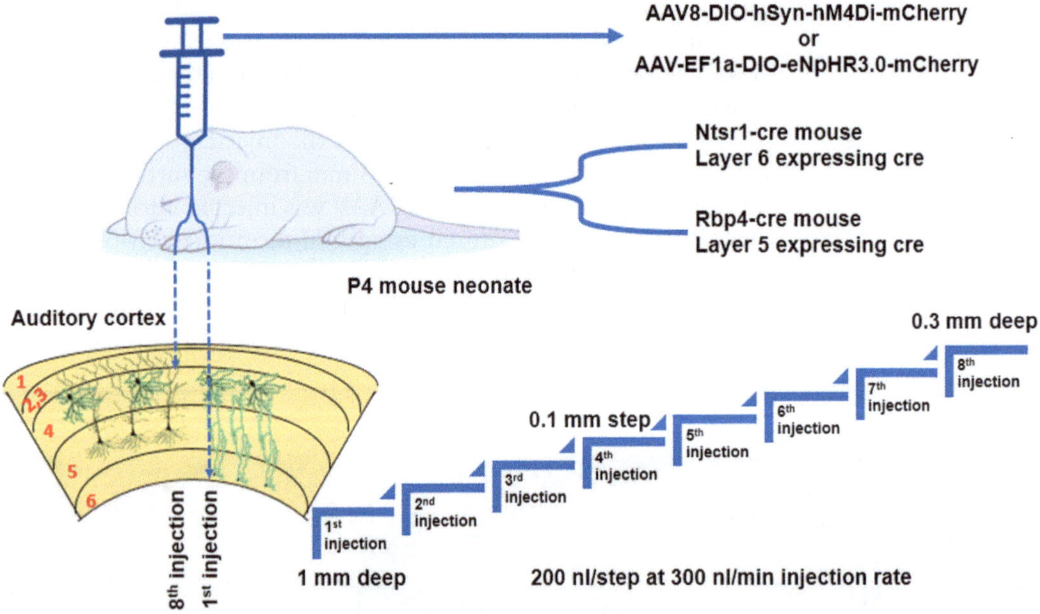

Fig. 10 Schematic cartoon showing the continuous stepwise injection of Ha-AAV and Di-AAV into Ntsr1-Cre or Rbp4-Cre neonatal mice

be some movement. At this point, they are returned back to the cage with the dam (*see* **Note 9**).

The brains were removed two weeks post-injection and sectioned for examination using fluorescence and confocal microscopy using the excitation and emission profiles of m-Cherry. For examining the cortical expression of Di-AAV and Ha-AAV in Ntsr1-Cre and Rbp4-Cre neonatal mice, colliculo-thalamocortical brain slices (developed by our laboratory), which retain the auditory cortical projections with the auditory thalamus, as well as projections from the inferior colliculus to the auditory thalamus, were used [30–32].

The injection of Di-AAV into Ntsr1-Cre neonatal mice showed a successful expression of DREADDs in auditory layer 6 cells transfected by Di-AAV as well as their projections to thalamic reticular nucleus (TRN) and auditory thalamus a.k.a. medial geniculate body (MGB) (Fig. 11a, b). In layer 6 cells, DREADDs expression was localized at the cell body (Fig. 11e, f-white arrows) as proven by nuclear stain (DAPI), their apical dendrites to upper layers (Fig. 11c, white arrows), and their axons to the subcortical brain structures (Fig. 11d, white arrows). As evidence of the ability of the viral vector to travel solely to the connected brain regions to cortical layer 6, the projections of layer 6 cells to TRN (Fig. 12a, b, white arrows) and MGB (Fig. 12c, d) were also shown. It is clear there were no projections to visual thalamus a.k.a. lateral geniculate nucleus (LGN). Also, the expression of DREADDs was seen in Di-AAV infected layer 5 cells in Rbp4-Cre neonatal mice (Fig. 13a–d). DREADDs expression was localized at the cell bodies of layer 5 cells (Fig. 13f, white arrows) and their apical dendrites to upper layers (Fig. 13d ,white arrows, and 13E-white arrows). Halorhodopsin expression was also observed in auditory cortex layer 5 cells transfected with Ha-AAV in Rpb4-Cre neonatal mice (Fig. 14a–d). Halorhodopsin expression was localized to the cell body of layer 5 cells (Fig. 14f-white arrows) and their nearby projections (Fig. 14e, white arrows) as well as their apical dendrites to upper cortical layers (Fig. 14d, white arrows).

3.1.3 Brain Slicing

At 12 days old age, Ntsr1-Cre and Rbp4-Cre mice were genotyped for Cre positivity. Cre-positive mice of both types were initially anesthetized with ketamine (100 mg/kg) and xylazine (3 mg/kg) intraperitoneally and transcardially perfused with chilled (4 °C) sucrose-based slicing solution containing the following (in mM): 234 sucrose, 11 glucose, 26 $NaHCO_3$, 2.5 KCl, 1.25 NaH_2PO_4, 10 $MgCl_2$, 0.5 $CaCl_2$. 600 μm thick auditory colliculo-thalamo-cortical brain slices were obtained [30–32]. All slices were incubated for 30 min in 33 °C incubation solution (in mM: 26 $NaHCO_3$, 2.5 KCl, 10 glucose, 126 NaCl, 1.25 NaH_2PO_4, 3 $MgCl_2$, and 1 $CaCl_2$). After incubation, slices were transferred to a perfusion chamber coupled to an upright Olympus BX51 microscope, perfused with artificial cerebrospinal fluid (ACSF) containing (in mM) 26 $NaHCO_3$, 2.5 KCl, 10 glucose, 126 NaCl, 1.25 NaH_2PO_4, 2 $MgCl_2$, and 2 $CaCl_2$, with 3–6 mL/min as a flow rate, and bubbled with 95% oxygen/5% carbon dioxide.

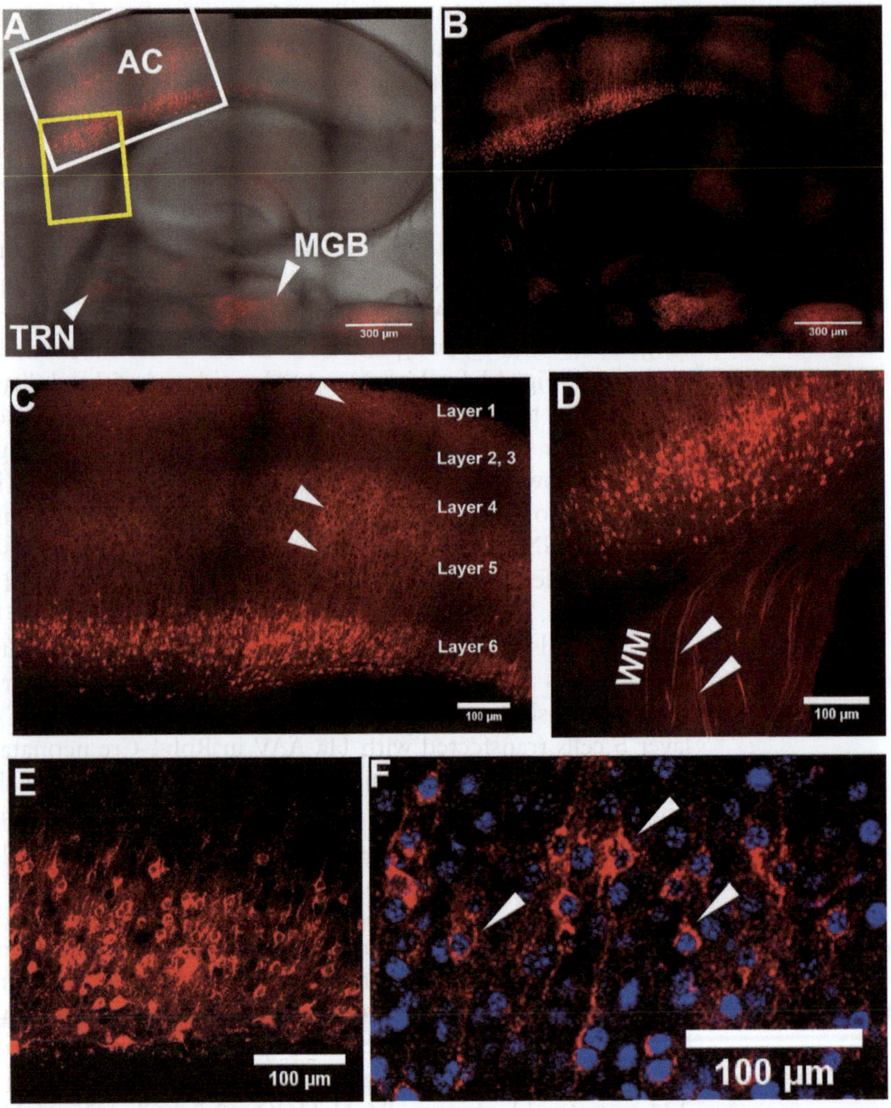

Fig. 11 DREADDs expression in auditory cortex layer 6 cells (**a**) a composite tile image [bright field (gray) and m-Cherry (red)] of a colliculothalamocortical brain slice showing DREADDs expression in layer 6 cells of the auditory cortex as well as its projections to TRN and MGB (**b**) the m-Cherry channel for A. (**c**) a tile image of the auditory cortex showing the apical dendrites (white arrows) of layer 6 cells to the upper cortical layers (**d**) layer 6 cells and their axons (white arrows) to the subcortical brain structures (**e**) High magnification image of layer 6 cells showing that DREADDs expression is localized at their cell body and their extended projections (**f**) a composite image [m-Cherry (red) and DAPI (blue)] showing the same as **e** (white arrows point to DREADDs expression in the cell membrane). *AC* auditory cortex, *MGB* medial geniculate body, *TRN* thalamic reticular nucleus, *WM* white matter

3.1.4 Electrophysiology

Whole-cell recording was performed using a visualized slice setup outfitted with infrared-differential interference contrast optics. Recording pipettes were pulled from borosilicate glass capillary tubes and had tip resistances of 2–5 MΩ when filled with

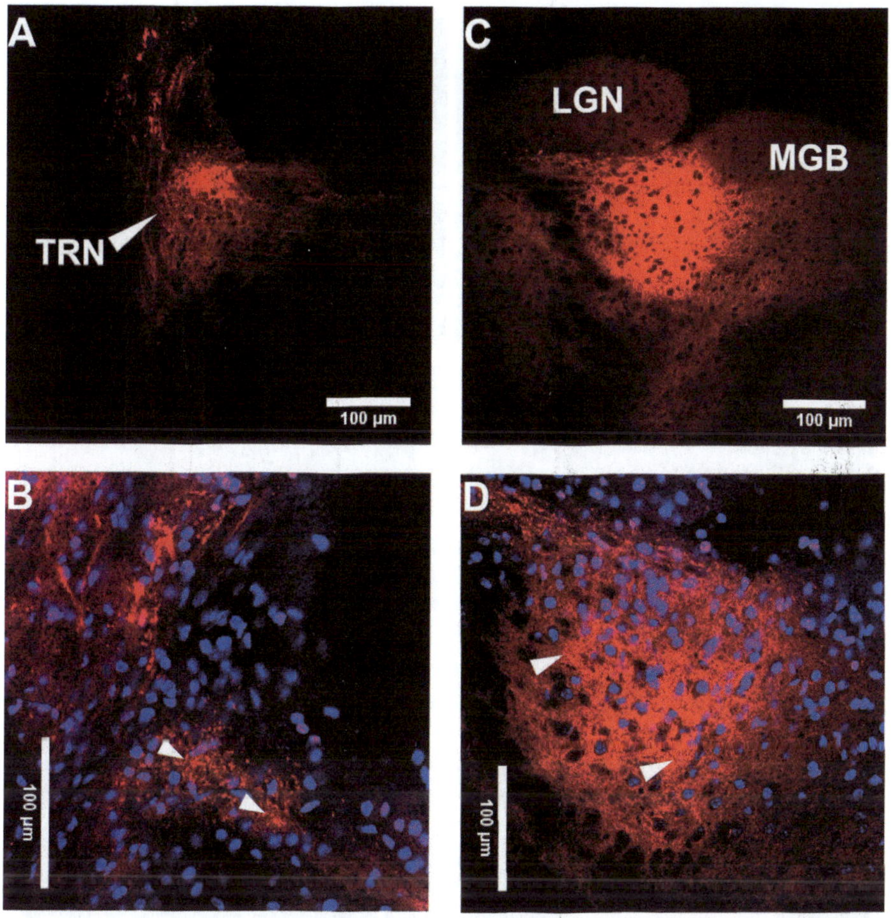

Fig. 12 DREADDs expression in auditory layer 6 projections to auditory thalamus (**a**) layer 6 projections to TRN (**b**) high magnification composite image [m-Cherry (red) and DAPI (blue)] of layer cells projections (white arrows) to TRN inhibitory neurons (**c**) layer 6 projections to MGB (**d**) high magnified composite image [m-Cherry (red) and DAPI (blue)] of layer cells projections (white arrows) to MGB cells. LGN: lateral geniculate nucleus, MGB: medial geniculate body, TRN: thalamic reticular nucleus

intracellular solution, which contained the following (in mM: 117 K-gluconate, 13 KCl, 1.0 $MgCl_2$, 0.07 $CaCl_2$, 0.1 ethyleneglycol-bis(2-aminoethylether)-N,N,N′,N′-tetra acetic acid, 10.0 4-(2-hydroxyethyl)-1-piperazineethanesulfonic acid, 2.0 Na-ATP, 0.4 Na-GTP, and 0.5% biocytin, pH 7.3). A Multiclamp 700B amplifier and pClamp software (Molecular Devices) were used for data acquisition (20 kHz sampling). Whole-cell patching configurations such as current clamp, −40 mV, and −50 mV voltage clamp were conducted.

The functionality of expressed DREADDs and halorhodopsin was tested. 10 mV hyperpolarization of the membrane potential of layer 5 or 6 cells transfected by Di-AAV was observed after the bath application of 5 μM CNO to the recording chamber (Fig. 15).

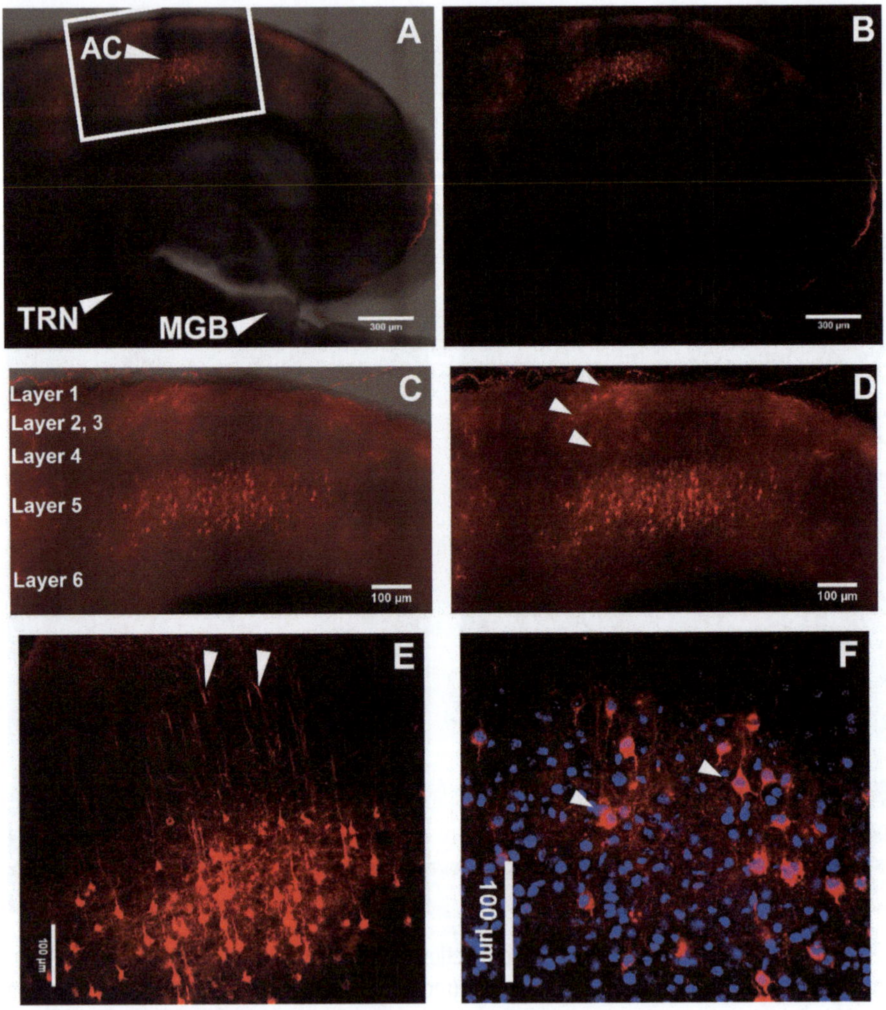

Fig. 13 DREADDs expression in auditory cortex layer 5 cells (**a**) a composite tile image [bright field (gray) and m-Cherry (red)] of a thalamocortical brain slice showing DREADDs expression in layer 5 cells of AC (**b**) The m-Cherry channel for A (**c**) a composite tile image [bright field (gray) and m-Cherry (red)] of AC showing the apical dendrites of layer 5 cells to the upper cortical layers (**d**) the m-Cherry channel for C (white arrows point to layer 5 apical dendrites) (**e**) high magnification image of layer 5 cells showing that DREADDs expression is localized at their cell body and their apical dendrites (white arrows) (**f**) a composite image [m-Cherry (red) and DAPI (blue)] showing the same as E (white arrows point to DREADDs expression in the cell membrane). AC: auditory cortex, MGB: medial geniculate body, TRN: thalamic reticular nucleus

Unlike chemogenetic inhibitory approach, which is characterized by a modest hyperpolarization, 590 nm light exposure to Ha-AAV transfected layer 5 cell was associated with a strong hyperpolarization that hyperpolarized the membrane potential from −60 mV to more negative than −80 mV (Fig. 16a). Also, 590 nm light exposure to Ha-AAV transfected layer 5 cell showed outward current at −40 mV and −50 mV voltage clamp configurations (Fig. 16b, c).

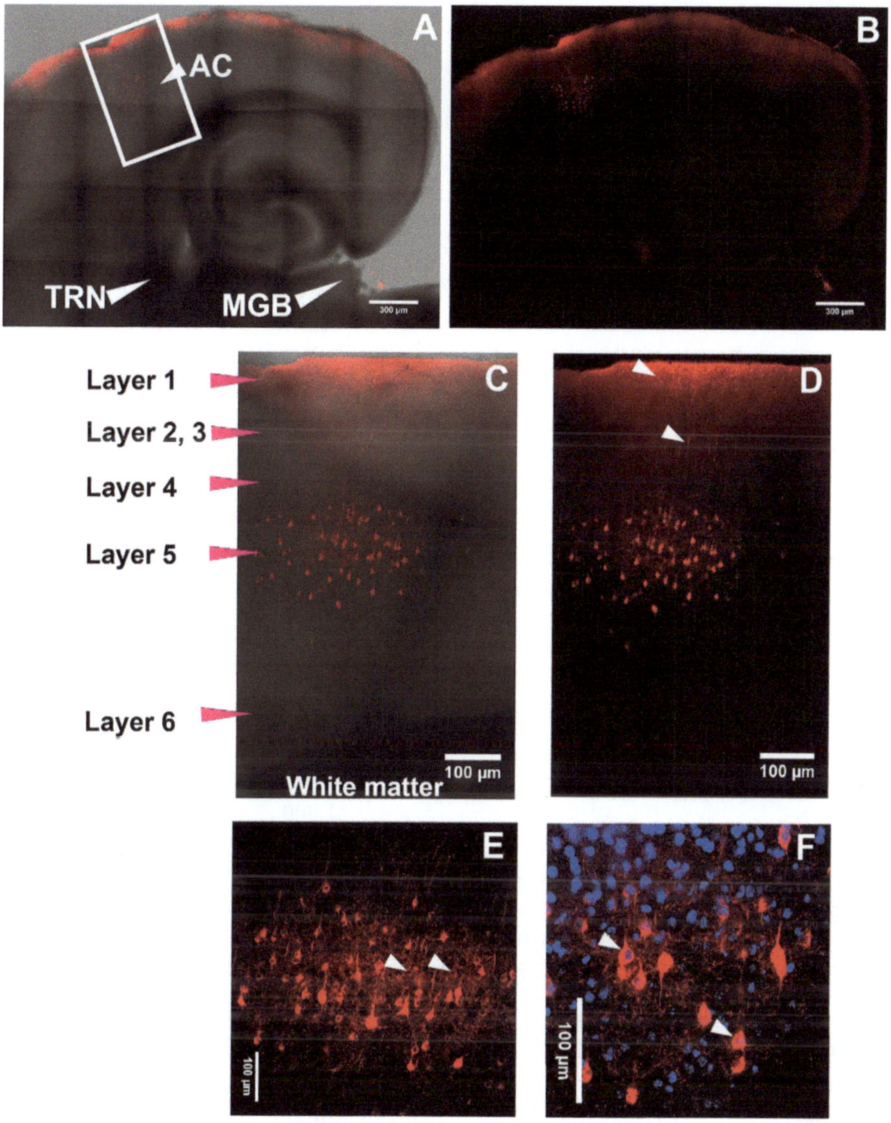

Fig. 14 Halorhodopsin expression in auditory cortex layer 5 cells (**a**) a composite tile image [bright field (gray) and m-Cherry (red)] of a colliculothalamocortical brain slice showing DREADDs expression in layer 5 cells of AC (**b**) The m-Cherry channel for A (**c**) a composite tile image [bright field (gray) and m-Cherry (red)] of AC showing the apical dendrites of layer 5 cells to the upper cortical layers (**d**) the m-Cherry channel for C (white arrows point to layer 5 apical dendrites) (**e**) enlarged view of layer 5 cells showing that DREADDs expression is localized at their cell body and their extended projections (white arrows) (**f**) a composite image [m-Cherry (red) and DAPI (blue)] showing the same as E (white arrows point to DREADDs expression in the cell membrane). AC: auditory cortex, MGB: medial geniculate body, TRN: thalamic reticular nucleus

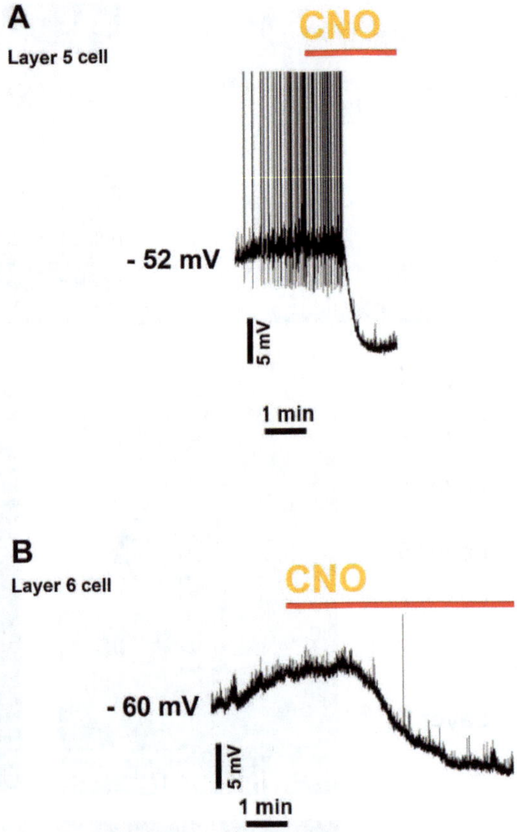

Fig. 15 Functional performance of DREADDs after Di-AAV expression. Bath application of 5 μM CNO hyperpolarized the membrane potential by 10 mV in (**a**) Di-AAV transfected layer 5 cell of Rbp4-Cre mouse brain slice and in (**b**) Di-AAV transfected layer 6 cell of Ntsr1-Cre mouse brain slice

3.2 Hippocampal Injection, Stereotaxic Procedure Using a Pressure Injector

Hippocampal injections were done with BALB/c neonatal mice. The stereotaxic setup follows the setup procedure for auditory cortical injections (*see* Subheading 3.1.1). For hippocampal injections, the micropipette tip was also injected 1–1.5 mm anterior to lambda, but only three-quarters of the way from the medial line to the edge of the skull's flat horizon. The tip was lowered to 1.2 mm from the brain surface, then it was pulled back to 1.0 mm. 500 nL of the Ch-AAV was injected into the hippocampus at 200 nL/min. After the injection was finished, the micropipette was left in the brain for 1 min before removing to allow the injectate to settle into the brain.

The incision was sutured using 6/0 thread size, nylon sutures (Fig. 10; CP Medical, Norcross, GA). The neonatal mice were transferred back onto a warming pad to recover. After around 20 min, their skin color should return to normal and there should be some movement. At this point, they are returned back to the cage with the dam (*see* **Note 9**).

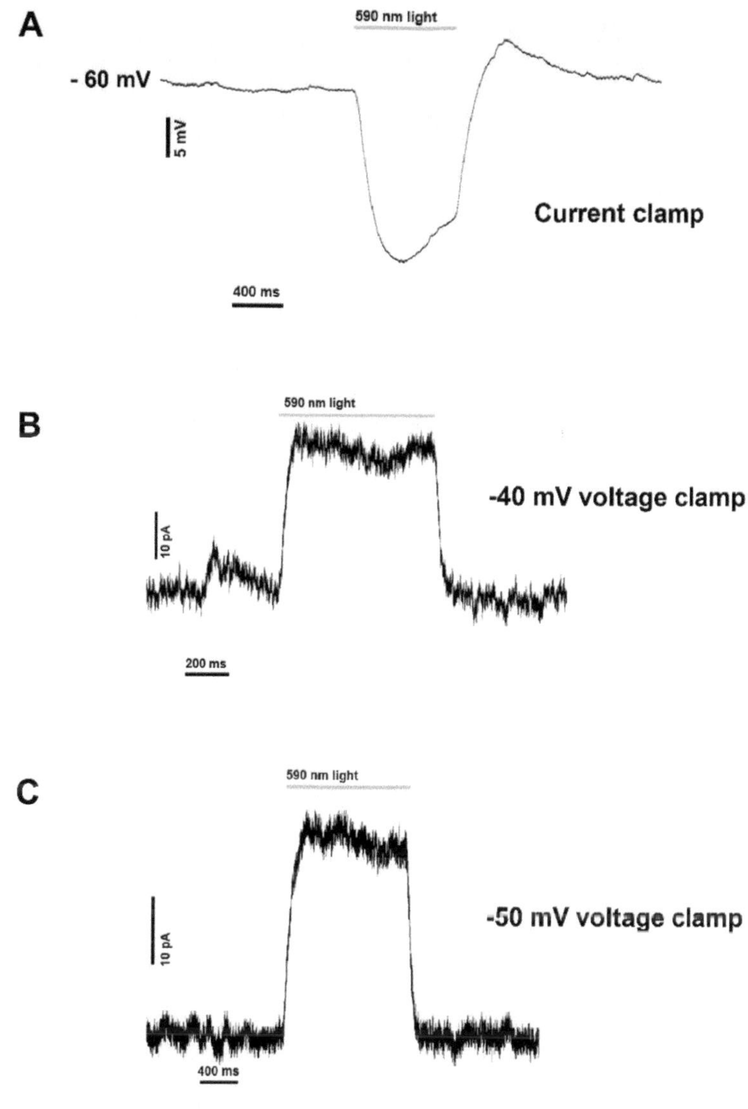

Fig. 16 Functional performance of halorhodopsin after Ha-AAV expression. The exposure of 590 nm light to Ha-AAV transfected layer 5 cell led to (**a**) a strong hyperpolarization as indicated by 20 mV hyperpolarization of the membrane potential, (**b**, **c**) outward current at −40 mV and 50 mV voltage clamp, respectively

The brains were removed 2 weeks post-injection and sectioned for examination using fluorescence and confocal microscopy using the excitation and emission profiles of m-Cherry. For examining the hippocampal expression of Ch-AAV, 50 microns thick coronal sections were used for imaging.

There was clear viral expression in the hippocampus (Fig. 17a) and hippocampal projections (Fig. 17b) that demonstrate successful injections of Ch-AAV containing optogenetic probes into neonatal mice.

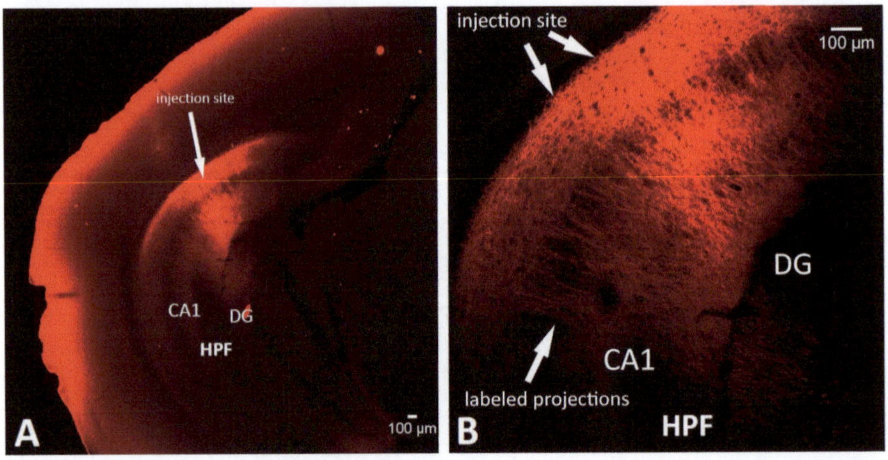

Fig. 17 Representative images for stereotaxic procedure using a pressure injector show transduction pattern of mCherry/Channelrhodopsin-2 at (**a**) the injection site in the hippocampus and (**b**) labeled projections to the dentate gyrus (DG)

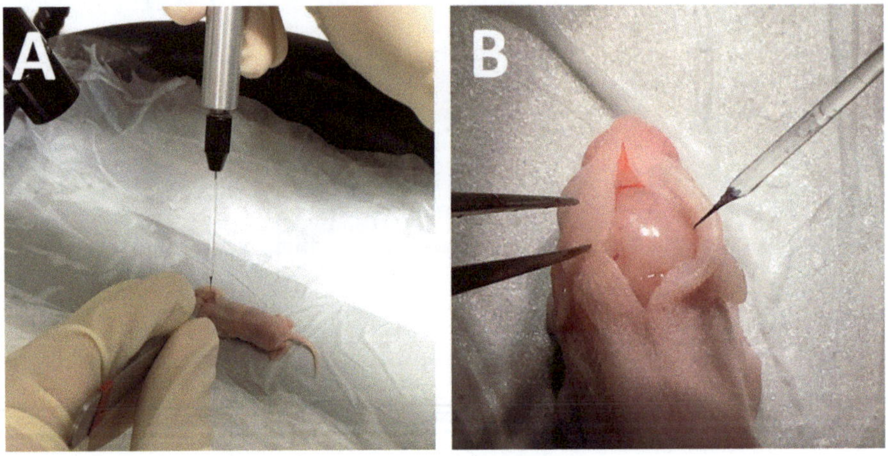

Fig. 18 Free-hand procedure using a pressure injector (**a**) Pressure injector with micropipette was directly injected through the skull into the hippocampus as forceps were used to hold back the skin (**b**) Tip injected up to the mark at 1.5 mm and pulled back slightly to an approximate depth of 1 mm. Pump controller was run to inject the AAV into the brain

3.3 Hippocampal Injection, Free-Hand Procedure Using a Pressure Injector

The free-hand procedure follows the stereotaxic procedure up until transferring the neonatal mouse to the stereotaxic adaptor stage. After making an incision at the midline between the ears, one-half of skin was moved away using the forceps to expose one-half of the skull. The pressure injector with the micropipette loaded with AAV was held in the right hand as the left hand held the forceps that was used to keep the region containing the hippocampus exposed. The micropipette was free-hand injected directly through the skull into the hippocampus (Fig. 18). The tip was injected up to two-thirds of

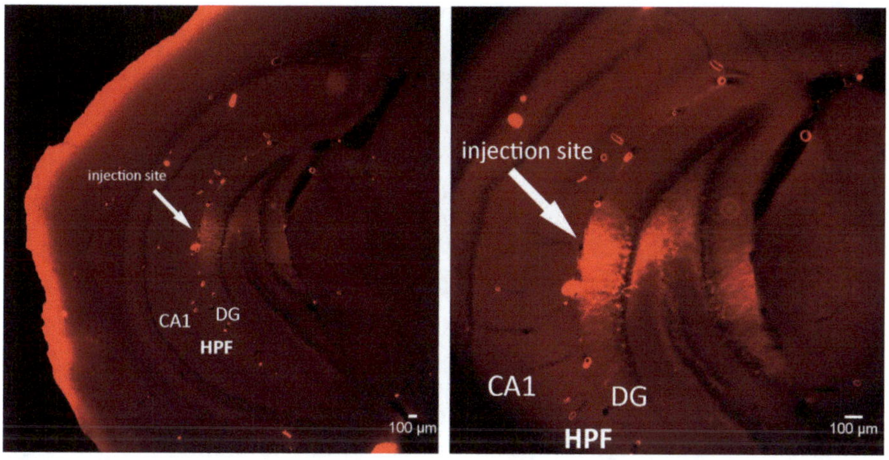

Fig. 19 Representative images for free-hand procedure using a pressure injector show transduction pattern of mCherry/Channelrhodopsin-2 at a deep site in the hippocampus

the way before the black mark, around 1.0 mm, and pulled back slightly to an approximate depth of 0.5 mm. 500 nL of the AAV was injected into the hippocampus at 200 nL/min for 2 min. After the injection was finished, the micropipette was left in the brain for 15 s before removing to allow the injectate to settle into the brain. The neonatal mouse scalp was sutured, allowed to recover, and analyzed as in the stereotaxic procedure.

There was clear viral expression in the brain slices, but it was frequently too deep (Fig. 19). There was difficulty with maintaining the top-heavy pressure injector at a steady and shallow depth during the injection period (*see* **Note 10**). Therefore, our laboratory recommends doing the AAV injections using a stereotaxic instrument.

3.4 Hippocampal Injection, Stereotaxic Procedure Using a Hamilton Syringe

Instead of using a pressure injector and glass micropipettes, a Hamilton syringe with needle could be used to inject the AAV into neonatal mice. This procedure follows the stereotaxic procedure using a pressure injector up until mounting the pressure injector to the universal syringe holder. Instead, a Hamilton Syringe (10 μL, Neuros Model 1701, Hamilton, Reno, NV) was loaded with 1000 nL of the AAV and attached to the universal syringe holder. The syringe was moved into position for the injection, the needle was injected to a depth of 1.0 mm, with reference to the digital display console readout, and pulled back slightly to an approximate depth of 0.5 mm. 500 nL of the AAV was injected into the hippocampus by slowly pushing the syringe plunger over a period of 2 min. After the injection was finished, the needle was left in the brain for 1 min before removing to allow the injectate to settle into the brain. The neonatal mouse was sutured, allowed to recover, and analyzed as in the stereotaxic procedure using a pressure injector.

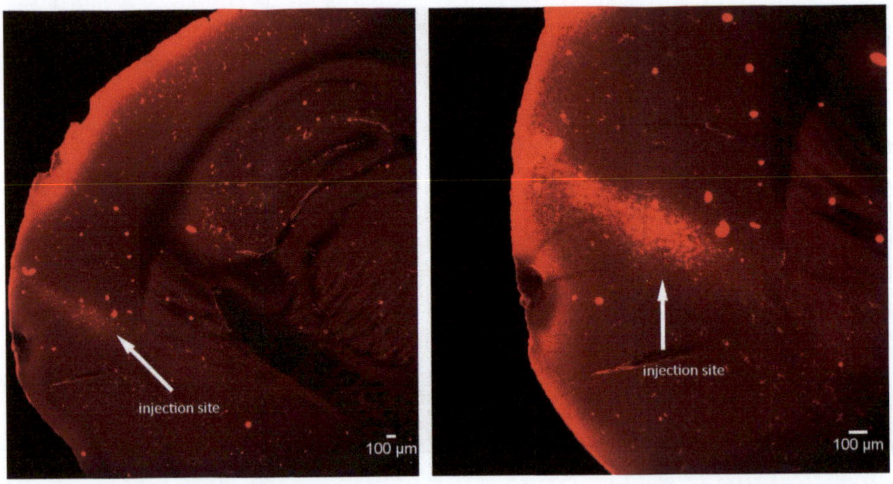

Fig. 20 Representative images for stereotaxic procedure using a pressure injector show transduction pattern of mCherry/Channelrhodopsin-2 in the cortex. Evidence of back flow at the injection site suggests the injection rate was too fast

There were fewer successfully injections using the Hamilton syringe. One neonatal mouse that had clear expression in the cortex displayed evidence of back flow of the injectate (Fig. 20). This finding suggests the rate of injection was too fast using manual depression of the syringe plunger (*see* **Note 11**). For these reasons, our laboratory recommends doing the neonatal AAV injections using a pressure injector.

4 Notes

1. Replace only half of the litter at a time to decrease the chances that the entire litter will be lost upon return after surgery due to cannibalism by the dam.
2. Lay the neonatal mice on their backs to prevent them from moving off the metal plate. After the neonatal mice become motionless, rotate them onto their ventral surface for the rest of the cryoanesthesia.
3. To use the aliquots, thaw them over ice for 20 min and refreeze any unused virus at −80 °C. The AAV can withstand 2–3 freeze-thaw cycles, but it is best to minimize the number of cycles [33].
4. Polybrene, a cationic polymer, is often used to increase the efficiency of viral transduction in cell cultures [34]. Pilot studies in our laboratory have found that there is no difference in the level of expression between AAV diluted with and without polybrene.

5. The use of a dark dye, like Trypan Blue, can be combined with the clear viral solution to produce a dark injectate that is visible upon injection beneath the thin neonatal mouse skin. A fluorescent tracer, like Fluro-Emerald, can also be mixed with the viral solution to produce confirmation of injection using a fluorescent scope by visualizing a luminescence at the injection site. However, pilot studies in our laboratory have found that the addition of a dye or a tracer resulted in reduced expression of the optogenetic probe.
6. The glass micropipette tips are broken back by grazing the tip over the edge of a piece of paper. The tips are examined using a light microscope (10X eyepiece and 10X objective) and an ocular micrometer to confirm they are between 50 and 150 μm (5–10 ocular divisions). This procedure can take several trials to get properly sized tips.
7. After each injection, the micropipette should be examined for the unmarked tip. If it is missing, the tip must have broken during the injection and a new micropipette should be used.
8. Remember to periodically add dry ice to the well to maintain the temperature between 1 °C and 8 °C. The pieces should be small to avoid inadvertently cooling the stage below 1 °C.
9. The litter should be wiped with used bedding or cage material and returned as a group to the cage. This approach decreases the chances that the entire litter will be lost due to cannibalism by the dam.
10. To increase the success of getting localized expression, do more than one injection in the same area. It can be difficult to hold the pressure injector steady for 1 min, so the 500 nL injection could be divided into several smaller injections with shorter time intervals.
11. A microsyringe pump injector (UMP3 UltraMicroPump, World Precision Instruments, Sarasota, FL) could be used to depress a syringe plunger at a slower rate to achieve a more controlled infusion. Using this injector would require adapting the pump injector to the stereotaxic stage. Our laboratory has not tested this device yet.

Acknowledgments

The research was supported by research grant DC013073 from the National Institutes of Health.

References

1. Mattis J, Tye KM, Ferenczi EA, Ramakrishnan C, O'Shea DJ, Prakash R, Gunaydin LA, Hyun M, Fenno LE, Gradinaru V, Yizhar O, Deisseroth D (2012) Principles for applying optogenetic tools derived from direct comparative analysis of microbial opsins. Nat Methods 9(2):159–172
2. Boyden ES, Zhang F, Bamberg E, Nagel G, Deisseroth K (2005) Millisecond-timescale, genetically targeted optical control of neural activity. Nat Neurosci 8(9):1263–1268
3. Zhang F, Wang LP, Brauner M, Liewald JF, Kay K, Watzke N, Wood PG, Bamberg E, Nagel G, Gottschalk A, Deisseroth K (2007) Multimodal fast optical interrogation of neural circuitry. Nature 446(7136):633–639
4. Klapoetke NC, Murata Y, Kim SS, Pulver SR, Birdsey-Benson A, Cho YK, Moriomoto TK, Chuong AS, Carpenter EJ, Tian Z, Wang J, Xie Y, Yan Z, Zhang Y, Chow BY, Surek B, Melkonian M, Jayaraman V, Constantine-Paton M, Wong GK, Boyden ES (2014) Independent optical excitation of distinct neural populations. Nat Methods 11(3):338–346
5. Cardin JA, Carlén M, Meletis K, Knoblich U, Zhang F, Deisseroth K, Tsai L, Moore CI (2010) Targeted optogenetic stimulation and recording of neurons in vivo using cell-type-specific expression of Channelrhodopsin-2. Nat Protoc 5(2):247–254
6. Beltramo R, D'Urso G, Dal Maschio M, Farisello P, Bovetti S, Clovis Y, Lassi G, Tucci V, Tonelli DDP, Fellin T (2013) Layer-specific excitatory circuits differentially control recurrent network dynamics in the neocortex. Nat Neurosci 16(2):227–234
7. Roth BL (2016) DREADDs for neuroscientists. Neuron 89(4):683–694
8. Armbruster BN, Li X, Pausch MH, Herlitze S, Roth BL (2007) Evolving the lock to fit the key to create a family of G protein-coupled receptors potently activated by an inert ligand. Proc Natl Acad Sci 104(12):5163–5168
9. Armbruster B, Roth B (2005) Creation of designer biogenic amine receptors via directed molecular evolution. In: Neuropsychopharmacology, vol 30. Nature Publishing Group, London, pp S265–S265
10. Alexander GM, Rogan SC, Abbas AI, Armbruster BN, Pei Y, Allen JA, Nonneman RJ, Hartmann J, Moy SS, Nicolelis MA, McNamara JO, Roth BL (2009) Remote control of neuronal activity in transgenic mice expressing evolved G protein-coupled receptors. Neuron 63(1):27–39
11. Vardy E, Robinson JE, Li C, Olsen RH, DiBerto JF, Giguere PM, Sassano FM, Huang X, Zhu H, Urban DJ, White KL, Rittiner JE, Crowley NA, Pleil KE, Mazzone CM, Mosier PD, Song J, Kash TL, Malanga CJ, Krashes MJ, Roth BL (2015) A new DREADD facilitates the multiplexed chemogenetic interrogation of behavior. Neuron 86(4):936–946
12. Stachniak TJ, Ghosh A, Sternson SM (2014) Chemogenetic synaptic silencing of neural circuits localizes a hypothalamus → midbrain pathway for feeding behavior. Neuron 82 (4):797–808
13. Allen JA, Roth BL (2011) Strategies to discover unexpected targets for drugs active at G protein–coupled receptors. Annu Rev Pharmacol Toxicol 51:117–144
14. Ferguson SM, Phillips PE, Roth BL, Wess J, Neumaier JF (2013) Direct-pathway striatal neurons regulate the retention of decision-making strategies. J Neurosci 33 (28):11668–11676
15. Roth BL, Craigo SC, Choudhary MS, Uluer A, Monsma FJ, Shen Y et al (1994) Binding of typical and atypical antipsychotic agents to 5-hydroxytryptamine-6 and 5-hydroxytryptamine-7 receptors. J Pharmacol Exp Ther 268 (3):1403–1410
16. Chavkin C, Sud S, Jin W, Stewart J, Zjawiony JK, Siebert DJ, Toth BA, Hufeisen SJ, Roth BL (2004) Salvinorin A, an active component of the hallucinogenic sage Salvia divinorum is a highly efficacious κ-opioid receptor agonist: structural and functional considerations. J Pharmacol Exp Ther 308(3):1197–1203
17. Jann MW, Lam YW, Chang WH (1994) Rapid formation of clozapine in guinea-pigs and man following clozapine-N-oxide administration. Archiv Int Pharmacodyn Ther 328 (2):243–250
18. Chen X, Choo H, Huang XP, Yang X, Stone O, Roth BL, Jin J (2015) The first structure–activity relationship studies for designer receptors exclusively activated by designer drugs. ACS Chem Nerosci 6(3):476–484
19. Zhu H, Pleil KE, Urban DJ, Moy SS, Kash TL, Roth BL (2014) Chemogenetic inactivation of ventral hippocampal glutamatergic neurons disrupts consolidation of contextual fear memory. Neuropsychopharmacology 39 (8):1880–1892
20. Scofield MD, Boger HA, Smith RJ, Li H, Haydon PG, Kalivas PW (2015) Gq-DREADD selectively initiates glial glutamate release and

inhibits cue-induced cocaine seeking. Biol Psychiatry 78(7):441–451

21. Peñagarikano O, Lázaro MT, Lu XH, Gordon A, Dong H, Lam HA, Peles E, Maidment NT, Murphy NP, Yang XW, Golshani P, Geschwind DH (2015) Exogenous and evoked oxytocin restores social behavior in the Cntnap2 mouse model of autism. Sci Transl Med 7(271):271ra8–271ra8
22. Roth BL, Marshall FH (2012) NOBEL 2012 chemistry: studies of a ubiquitous receptor family. Nature 492(7427):57–57
23. Samama P, Cotecchia S, Costa T, Lefkowitz RJ (1993) A mutation-induced activated state of the beta 2-adrenergic receptor. Extending the ternary complex model. J Biol Chem 268(7):4625–4636
24. Krashes MJ, Koda S, Ye C, Rogan SC, Adams AC, Cusher DS, Maratos-Flier E, Roth BL, Lowell BB (2011) Rapid, reversible activation of AgRP neurons drives feeding behavior in mice. J Clin Invest 121(4):1424
25. Phifer CB, Terry LM (1986) Use of hypothermia for general anesthesia in preweanling rodents. Physiol Behav 38(6):887–890
26. National Research Council (2003) Guidelines for the care and use of mammals in neuroscience and behavioral research. National Academies Press, Washington, DC
27. Kim JY, Grunke SD, Levites Y, Golde TE, Jankowsky JL (2013) Intracerebroventricular viral injection of the neonatal mouse brain for persistent and widespread neuronal transduction. J Vis Exp (91):51863–51863
28. Watson DJ, Passini MA, Wolfe JH (2005) Transduction of the choroid plexus and ependyma in neonatal mouse brain by vesicular stomatitis virus glycoprotein-pseudotyped lentivirus and adeno-associated virus type 5 vectors. Hum Gene Ther 16(1):49–56
29. Chakrabarty P, Rosario A, Cruz P, Siemienski Z, Ceballos-Diaz C, Crosby K, Jansen K, Borchelt DR, Kim J, Jankowsky JL, Golde TE (2013) Capsid serotype and timing of injection determines AAV transduction in the neonatal mice brain. PLoS One 8(6):e67680
30. Ibrahim BA, Wang H, Lesicko AM, Bucci B, Paul K, Llano DA (2017) Effect of temperature on FAD and NADH-derived signals and neurometabolic coupling in the mouse auditory and motor cortex. Pflügers Arch 469(12):1631–1649
31. Cruikshank SJ, Rose HJ, Metherate R (2002) Auditory thalamocortical synaptic transmission in vitro. J Neurophysiol 87(1):361–384
32. Stebbings KA, Choi HW, Ravindra A, Caspary DM, Turner JG, Llano DA (2016) Ageing-related changes in GABAergic inhibition in mouse auditory cortex, measured using in vitro flavoprotein autofluorescence imaging. J Physiol 594(1):207–221
33. Optogenetics Material Request/FAQ (n.d.) Optogenetics Resource Center. https://web.stanford.edu/group/dlab/optogenetics/request_dna.html
34. Davis HE, Morgan JR, Yarmush ML (2002) Polybrene increases retrovirus gene transfer efficiency by enhancing receptor-independent virus adsorption on target cell membranes. Biophys Chem 97(2):159–172

Chapter 3

Guide to Transcranial Imaging of Sound-Evoked Activity in the Auditory Cortex of GCaMP6s Mice In Vivo

Georgiy Yudintsev, Christopher M. Lee, Alexander R. Asilador, and Daniel A. Llano

Abstract

Transcranial in vivo imaging of rodent cortical activity during sensory events is a fast, minimally invasive and reliable way to measure the nature and location of neuronal responses, allowing the creation of topographic maps and assessment of how the maps change over different time scales. In this chapter, we describe a straightforward and robust method of transcranial in vivo imaging of auditory cortical responses after sound stimulation in C57BL6J GCaMP6s transgenic mice. We aim to describe to the reader some theoretical background of this imaging method and, most importantly, practical guidelines on how to apply it in their own laboratory setting, including the necessary equipment along with common qualitative and quantitative methods of analysis. Although this chapter specifically focuses on the use of C57BL6J-Tg(Thy1-GCaMP6s)GP4.3 transgenic mouse line in auditory research, the method can be further modified for visual, motor, and somatosensory research, as well as the areas of clinical and translational neuroscience for studying mouse models of neurological diseases.

Key words GCaMP6s, GCaMP, In vivo imaging, Mouse, Auditory cortex, Tonotopic organization, Transcranial imaging

1 Introduction

The ability to record and measure neuronal activity in vivo is essential for understanding brain function, as neurons are organized in networks that are necessary for behavioral and cognitive functions [1–3]. Therefore, the development of new methods which allow recording and/or imaging activity from the intact brain is crucial to advancing the field of neuroscience. Particularly, the use of the mouse and the rat as model organisms is instrumental in this process. The neural activity of rodents has been investigated by various in vivo imaging techniques, including optical imaging of the hemodynamic response, flavoprotein autofluorescence, and various types of applied chemical agents, such as voltage-sensitive

Nicholas J. D. Wright (ed.), *Basic Neurobiology Techniques*, Neuromethods, vol. 152,
https://doi.org/10.1007/978-1-4939-9944-6_3,

dyes or calcium indicators [4–10]. However, the search for improved neuroimaging approaches continues as each of these techniques comes with some certain shortcomings that are important to consider in experimental designs. The descriptions of major and commonly used neuroimaging and recording techniques for the study of the rodent brain, with particular emphasis on the auditory cortex, are provided below. Some key experimental and analytical advantages and limitations of these methods are discussed, and, specifically, recent developments and important considerations for genetically encoded calcium imaging are outlined.

The presence of topographic maps is one of the commonly found features of cerebral cortex organization. Indeed, many studies found consistent tonotopic gradients when mapping the primary auditory cortex, as well as topographic maps of other sensory systems in the rodent brain, and the brains of many other mammalian species, including the primate and the human [11–21]. In earlier periods, traditional microelectrode recordings were widely used to describe and study sensory representations by the cerebral cortex. Although precise cortical maps could be obtained with this method, the experiments were rather invasive, which in turn limited the number of manipulations that could be performed—the brain tissue would become damaged due to multiple electrode penetrations required to obtain detailed maps. Often, this approach made it difficult to trace plastic changes over time where the neural activity with regards to topography needed to be measured at several time points, sometimes over the course of a few months.

As an alternative to traditional recordings, the use of intrinsic signal and flavoprotein fluorescence imaging became well established for large-scale brain imaging [4, 22]. In mouse, both imaging modalities can be used transcranially, although the surgical thinning of the skull may be required in some cases, for example, when imaging the cortical activity of adult animals. As such, imaging the neural activity rather than recording it using microelectrodes circumvents the need for most invasive surgical procedures that often cause some damage to the brain tissue. However, although minimally invasive, these methods have two major limitations: a low spatiotemporal resolution and a small signal-to-noise ratio. Such method characteristics pose a set of constraints on the types of experiments that can be conducted, with a higher number of repetitions being required to capture the signal, and more sophisticated methods of image analyses are often required to generate functional maps [23]. Additionally, prolonged exposure to blue light used for flavoprotein fluorescence imaging may temporarily inactivate neurons in the illuminated brain regions [24]. Therefore, in practice, the amount of time that this imaging may be conducted for an individual experiment is limited. Unlike intrinsic signal and flavoprotein autofluorescence imaging, voltage and calcium sensitive indicators have faster temporal responses and sometimes improved signal-to-noise, but these methods often require invasive surgical preparations, in which the photosensitive compounds are loaded into the brain [25–27]. The addition of these dyes also creates a concentration gradient in the live brain tissue due to diffusion, which establishes regional inhomogeneity, consequently making the results more difficult to interpret.

Although the above-listed neuroimaging techniques are all readily available to neuroscientists, calcium imaging relying on genetically encoded indicators is becoming particularly popular [28–32]. This method offers higher spatial resolution and more precise temporal dynamics, and single-cell resolution can be achieved in vivo with multiphoton imaging [33]. Large-scale imaging of global cortical responses can also be performed using this type of imaging. Genetically encoded calcium indicators (GECIs) are specifically suited for in vivo studies of brain activity, as they provide experimenters with the opportunity to image the brain in a noninvasive fashion and over chronic timescales, avoiding highly invasive procedures or side effects of dye loading, yet offering improved temporal resolution and higher signal-to-noise ratio.

Today, there exists a family of GECIs, each with its own temporal dynamics and signal-to-noise ratio [34], among them are GCaMP proteins. These proteins consist of three main parts: a modified green fluorescent protein (cpGFP), calmodulin—a calcium-binding protein [35], and M13 peptide sequence of myosin light chain kinase, which interacts with calmodulin [36]. Such a bioengineered arrangement forms an indicator molecule sensitive to calcium concentration: calcium binding in this protein induces a conformational change leading to increased fluorescence by the GFP component. More recently, a class of GCaMP6 proteins has been developed which has superior temporal dynamics and ultra-sensitivity to calcium ions when compared to previous GCaMPs or calcium dyes [37]. The following year, several transgenic mouse lines expressing slow and fast variants of GCaMP6 were created and used successfully for a number of in vivo applications [38, 35].

In the following sections of this chapter, we describe a method of large-scale, transcranial in vivo imaging in the anesthetized C57BL6J-Tg(Thy1-GCaMP6s)GP4.3DkimJ mice. We particularly offer insight into practical considerations on the use of these transgenic animals for fast generation of auditory cortical maps. The equipment, surgical procedures, optimal imaging parameters and methods of analysis are described in detail. Although our focus is on auditory applications, these procedures can be adjusted for application in other sensory systems or in disease models.

2 Materials

2.1 Macroscope and Imaging Setup

Imager 3001 Integrated data acquisition and analysis system (Optical Imaging Ltd., Israel) was used to image the cortical responses to sound in mice. C57BL6J-Tg(Thy1-GCaMP6s)GP4.3DkimJ mouse line was obtained from Jackson Laboratories. A macroscope consisting of 85 mm f/1.4 and 50 mm f/1.2 Nikon lenses was mounted to an Adimec 1000m high-end CCD camera (7.4 × 7.4 μm pixel size, 1004 × 1004 resolution), and centered above the left auditory cortex, focused approximately 0.5 mm below the surface of the exposed skull. The macroscope system has two main significant advantages over microscope objectives for large-scale imaging of the cortex: it provides a longer working distance between the specimen and the lens and 100–700 fold increase in brightness [39]. Blue excitation (450 nm, 30 nm band-pass), green emission (515 nm, long-pass) filters and a 495 DRLP dichroic mirror were used. Imager 3001 VDAQ software controlled the acquisition and stimulus trigger.

2.2 Acoustic Stimulation

Acoustic stimuli were generated using a TDT system 3 with an RP 2.1 Enhanced Real-Time Processor and delivered via an ES1 free field electrostatic speaker (Tucker-Davis Technologies, FL, USA), located approximately 2 cm away from the contralateral ear. All imaging experiments were conducted in a sound-proof chamber. 500 ms pure tones of 5, 10, 20, and 30 kHz were used, 100% amplitude-modulated at 20 Hz.

2.3 Animal Preparations

All surgical procedures were approved by the Institutional Animal Care and Use Committee at the University of Illinois at Urbana-Champaign, and animals were housed in animal care facilities approved by the American Association for Accreditation of Laboratory Animal Care (AAALAC). Every attempt was made to minimize the number of animals used and to reduce suffering at all stages of the experiments.

C57BL6J-Tg(Thy1-GCaMP6s)GP4.3DkimJ mice were purchased from Jackson Laboratories, bred as heterozygous in the local animal facility. Prior to the experiments, mice were genotyped for the presence of Tg(Thy1-GCaMP) transgene (forward PCR primer CAT CAG TGC AGC AGA GCT TC, reverse PCR primer CAG CGT ATC CAC ATA GCG TA). Before the start of the surgery, mice were anesthetized with a mixture of ketamine and xylazine (100 mg/kg and 3 mg/kg respectively) delivered intraperitoneally using a 27 gauge needle, followed by an intraperitoneal injection of acepromazine (2–3 mg/kg). We found that the addition of acepromazine provides deeper and longer duration anesthesia (*see* **Note 1**). After the animals were anesthetized, proper care was taken to maintain the body temperature within the range of 35.5 and 37 °C, using a DC temperature controller (FHC, ME, USA). See Table 1 for a list of materials.

Table 1
List of imaging and auditory stimulation equipment, mouse strains, anesthesia, surgical tools and supplies

#	Item	Vendor
Imaging equipment		
1.	Imager 3001 data acquisition and analysis system	Optical Imaging, Ltd., Israel
2.	Camera/lens holder	Optical Imaging, Ltd., Israel
3.	Tandem macroscope lens (85 mm f/1.4 and 50 mm f/1.2)	Nikon, Purchased from Optical Imaging, Ltd., Israel
4.	VSD illumination system with light guides	Optical Imaging, Ltd., Israel
5.	Adimec 1000m CCD camera	Adimec, Purchased from Optical Imaging, Ltd., Israel
6.	Blue excitation filter (450 nm, 30 nm bandpass)	Optical Imaging, Ltd., Israel
7.	Green emission filter (515 nm, longpass)	Optical Imaging, Ltd., Israel
8.	495 DRLP dichroic mirror	Optical Imaging, Ltd., Israel
Auditory stimulation system		
9.	PA5 programmable attenuator	Tucker-Davis Technologies, FL, USA
10.	RP 2.1 real-time processor	Tucker-Davis Technologies, FL, USA
11.	ED1 electrostatic speaker driver	Tucker-Davis Technologies, FL, USA
12.	ES1 free electrostatic speaker	Tucker-Davis Technologies, FL, USA
Mouse strain		
13.	C57BL6J-Tg(Thy1-GCaMP6s)GP4.3DkimJ	Jackson Laboratories, ME, USA
Anesthesia		
14.	Ketamine (Ketaset)	Zoetis
15.	Xylazine (AnaSed)	Lloyd Laboratories
16.	Acepromazine (PromAce)	Boehringer Ingelheim
Surgical tools, equipment and supplies		
17.	DC temperature controller	FHC, ME, USA
18.	Standard pattern forceps	Fine Science Tools, CA, USA
19.	Vannas scissors (5 mm cutting edge)	Fine Science Tools, CA, USA
20.	Scalpel handle #3	Fine Science Tools, CA, USA
21.	Scalpel blades #15	Fine Science Tools, CA, USA
22.	Dumont #5 forceps—standard tips/straight/ Dumoxel/11 cm	Fine Science Tools, CA, USA
23.	Crile hemostat—serrated/straight/14 cm	Fine Science Tools, CA, USA
24.	Mineral oil	Bio-Rad Laboratories, Inc., USA

(continued)

Table 1 (continued)

#	Item	Vendor
25.	Cotton tipped applicators	Covidien, USA
26.	Povidone-iodine Swabstick	Dynarex, NY, USA
27.	Sterile latex surgical gloves	Dynarex, NY, USA
28.	Stereotaxic apparatus with head-holder and non-puncture ear bars for the mouse	Kopf Instruments, CA, USA
29.	Model 1474 High Speed Stereotaxic Drill	Kopf Instruments, CA, USA
30.	Small grooming clippers	Generic brand
31.	Sterile saline	
32.	1 cc syringes	BD, NJ, USA
33.	27 gauge syringe needles	BD, NJ, USA
34.	Light metal bolt	Custom-made
35.	Stick head holder	Custom-made
36.	Dental cement	3 M ESPE KETAC
37.	Razor blade (one-sided)	
38.	Micromanipulator	Custom-made from separately purchased Kopf Instruments parts
39.	Monofilament nylon nonabsorbable suture, 13 mm, 3/8, cutting	CP Medical, GA, USA
40.	Triple antibiotic ointment	Sheffield Pharmaceuticals

3 Methods

3.1 Surgical Preparation

Upon the induction of deep anesthesia, as indicated by the lack of a response to a hind paw pinch, the mouse was placed in a stereotaxic apparatus (Kopf Instruments, CA, USA) (Fig. 1a).

1. The fur of the dorsal and lateral (left) parts of the scalp was trimmed with small animal clippers (Fig. 1b), and the skin was first cleaned with ethanol to remove any remaining hairs, as they may get into the surgical site and obstruct the field of view for the imaging.
2. Then, povidone-iodine solution was applied to the skin (Fig. 1c), and mid-sagittal and mid-lateral incisions were made to expose the dorsal and lateral aspects of the skull along with the temporalis muscle (Fig. 1d, e).

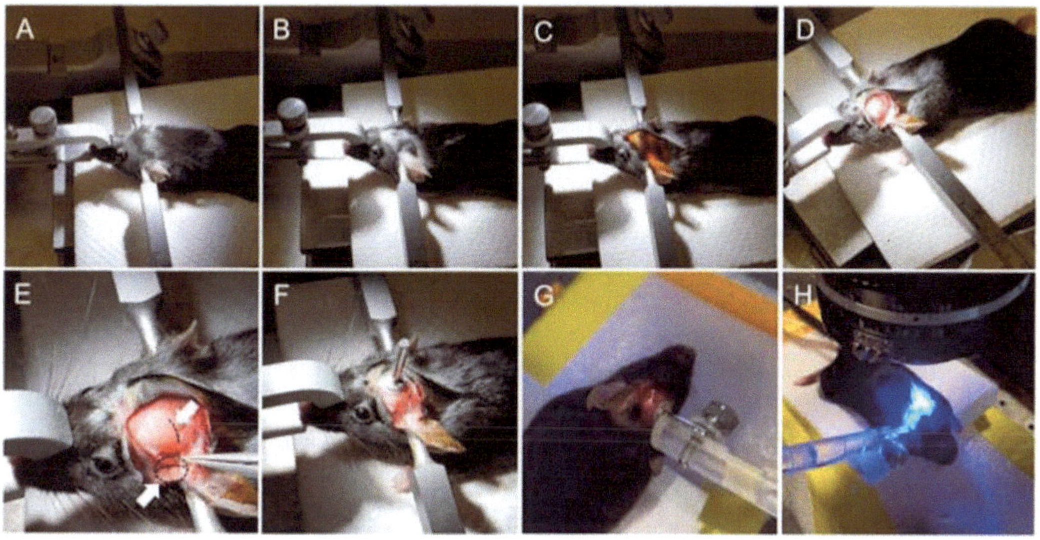

Fig. 1 Step-by-step description of the surgical procedure for head bolt placement and preparation of the site for the auditory cortex imaging. (**a**) The mouse is positioned in a stereotaxic and on a heating pad covered by a paper towel, and secured using non-puncture ear bars placed into the ear canal openings. (**b**) The fur on the scalp is trimmed, and povidone iodine solution is applied to the area (**c**). (**d**) A mid-sagittal incision is made, followed by another dorsoventral incision rostral to the left ear, to expose the skull and left temporalis muscle. (**e**) The top white arrow in points to lambda suture, and the bottom white arrow points to the exposed imaging site along with deflected temporalis muscle. (**f**) A light metal bolt is attached to the top of the skull with dental cement. (**g**) Once the cement solidifies, and the head bolt is secured, a custom-made head holder is attached to the bolt, and the animal is brought to the imaging chamber and placed on a covered heating pad under the macroscope objective (**h**)

3. The temporalis muscle was separated from the skull by an incision made along its insertion site and deflected to expose the ventral parts of the underlying auditory cortex (Fig. 1e). Cutting off the muscle completely is not recommended, as even partial removal of the muscle usually leads to excessive bleeding which may cover the site of the imaging. Rather, deflecting the muscle to the side is sufficient (*see* **Note 2**).
4. The site was cleaned with sterile saline. At this point, it is imperative to slowly thin the skull surface over the auditory cortex with a drill (*see* **Note 3**). The former insertion site of the muscle temporalis is evened out, and the dorsal aspects of the skull over the auditory cortex are thinned. The ventral and lateral portion of the skull, which was previously covered by the muscle, is significantly thinner than its dorsomedial aspect, therefore less skull thinning, if any, is required in this area. We found that the GCaMP6 signal is strong enough to easily pass through even the unthinned parts of the mouse skull. However, removing some curvature and surface inhomogeneities of the bone covering the auditory cortex helps to create a cleaner

site for improved imaging. If the procedure is performed as described, there should be no bleeding in the site.

5. Next, we attached a light metal bolt to the dorsal portion of the skull (Fig. 1f). The deep fascia covering the dorsal portion of the skull was cut, and the bone was slightly shaved down (*see* **Note 4**), rinsed with sterile saline and dried with a sterile cotton swab. This procedure helps to increase the surface area for stronger bolt attachment.
6. A small amount of dental cement (3 M ESPE KETAC) was mixed to a medium level of viscosity, and added to the head of the bolt just enough to cover it. It is important to make sure that the skull surface is clean and dry.
7. The bolt was bonded to the top of the skull, and the dental cement was allowed to set. This process usually requires approximately 10 min.
8. Once the bolt was securely in place, a holding pole was attached to the bolt, and the mouse was moved to the imaging chamber (Fig. 1g, h).

3.2 Imaging

Following the surgery and the bolt placement, the mouse was moved to the imaging chamber and placed on a heating pad under the macroscope, with the holding pole attached to the head bolt and secured in a micromanipulator. The core body temperature was monitored by a rectal thermometer and maintained approximately at 36 °C by a temperature controller and a heating pad. The state of the anesthesia was periodically monitored by testing a toe pinch reflex every 15–30 min, and supplemental doses of ketamine and xylazine mixture were delivered intraperitoneally as needed. For survival experiments, extending the imaging sessions for longer than 2.5–3 h is not suggested, as the chances of mortality increase following this time frame (*see* **Note 5**). Periodic, hourly intraperitoneal injections of sterile saline (0.3–0.35 mL for a 20–24 g mouse) also increase the survival rate and recovery of the animal following the surgery (*see* **Note 6**).

The mouse auditory cortical areas are located in the temporal cortex, usually between two temporal branches of the rhinal vein and the middle cerebral artery [40, 41]. For optimal localization of the auditory areas under the macroscope for the initial imaging session, it is typically best to position the mouse's head at a 45° angle to the horizontal plane (*see* **Note 7**). A small amount of sterile saline or mineral oil was applied to the skull over the temporal cortex, and blue light was used to find the blood vessels supplying the area (Fig. 2a) (*see* **Note 8**). To first locate auditory-evoked activity in the cortex, either a broadband white noise stimulus or a pure tone at 55–60 dB SPL was used. The optimal duration of the sound stimulus was between 200 ms and 500 ms, arriving at 5 s

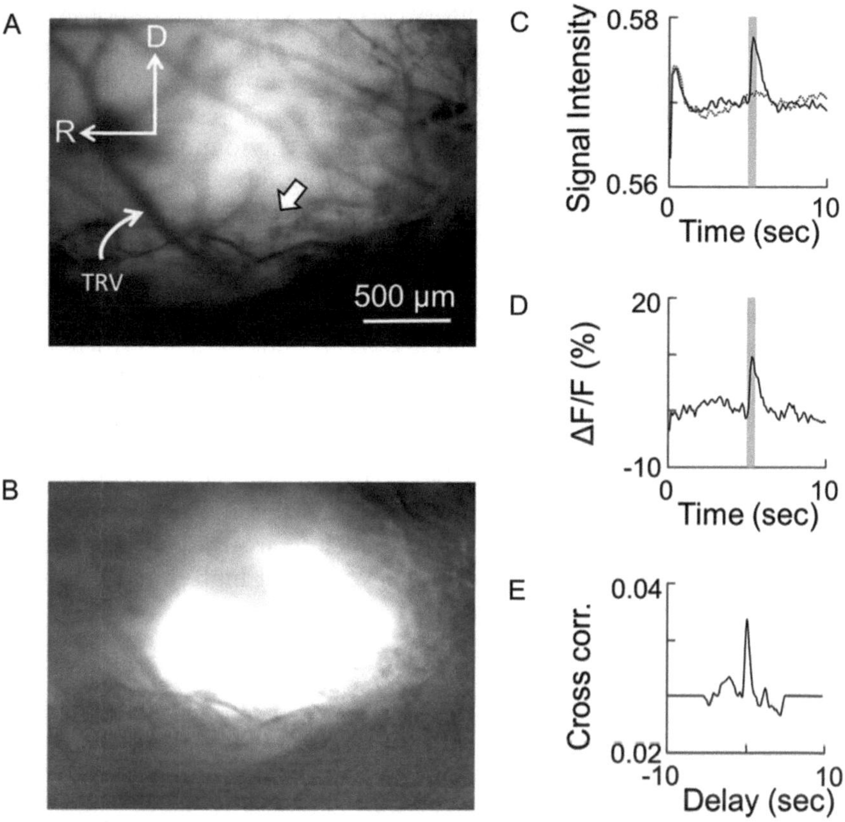

Fig. 2 Analysis of fluorescence imaging signals. (**a**) The temporal cortex was imaged with a CCD camera at 10 frames per second, in 10-s trials. Curved arrow marks a temporal branch of the rhinal vein (TRV), used as a landmark to locate the auditory cortex. We illustrate the fluorescence signal recorded from a single pixel, indicated by the white box arrow. (**b**) The imaging of responses to broadband white noise reveals the location of auditory cortices. (**c**) We plot the response during sound presentation trials (500 ms SAM tone, timing indicated by gray bar) averaged across 10 trials (solid line). A baseline signal was established from 10 silent trials (dotted line). (**d**) The $\Delta F/F$ signal yields a peak that follows the sound presentation timing in regions that are driven by sound stimuli. (**e**) The cross-correlation of the $\Delta F/F$ signal with the sound envelope has a prominent peak with a small positive delay (100–500 ms) in auditory cortices

after the onset of imaging, followed by another 5 s of acquisition. The frames were collected at 10 frames/second, equating to 100 frames total per one stimulus presentation. Reliable imaging of sound-driven activity requires averaging responses across multiple trials, due to inherent variability between auditory cortical responses to individual stimulus presentations. However, we found that a smaller number of trials were required for obtaining reliable maps when compared to flavoprotein autofluorescence imaging. Generally, the averaging of 10 trials was sufficient to map the cortical responses. The sound stimulus activated areas of the auditory cortex, but did not activate nearby visual and somatosensory cortices.

Once the auditory cortex was localized, responsiveness to pure tones or other stimuli could be mapped. Regions that were responsive to sound typically produced a peak in the fluorescence signal that began within a few hundred milliseconds of the sound onset, and decayed within a second of the sound offset (Fig. 2b, c). The signal was either quantified using the $\Delta F/F$ metric (Fig. 2d), or by its correlation with the sound envelope (Fig. 2e).

3.3 Delineation of Individual Auditory Cortical Areas Using Neural Responses to Pure Tones

Four areas of the mouse auditory cortex could be reliably delineated using C57BL6J-Tg(Thy1-GCaMP6s)GP4.3DkimJ transgenic mouse line (Figs. 3a, b and 7). The presentation of pure tones whose frequencies fell under 20 kHz would lead to activation of three main auditory areas: the primary auditory cortex (A1), the anterior auditory field (AAF) and the secondary auditory cortex (A2). A1 and AAF both have a tonotopic gradient and receive their primary input from the ventral division of the medial geniculate body (MGBv). These gradients are roughly arranged as mirror images of each other, converging at high sound frequencies, while

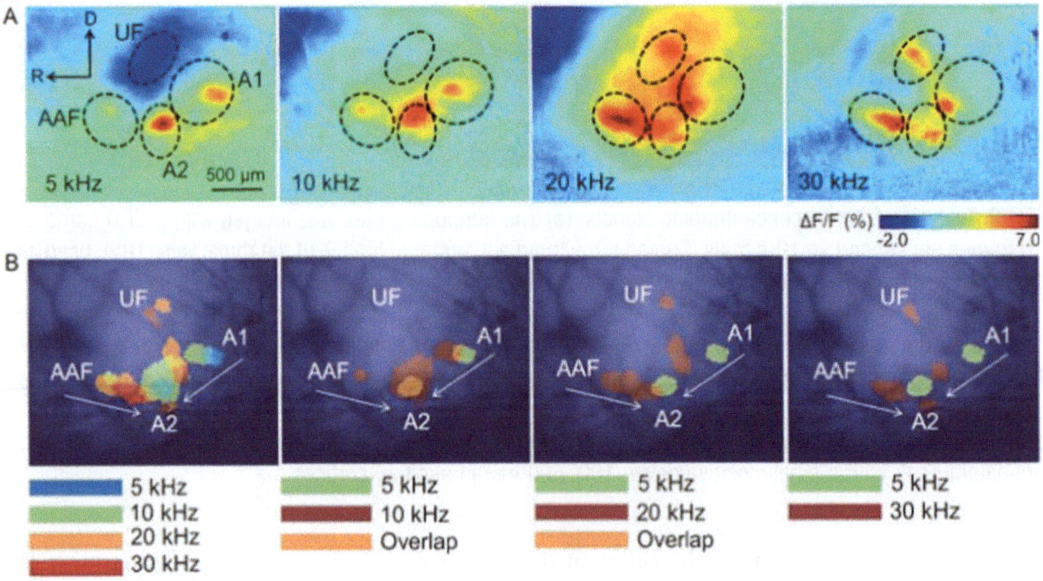

Fig. 3 Pure tone evoked activity in the auditory cortex of Thy1-GCaMP6s mouse. (**a**) In separate sessions, 500 ms sinusoidally 100% amplitude modulated (SAM) pure tones of 5, 10, 20, and 30 kHz were each presented for 10 times. The responses were analyzed using a custom-written software in MatLab for $\Delta F/F$ analysis. Additionally, prior to each single pure tone stimulus presentation, baseline activity was collected for 10 s, and later subtracted from the baseline activity of stimulus runs. AAF—anterior auditory field. UF—ultrasonic field. A1—primary auditory cortex. A2—secondary auditory cortex. (**b**) Four auditory fields mapped onto the surface of the mouse skull as thresholded responses from (**a**). AAF is revealed by its low-to-high rostral-to-caudal tonotopic gradient. A1 is identified by its low-to-high caudal-to-rostral tonotopic organization. A2 lies ventral to the high-frequency regions of A1 and AAF, and is activated by all frequencies. UF is only activated by the frequencies of the ultrasonic range, and is located dorsally with respect to AAF, A2, and A1. Scale bar = 500 μm, dorsal (D) is up, rostral (R) is left

the activations in response to lower sound frequencies are found more rostrally in the AAF and more caudally in A1. A2 is located ventrally relative to where A1 and AAF converge at high frequencies, and is activated by all frequencies we used. When pure tones with frequencies of 20 kHz and above are presented, in addition to A1, AAF and A2, the ultrasonic field (UF) should be activated. This activation lies dorsally to A1 and AAF.

$\Delta F/F$ of responses to pure tones (5, 10, 20, and 30 kHz) were obtained for multiple auditory cortical regions (Fig. 3a). The topography of the frequency sensitivity is reconstructed in Fig. 3b. Here, regions sensitive to 5, 10, 20, and 30 kHz are delineated by thresholding the $\Delta F/F$ images to keep the peaks of the signal. Alternatively, topography of frequency sensitivity can also be reconstructed by identifying the stimulus frequency yielding the largest signal at each pixel (Fig. 7).

To optimize the imaging parameters, we explored the effects of sound pressure level and the number of stimulus presentations on the kinds of cortical maps that can be obtained in the GCaMP6s transgenic mice. Responses to individual presentations of pure tones were found to be quite variable, from robust activations to complete absence of responses. Therefore, we sought out the optimal number of repeated sound stimuli, which would be required for effective mapping of the tonotopically and non-tonotopically organized auditory subfields. Averaging of the first 10 presentations of the stimulus generally resulted in reliable maps (Fig. 3a, b). Occasionally, accurate maps of sound-evoked activity ensued from the averaging of the first five stimulus repetitions (Fig. 4a, b). These maps changed little with increased amount of averaging, with virtually no further improvement beyond 25 repetitions. A relatively small number of stimulus repetitions required for $\Delta F/F$ analysis is a considerable advantage of GCaMP6s mice for in vivo applications. Cortical responses become progressively more variable after 2–2.5 h of imaging, and delineation of auditory fields may not be achieved at this point. Because we did not observe significant decreases in GCaMP6 fluorescence itself, it is possible that prolonged exposure of the cortex to light suppressed cortical activity, changing its dynamics and responsiveness to sensory events. Mechanistically, this phenomenon could happen via blue light's interference with mitochondrial flavoproteins, which are critical for energy metabolism and ATP generation [24]. Photo-bleaching of flavoproteins may lead to suppression of neuronal activity (*see* **Note 9**).

The intensity of the GCaMP6 fluorescent signal is proportional to sound pressure level (Figs. 4 and 5). With increasing sound amplitude, the GCaMP6 signal increases at different rates in A1 and AAF, generally with smaller and more diffuse activations in the AAF and larger activations in A1 (Fig. 5b). For example, in Fig. 4a, responses to a 100% sinusoidal amplitude modulated 5 kHz pure

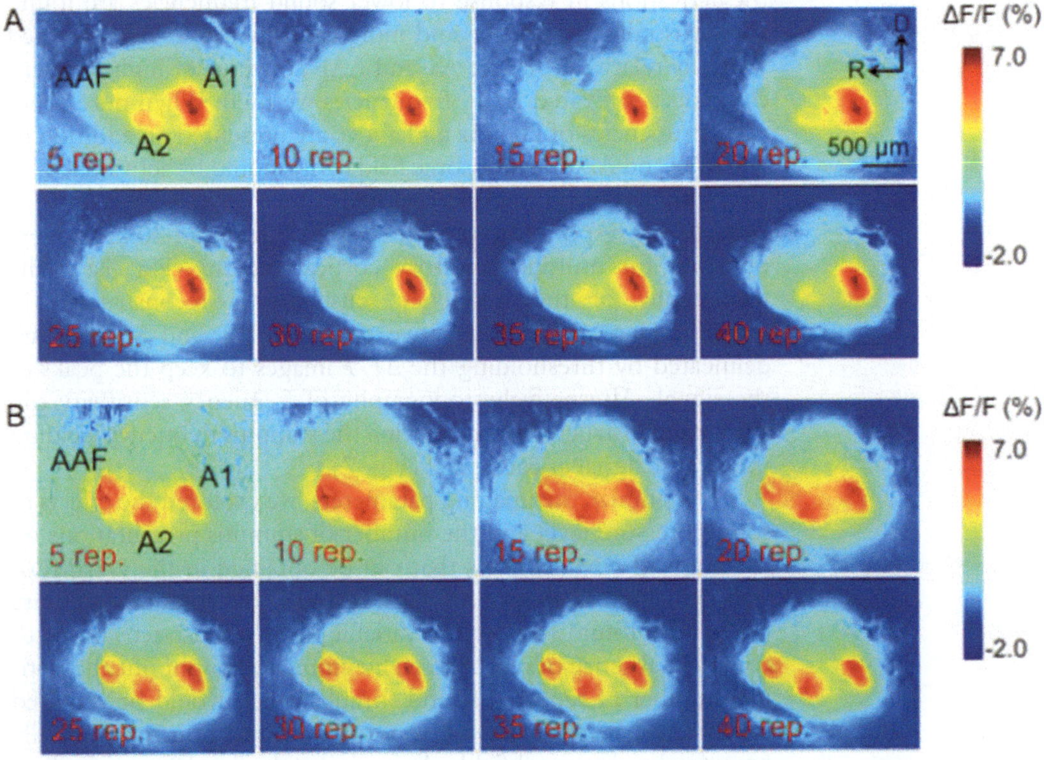

Fig. 4 The effects of the number of stimulus repetitions and sound level on frequency-specific maps. (**a**) Responses to a 5 kHz SAM pure tone at 37 dB SPL. The sound stimulus was played every 10 s, and the average responses to 5, 10, 15, 20, 25, 30, 35 and 40 presentations of the same stimulus were computed using a custom-written software in MATLAB. Primary auditory cortex (A1) is robustly activated, and its area of activation is distinct even after averaging the responses to the first five stimulus presentations. However, the activity in the anterior auditory field (AAF) and secondary auditory cortex (A2) is less powerful at this amplitude. Increasing the number of stimulus presentations with subsequent averaging of these responses, make A2 and AAF slightly more defined. Compare 10 repetitions with 40 repetitions. (**b**) Responses to a 5 kHz SAM pure tone at 55 dB SPL. Increasing the tone's amplitude further above the hearing threshold level led to more robust activation of A1, AAF, and A2. Generally, the optimal number of stimulus presentations for time-effective mapping is between 5 and 10 repetitions. Although the sharpness of responses was improved with the number of averaged responses (compare 5 repetitions window with 30 repetitions), the changes were minor, and the first 5–10 responses provided a rather clear frequency-specific representation in cortical fields

tone in AAF and A2 were not as conspicuous at 37 dB SPL for this particular animal. However, increasing the amplitude to 55 dB SPL revealed more complete responses, and led to full activations of A1, AAF and A2 (Fig. 4b). Playing sounds to the anesthetized mouse even at higher levels generally leads to very robust activation of all auditory fields activated by that sound, and in that case delineation between the fields becomes difficult using *ΔF/F*-type analysis. Therefore, it is important to consider mouse hearing thresholds for each frequency at the beginning of the experiment. Presenting tones near or slightly above this threshold generally provides the

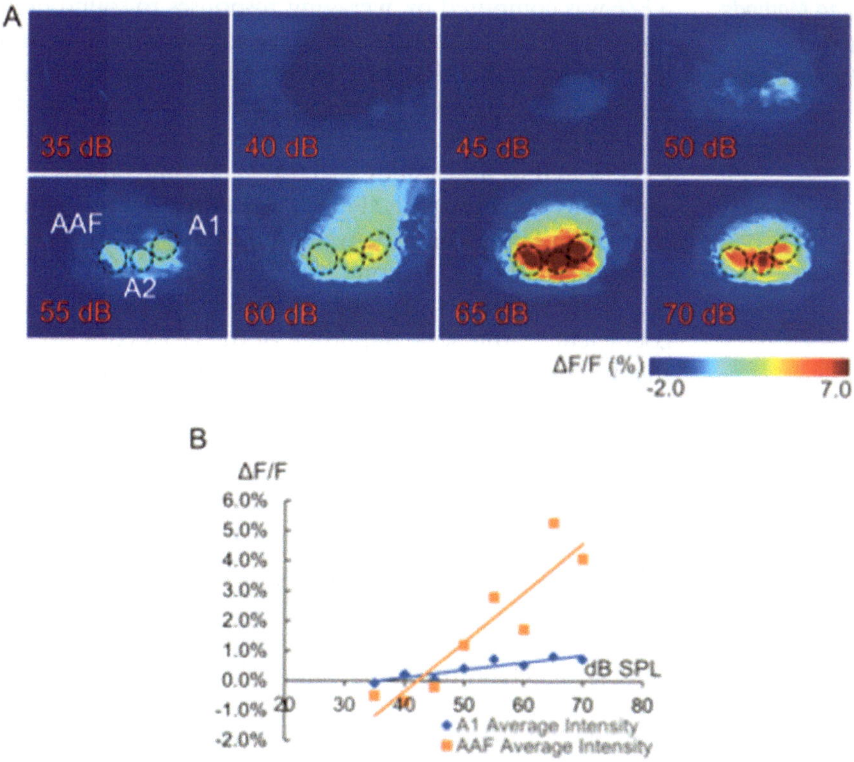

Fig. 5 The frequency-specificity of GCaMP6 signal decreases with increasing sound pressure level. (**a**) The average response to ten 5 kHz SAM tone presentations depicted as a function of increasing sound pressure level, starting at 35 dB SPL and going up to 70 dB SPL. In this particular example, the stimulus-evoked activity in the primary auditory cortex, A1, became apparent at 45–50 dB SPL. All three areas which had been shown to respond to low-frequency tones—anterior auditory field (AAF), primary (A1) and secondary (A2) auditory cortices—display stimulus-evoked activation at 55 dB SPL. Importantly, frequency-specific regions of A1, AAF and A2 could be discerned. Further increases in the sound amplitude were usually followed by the spread of activation outside of the frequency-specific regions. This phenomenon was observed at 65 dB SPL. (**b**) A plot of the average ΔF/F signal (%) as a function of sound pressure level for a 5 kHz tone. On average, responses in A1 were stronger than in AAF

experimenter with fine auditory cortical maps after averaging the activity from 10 repetitions of the same sound stimulus. Biological variability and the state of anesthesia also affect the quality of the maps, and some of this variability can be reduced by closely monitoring the anesthesia and administering small supplemental doses of the anesthetic (100 mg/kg/3 mg/kg of ketamine and xylazine, respectively) when the animal's breathing rate increases by nearly two times from the deeply anesthetized state's breathing rate. It is also important to monitor the temperature of the animal and keep it within the range of 35–37 °C.

3.4 Applied Methods of Quantitative and Qualitative Image Analysis

$\Delta F/F$ was computed by averaging responses to sound presentation across 10 trials to capture sound-evoked activity. The mean signal level from 0 to 5 s (frames 1 to 50) was subtracted from the mean signal level from 5 to 6 s (frames 51 to 60) to determine the change in fluorescence. This change was normalized to the mean signal level from 0 to 5 s (frames 1 to 50). The same procedure was then applied to 10 silent trials. Finally, the difference in the normalized changes in fluorescence between the sound presentation trials and silent trials was used to compute the $\Delta F/F$.

As alternatives to the $\Delta F/F$ metric, the Pearson's correlation coefficient or the peak amplitude of the cross-correlation between the fluorescence signal and the stimulus amplitude envelope can be used to quantify and image responses (Fig. 2e). One of the advantages of the cross-correlation metric is that it does not depend on an analysis time window to estimate the timing of the cortical response. Imaging based on either the *Pearson's correlation* coefficient and the cross-correlation peak amplitude could localize regions sensitive to individual frequencies (Fig. 6a, b).

First, the correlation between the stimulus and calcium response was quantified by computing the Pearson's correlation coefficient. Here, the stimulus signal was simplified as a rectangle function from $t = 5.1$ to $t = 5.5$ s, and the response signal was estimated as the difference between the averaged response across 10 silent trials subtracted from the averaged response across

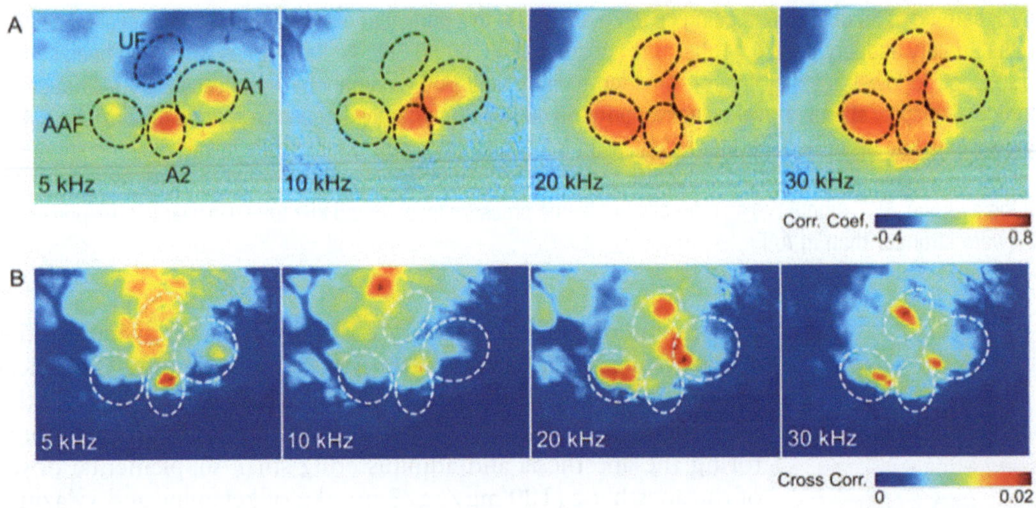

Fig. 6 Imaging of sound evoked responses. (**a**) Responses to four SAM tones were mapped based on the Pearson's correlation coefficient between the fluorescence signal and the sound amplitude envelope. Localization of responses can be seen in A1, A2, and AAF at all tested frequencies, and can be seen in UF at 20 and 30 kHz. Overall, these responses appeared to be frequency-specific, similar to the maps obtained with the spatial and temporal averaging method. (**b**) Mapping of cross correlation peak amplitude of SAM tone responses

10 sound presentation trials. Thus, regions that increase in the response signal between $t = 5.0$ and $t = 5.1$ and decrease in the response signal between $t = 5.5$ to $t = 5.6$ yielded a positive correlation, and regions with larger magnitudes of these changes yielded larger correlation coefficients (illustrated in Fig. 6a).

The main shortcoming of the Pearson's correlation coefficient is that it captures only the covariance of simultaneous fluctuations in the stimulus and the imaging signal. However, cortical GCaMP6s fluorescence responses may follow acoustic stimuli on the order of hundreds of milliseconds [36, 37]. Thus, to characterize the similarity between the stimulus and imaging signals while accounting for delays between the two signals, the cross-correlation of the stimulus and response signals was computed.

The cross-correlation measures the correlation between two signals as a function of time delay between the signals. As it is illustrated in Fig. 2e, the cross-correlation between the stimulus and the example response in Fig. 2. At a time delay of zero, the cross-correlation was the dot product of the stimulus and response signals and was directly proportional to the Pearson's correlation coefficient. Next, the cross-correlation was recomputed as the dot product between the stimulus and a time-delayed version of the response signal, for each possible delay. For the example response signal in Fig. 2, the cross-correlation yielded a single prominent peak that reaches a maximum value with a delay of 300 ms. Across our imaged auditory-responsive regions, the cross-correlation was usually maximal with a delay of 100–500 ms.

Tonotopic maps were reconstructed from the Pearson's correlation coefficient and cross correlation of responses to 5, 10, 20, and 30 kHz pure tones (Fig. 7). Notably, these maps are comparable to the tonotopic map reconstructed from the $\Delta F/F$ metric. In A1, a low to high frequency gradient is oriented along the dorsal-caudal to ventral-rostral axis. In AAF a low to high frequency is oriented along the dorsal-rostral to ventral-caudal axis. The gradients converge and reverse near A2, which is consistent with the tonotopy reconstructed in Fig. 3b.

3.5 Biological Variability of Individual Responses

It was noted that significant variability of global cortical responses to acoustic stimuli was present. Such variability could not be related to temperature changes, as the temperature was kept at a normal physiological level and monitored throughout the experiment. One possibility is that responses were more robust and representative of a given stimulus at the beginning of trials, and became progressively more variable as the trials continued. Another explanation could lie in different sound levels of the presented stimulus, with more variability of cortical responses at lower sound levels and more constant responses to louder sounds. To look at these two possibilities, rather than averaging the $\Delta F/F$ metric across trials, it was calculated for individual stimulus presentations (Figs. 8 and 9). The

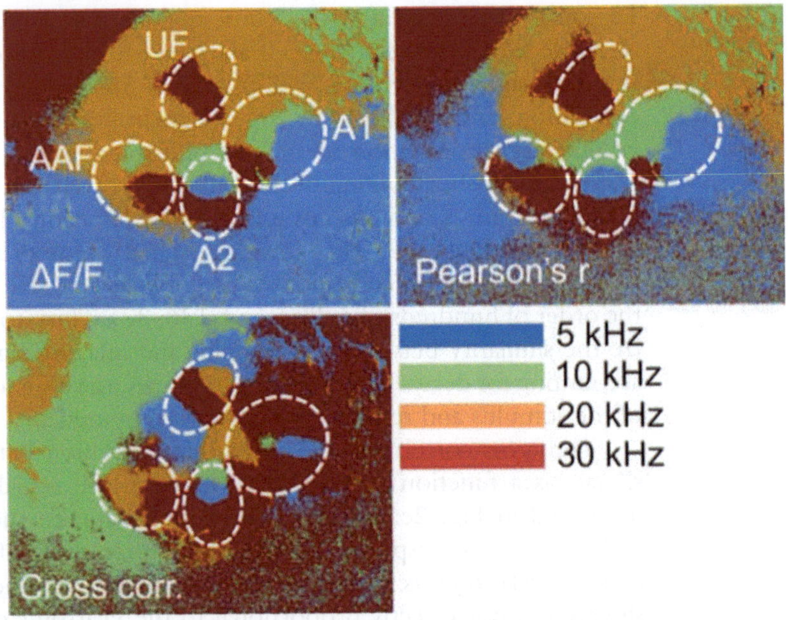

Fig. 7 Tonotopic organization of auditory cortices. Maps of best frequency were constructed based on the $\Delta F/F$ (top left), the *Pearson's correlation* coefficient (top right), or the cross-correlation peak amplitude (bottom left). For each pixel, the frequency eliciting the largest magnitude of the metric is represented by color. Using this approach quickly produces a topographic map of the entire imaged area. In this example, the orientations of the topographic gradients are relatively preserved when using any of the three metrics in A1, A2, AAF, and UF

same data set which was used to generate Fig. 4a and b was also used for this analysis. Each map is the $\Delta F/F$ response to individual stimulus presentation, with the total time window being 10 s. Although the responses appeared stochastic across 40 consecutive presentations of the stimulus, clear tonotopic maps became evident upon averaging across trials (*see* Fig. 4). Similar levels of variability were observed at near-threshold levels (37 dB SPL, 5 kHz SAM pure-tone, Fig. 8) as well as levels above threshold (55 dB SPL, 5 kHz SAM pure-tone, Fig. 9). This variability was unlikely the results of changes in the level of anesthesia, as the time between the start and end of each run was equal to 400 s, which was well below the duration of action of ketamine and xylazine anesthesia.

4 Conclusions

Transcranial in vivo imaging of acoustically evoked responses in the auditory cortex of GCaMP6s mice is a fast and robust method for visualizing the organization of receptive fields. The GCaMP6s fluorescence signal allows auditory evoked responses to be captured efficiently, requiring fewer repetitions when compared to other established modes of in vivo imaging of brain activity, such as

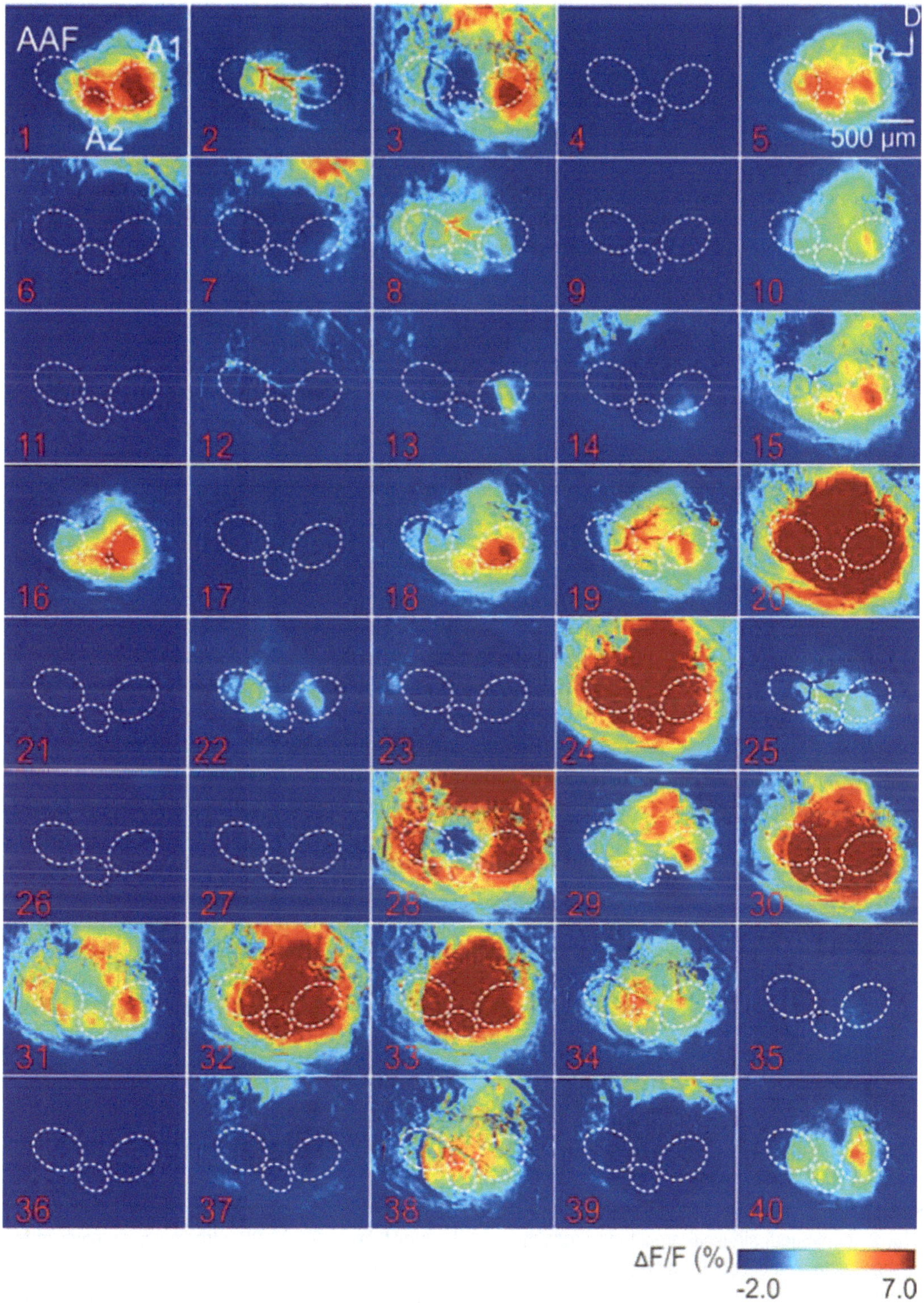

Fig. 8 Variability of the magnitude of individual responses at near-threshold sound pressure level. The same imaging data set that was used to generate Fig. 4a, was subdivided into individual response analysis windows. Each map (1–40) represents the magnitude and location of GCaMP6s cortical response to a 500 ms 100%

intrinsic signal or flavoprotein autofluorescence. The surgical procedures required for this type of imaging are fast and minimally invasive, giving the experimenter an opportunity to conduct multiple experiments on the same animal, for example, in the assessment of the effects of a pharmacological treatment or other types of therapeutic interventions. Additionally, because the imaging experiments can be significantly shorter in order to obtain high-quality cortical maps, the animal survival rate 100% can be achieved without compromising the amount of data collection, which is of particular value when more than one experiment needs to be conducted in one animal. Finally, although there is significant intrinsic variability of responses among individual trials, the fluorescence signal can be easily quantified with the traditional $\Delta F/F$ metric, or alternative metrics, such as correlation-based measures, to produce detailed response maps.

5 Notes

1. It is imperative to carefully select the type of anesthesia when performing in vivo imaging, as various types of general anesthesia may differentially affect the neural activity. For example, it has been reported that isoflurane—a commonly used type of gas anesthesia for many in vivo applications—completely abolished calcium responses in GCaMP mice [29]. For shorter imaging experiments (45 min–1 h), intraperitoneal administration of a mixture of ketamine and xylazine is generally sufficient, although a supplemental dose could be required. However, when longer experiments are needed, we found that addition of acepromazine prolonged the duration of the anesthetized state in GCaMP6s mice without noticeably affecting the quality of calcium signals in response to sound stimuli. The administration of acepromazine also decreased the number of times supplemental doses of ketamine and xylazine were needed, together reducing the mortality rate of GCaMP6s mice during the experiments, with virtually all mice successfully tolerating the longer (sometimes more than 3 h) imaging protocol.

Fig. 8 (continued) sinusoid amplitude-modulated 5 kHz pure tone at 37 dB SPL. Interestingly, each of the responses to a single stimulus presentation was quite variable and random. For some stimulus presentations, such as in 4 and 9, there was no evoked neural activity observed. For other stimulus presentations, frequency-specific areas in A1, AAF, and A2 could be delineated, such as in window 22. At other stimulus presentations, very strong and nonspecific responses were observed, for example in windows 20, 24, and 30. The presence of tonotopy became clear when these data were averaged together (refer to Fig. 4a)

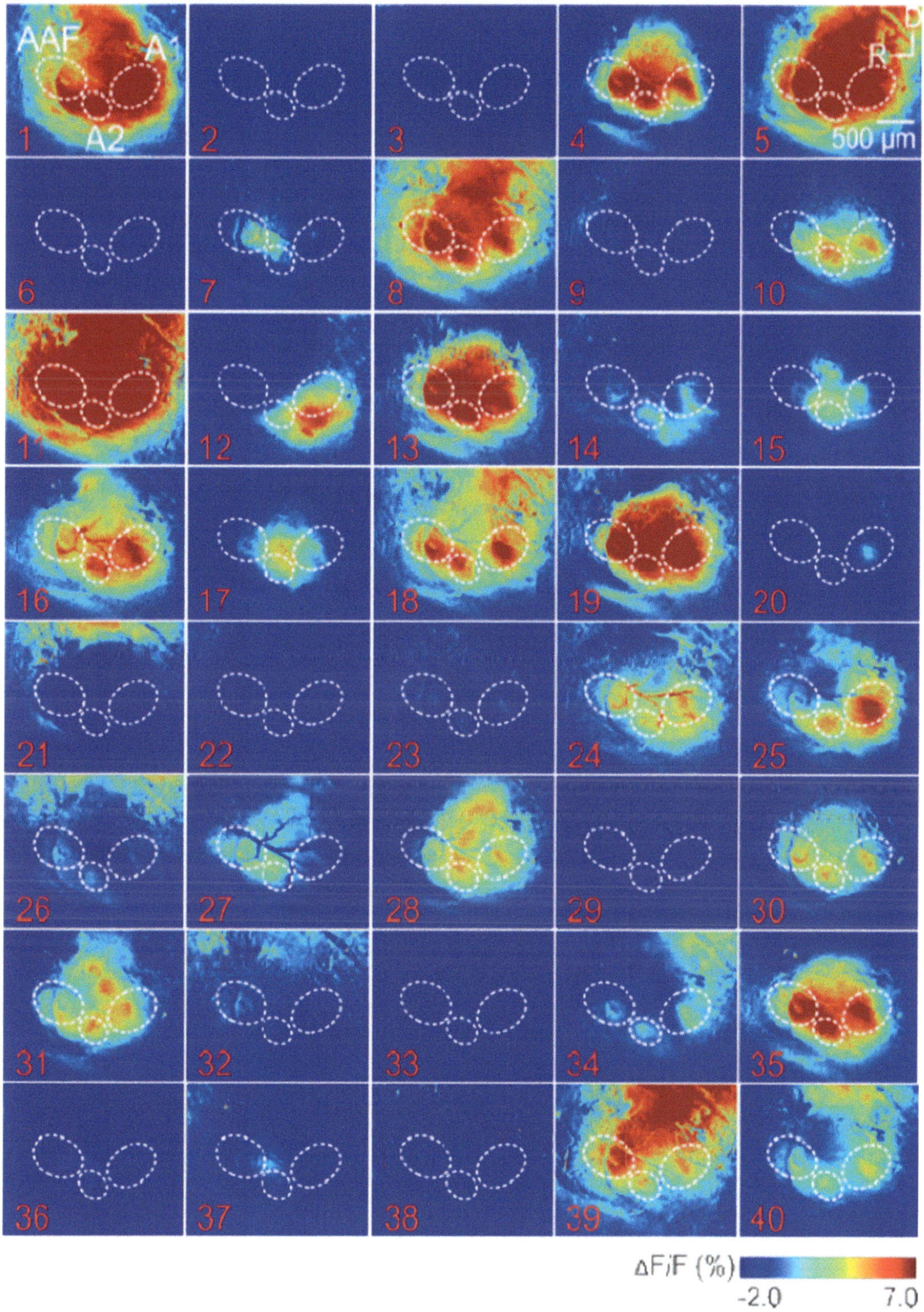

Fig. 9 Variability of the magnitude of individual responses at above-threshold sound pressure level. The same imaging data set that was used to generate Fig. 4b, was subdivided into individual response analysis windows. Each map (1–40) represents the magnitude and location of GCaMP6s cortical response to a 500 ms 100%

2. The temporalis muscle inserts into the portion of the mouse skull that protects the ventral aspects of the auditory cortex. Therefore, the muscle is either cut along its insertion site or completely removed. We suggest that removing large parts of the muscle is not necessary: it often leads to excessive bleeding, which may obscure the imaging field, or even lead to the death of the animal when major arteries supplying the muscle are cut. Simply separating the muscle from the insertion site and deflecting it further down to the ventral side causes minimal bleeding, if any, and is sufficient to uncover the ventral aspects of the auditory cortex for imaging.
3. Unlike the somatosensory and visual cortices, the mouse auditory cortex is protected by the skull with more pronounced curvature. Moreover, additional irregularities of the skull's surface are introduced by the insertion site of the temporalis muscle. These surface inhomogeneities may increase light scattering and thus perturb the quality of imaging. Improved imaging of auditory cortical activity can be attained by flattening out the surface of the overlying skull. This can be done by shaving down and thinning the skull with a burr drill. This process needs to be done slowly, with repeated saline washes, to reduce overheating of the skull and the underlying cortical areas.
4. We found that without shaving down the surface of the skull for bolt attachment, the bolt often detaches at some point during the experiment. Even animal movement due to breathing may cause the bolt dismount. As described above in section 3.1.5, the surface of the skulls is shaved down, rinsed with sterile saline and dried out before the placement of the bolt.
5. One of the main advantages of transcranial in vivo imaging, as introduced earlier, is the possibility of conducting a longitudinal study, or a pre- and post-intervention imaging experiment in the same animal. If such experiments are planned, we suggest not to exceed the imaging sessions beyond 3 h, which may increase the likelihood of mortality over the subsequent 24 h.
6. The animals may experience significant dehydration due to the loss of water with excretions as well as evaporation from the surgical site. This water loss decreases the survival rate of the animals, especially when the imaging experiments are prolonged. Additional periodic intraperitoneal injections of sterile

Fig. 9 (continued) sinusoid amplitude-modulated 5 kHz pure tone at 55 dB SPL. A similar phenomenon of random activations in the auditory cortex was observed above the hearing threshold level. Although the individual responses to sound were quite variable, when the data were averaged together in spatial and temporal domain, clear tonotopic organization became apparent (compare to Fig. 4b) saline (0.30–0.35 mL) improve the survival rate of the animals and their overall recovery following the experiment. We suggest delivering at least one such injection of sterile saline at the midpoint of the imaging session. More injections may be beneficial for the animals undergoing experiments that are more than 2.5–3 h in duration.

7. The angle of rotation of the mouse's head with respect to the horizontal axis is another source of variability for the obtained auditory maps, and may need to be adjusted from animal to animal. In general, a 45° angle serves as a reasonable starting point, and may need to be adjusted for each animal, and depending on which portions of the auditory cortex need to be captured the best. For example, for clearer imaging of the UF, which is located more dorsally, the angle would need to be decreased. However, if it is desired to focus more on the ventral aspects of the auditory cortex, such as secondary auditory area A2, the experimenter may benefit from slightly increasing the angle.
8. It is suggested to periodically reapply sterile saline or mineral oil to the surgical site and the site of imaging. This repeated application of sterile saline or the oil keep the surgical site moist and prevent additional inflammation of the soft tissues; keeping the imaging site moist also improves the quality of the imaging, as it should decrease the scattering of the fluorescent light signal.
9. It is imperative to minimize or avoid unnecessary blue light exposure when data are not being collected. Minimizing exposure of brain tissue to blue light decreases the photobleaching of the GCaMP6s molecules in neurons, and likely minimizes deactivation of endogenously expressed mitochondrial flavoproteins, which together improve the GCaMP6s signal-to-noise ratio and diminish the frequency of aberrant maps.

6 Notes Summary

It is important to pay special attention to several aspects of the surgical and imaging procedure for a successful in vivo imaging experiment. First, various anesthetics may differentially affect neural activity and particularly calcium signaling in GCaMP6s mice. We suggest the use of ketamine and xylazine anesthesia, supplemented with acepromazine for longer experiments. Second, it is important to keep the procedure as minimally invasive as possible. Although the temporalis muscle needs to be cut to expose the underlying auditory cortex, and the shaving down of the surface of the skull is usually needed to slightly flatten the curvature of the skull, other procedures such as completely removing the temporalis muscle are best to be avoided due to higher risks of excessive bleeding. Additionally, minimizing unnecessary light exposure, finding an optimal angle of the animal's head during imaging, keeping the imaging site clean and moisturized, as well as periodically administering intraperitoneal sterile saline injections will all help to ensure obtaining the best results. Lastly, individual variability among animals is also a factor to keep in mind, remembering that not every in vivo experiment could yield ideal cortical maps of sensory representation.

Acknowledgments

The research was supported by research grant DC013073 from the National Institutes of Health.

References

1. Harris-Warrick RM, Marder E (1991) Modulation of neural networks for behavior. Annu Rev Neurosci 14(1):39–57
2. Enquist M, Ghirlanda S (2013) Neural networks and animal behavior. Princeton University Press, Princeton, NJ
3. Hogan JA (2015) A framework for the study of behavior. Behav Processes 117:105–113. https://doi.org/10.1016/j.beproc.2014.05.003
4. Reinert KC, Gao W, Chen G, Ebner TJ (2007) Flavoprotein autofluorescence imaging in the cerebellar cortex in vivo. J Neurosci Res 85(15):3221–3232
5. Husson TR, Mallik AK, Zhang JX, Issa NP (2007) Functional imaging of primary visual cortex using flavoprotein autofluorescence. J Neurosci 27(32):8665–8675
6. Bandyopadhyay S, Shamma SA, Kanold PO (2010) Dichotomy of functional organization in the mouse auditory cortex. Nat Neurosci 13(3):361–368. https://doi.org/10.1038/nn.2490
7. Lippert MT, Takagaki K, Xu W, Huang X, Wu JY (2007) Methods for voltage-sensitive dye imaging of rat cortical activity with high signal-to-noise ratio. J Neurophysiol 98(1):502–512. https://doi.org/10.1152/jn.01169.2006
8. Masino S, Kwon M, Dory Y, Frostig R (1993) Characterization of functional organization within rat barrel cortex using intrinsic signal optical imaging through a thinned skull. Proc Natl Acad Sci U S A 90(21):9998–10002
9. Kalatsky VA, Stryker MP (2003) New paradigm for optical imaging: temporally encoded maps of intrinsic signal. Neuron 38(4):529–545
10. Tohmi M, Takahashi K, Kubota Y, Hishida R, Shibuki K (2009) Transcranial flavoprotein fluorescence imaging of mouse cortical activity and plasticity. J Neurochem 109(Suppl 1):3–9. https://doi.org/10.1111/j.1471-4159.2009.05926.x
11. Hackett TA, Barkat TR, O'Brien BM, Hensch TK, Polley DB (2011) Linking topography to tonotopy in the mouse auditory thalamocortical circuit. J Neurosci 31(8):2983–2995
12. Kanold PO, Nelken I, Polley DB (2014) Local versus global scales of organization in auditory cortex. Trends Neurosci 37(9):502–510. https://doi.org/10.1016/j.tins.2014.06.003
13. Gardner JL, Merriam EP, Movshon JA, Heeger DJ (2008) Maps of visual space in human occipital cortex are retinotopic, not spatiotopic. J Neurosci 28(15):3988–3999
14. Wang Q, Burkhalter A (2007) Area map of mouse visual cortex. J Comp Neurol 502(3):339–357
15. Kaas JH, Nelson RJ, Sur M, Lin C-S, Merzenich MM (1979) Multiple representations of the body within the primary somatosensory cortex of primates. Science 204(4392):521–523
16. Krubitzer L, Sesma M, Kaas J (1986) Microelectrode maps, myeloarchitecture, and cortical connections of three somatotopically organized representations of the body surface in the parietal cortex of squirrels. J Comp Neurol 250(4):403–430
17. Merzenich MM, Kaas JH, Roth GL (1976) Auditory cortex in the grey squirrel: tonotopic

organization and architectonic fields. J Comp Neurol 166(4):387–401
18. Hellweg FC, Koch R, Vollrath M (1977) Representation of the cochlea in the neocortex of guinea pigs. Exp Brain Res 29(3):467–474. https://doi.org/10.1007/bf00236184
19. Merzenich MM, Knight PL, Roth GL (1975) Representation of cochlea within primary auditory cortex in the cat. J Neurophysiol 38 (2):231–249
20. Suga N, Jen P (1976) Disproportionate tonotopic representation for processing CF-FM sonar signals in the mustache bat auditory cortex. Science 194(4264):542–544
21. Morel A, Kaas JH (1992) Subdivisions and connections of auditory cortex in owl monkeys. J Comp Neurol 318(1):27–63. https://doi.org/10.1002/cne.903180104
22. Frostig RD (2009) In vivo optical imaging of brain function. CRC Press, Boca Raton, FL
23. Turley J, Zalewska K, Nilsson M, Walker F, Johnson S (2017) An analysis of signal processing algorithm performance for cortical intrinsic optical signal imaging and strategies for algorithm selection. Sci Rep 7:7198
24. Kubota Y, Kamatani D, Tsukano H, Ohshima S, Takahashi K, Hishida R, Kudoh M, Takahashi S, Shibuki K (2008) Transcranial photo-inactivation of neural activities in the mouse auditory cortex. Neurosci Res 60(4):422–430. https://doi.org/10.1016/j.neures.2007.12.013
25. Kerr JN, Greenberg D, Helmchen F (2005) Imaging input and output of neocortical networks in vivo. Proc Natl Acad Sci U S A 102 (39):14063–14068
26. Sato TR, Gray NW, Mainen ZF, Svoboda K (2007) The functional microarchitecture of the mouse barrel cortex. PLoS Biol 5(7):e189
27. Peterka DS, Takahashi H, Yuste R (2011) Imaging voltage in neurons. Neuron 69 (1):9–21. https://doi.org/10.1016/j.neuron.2010.12.010
28. Rickgauer JP, Deisseroth K, Tank DW (2014) Simultaneous cellular-resolution optical perturbation and imaging of place cell firing fields. Nat Neurosci 17(12):1816–1824. https://doi.org/10.1038/nn.3866
29. Issa JB, Haeffele BD, Agarwal A, Bergles DE, Young ED, Yue DT (2014) Multiscale optical Ca 2+ imaging of tonal organization in mouse auditory cortex. Neuron 83(4):944–959
30. Akerboom J, Chen TW, Wardill TJ, Tian L, Marvin JS, Mutlu S, Calderon NC, Esposti F, Borghuis BG, Sun XR, Gordus A, Orger MB, Portugues R, Engert F, Macklin JJ, Filosa A, Aggarwal A, Kerr RA, Takagi R, Kracun S, Shigetomi E, Khakh BS, Baier H, Lagnado L, Wang SS, Bargmann CI, Kimmel BE, Jayaraman V, Svoboda K, Kim DS, Schreiter ER, Looger LL (2012) Optimization of a GCaMP calcium indicator for neural activity imaging. J Neurosci 32(40):13819–13840. https://doi.org/10.1523/JNEUROSCI.2601-12.2012
31. Zhao Y, Araki S, Wu J, Teramoto T, Chang Y-F, Nakano M, Abdelfattah AS, Fujiwara M, Ishihara T, Nagai T (2011) An expanded palette of genetically encoded Ca2+ indicators. Science 333(6051):1888–1891
32. Looger LL, Griesbeck O (2012) Genetically encoded neural activity indicators. Curr Opin Neurobiol 22(1):18–23. https://doi.org/10.1016/j.conb.2011.10.024
33. Grewe BF, Langer D, Kasper H, Kampa BM, Helmchen F (2010) High-speed in vivo calcium imaging reveals neuronal network activity with near-millisecond precision. Nat Methods 7(5):399–405
34. Rose T, Goltstein PM, Portugues R, Griesbeck O (2014) Putting a finishing touch on GECIs. Front Mol Neurosci 7:88. https://doi.org/10.3389/fnmol.2014.00088
35. Chen Q, Cichon J, Wang W, Qiu L, Lee SJ, Campbell NR, Destefino N, Goard MJ, Fu Z, Yasuda R, Looger LL, Arenkiel BR, Gan WB, Feng G (2012) Imaging neural activity using Thy1-GCaMP transgenic mice. Neuron 76 (2):297–308. https://doi.org/10.1016/j.neuron.2012.07.011
36. Grienberger C, Konnerth A (2012) Imaging calcium in neurons. Neuron 73(5):862–885. https://doi.org/10.1016/j.neuron.2012.02.011
37. Chen TW, Wardill TJ, Sun Y, Pulver SR, Renninger SL, Baohan A, Schreiter ER, Kerr RA, Orger MB, Jayaraman V, Looger LL, Svoboda K, Kim DS (2013) Ultrasensitive fluorescent proteins for imaging neuronal activity. Nature 499(7458):295–300. https://doi.org/10.1038/nature12354
38. Dana H, Chen T-W, Hu A, Shields BC, Guo C, Looger LL, Kim DS, Svoboda K (2014) Thy1-GCaMP6 transgenic mice for neuronal

population imaging in vivo. PLoS One 9(9): e108697

39. Ratzlaff EH, Grinvald A (1991) A tandem-lens epifluorescence macroscope: hundred-fold brightness advantage for wide-field imaging. J Neurosci Methods 36(2):127–137

40. Polley DB, Read HL, Storace DA, Merzenich MM (2007) Multiparametric auditory receptive field organization across five cortical fields in the albino rat. J Neurophysiol 97 (5):3621–3638. https://doi.org/10.1152/jn.01298.2006

41. Joachimsthaler B, Uhlmann M, Miller F, Ehret G, Kurt S (2014) Quantitative analysis of neuronal response properties in primary and higher-order auditory cortical fields of awake house mice (Mus musculus). Eur J Neurosci 39 (6):904–918. https://doi.org/10.1111/ejn.12478

Chapter 4

A Novel Preclinical Rat Model of Alzheimer's Disease

Karim A. Alkadhi

Abstract

We have developed a novel rat model that correlates with seemingly normal individuals who are predisposed (at-risk) to developing Alzheimer's disease (AD). This work summarizes the findings we have reported on the effect of chronic psychosocial stress in this at-risk rat model of AD. Behavioral (learning and memory tests), electrophysiological (evoked LTP) and molecular (determining protein levels of signaling molecules) studies suggest that even mild chronic psychosocial stress can converts this seemingly normal rat into one showing clear AD phenotype. It is well known that vast individual variations exist in the time of onset and severity of the sporadic type of AD. Therefore, a patient-related external factor must be assumed to play a significant role in the development of the sporadic type of the disease. Since stress is increasingly recognized as an external factor in the development of AD, we tested the effect of mild psychosocial stress on our at-risk model.

Key words Rat subclinical AD model, Amyloid-beta, Learning and memory, Signaling molecules, Synaptic plasticity

1 Introduction: Stress and High Aβ Levels in the Hippocampus

Alzheimer's disease (AD) develops as a steady deterioration of intellectual and emotional health. The disease is characterized by extracellular amassing of high levels of neurotoxic amyloid-beta (Aβ) peptides, followed later by intracellular accumulation of hyper-phosphorylated tau protein, leading to neuronal death and gradual loss of mental abilities [1, 2]. The early symptoms of the disease are slow and subtle personality changes with failure of memory and cognitive skills.

Two forms of AD have been identified an early-onset and a sporadic types. While the early-onset, familial form of AD, which is due to mutations in the genes for amyloid precursor protein (APP), is rare, the overwhelming majority of AD cases are of the late onset, sporadic type. The sporadic nature of the disease may suggest an environmental link that speeds the appearance of AD symptoms. The sporadic AD shows highly variable time of onset, which

Nicholas J. D. Wright (ed.), *Basic Neurobiology Techniques*, Neuromethods, vol. 152,
https://doi.org/10.1007/978-1-4939-9944-6_4, © Springer Science+Business Media, LLC, part of Springer Nature 2020

suggests that environmental determinants, particularly chronic stress, may play a significant role in the development of the disease [3–5].

It is generally accepted that the cognitive deficits resulting from progressive synaptic dysfunction and neurodegeneration are probably started by soluble small oligomeric Aβ peptides. The amyloid cascade hypothesis suggests that soluble Aβ can cause cognitive impairment in the absence of neurodegeneration [6] where soluble Aβ species disrupts all forms of synaptic plasticity, including long-term potentiation (LTP), the cellular correlate of learning and memory [7–10].

Chronic stress is known to worsen the negative effects of different brain disorders including schizophrenia [11], Cushing's disease [12], hypothyroidism [13–15], and AD [3–5]. The involvement of the stress hormones in the pathogenesis of AD is implied by findings of high levels of cortisol in AD patients [16, 17] and cortisone in animal models of the disease [18, 19]. This suggestion is supported by epidemiological findings showing that individuals under chronic stress are more likely to develop mild cognitive deficits, or even AD, than unstressed individuals [20–22]. Exposure to severe and/or prolonged mental stress can cause disruption of the hypothalamic-pituitary-adrenal (HPA) axis, which negatively impacts brain morphology and function [23, 24].

Among the first brain regions affected by AD is the hippocampal formation, which is also particularly vulnerable to the harmful effects of stress. Reports from my laboratory have shown that 2-week intracerebroventricular infusion of pathogenic doses of Aβ peptides in chronically stressed rats exacerbates the already impaired learning and memory and negatively impacts long-term potentiation (LTP) in the hippocampal CA1 and dentate gyrus areas [3, 4, 25, 26]. Thus, we have shown that the combination of chronic stress and Aβ infusion is decidedly more harmful than either condition alone. We then wanted to establish that chronic stress accelerates or reveals symptoms of AD in normal individuals who are at risk for developing the disease. Therefore, we have conceived a novel rat model aimed to exemplify normal individuals who are predisposed to developing AD but are not yet showing AD symptoms. We also wanted to test the hypothesis that in at-risk animal (equivalent to preclinical AD), chronic stress hastens or unveils AD phenotype.

2 Materials

2.1 The Radial Arm Water Maze

The radial arm water maze (RAWM) is a water-filled black circular trough with six stainless steel V-shaped structures positioned such that six swim paths (arms) radiating from one open central area is constructed as described [14]. The RAWM is a reliable and sensitive behavioral test for analyzing hippocampus dependent spatial learning and memory. It is a hybrid of the radial arm maze (RAM) and the Morris water maze, combining the spatial difficulty of the RAM with the driven quick learning of the Morris water maze while reducing their deficiencies (Fig. 1a) [27–30].

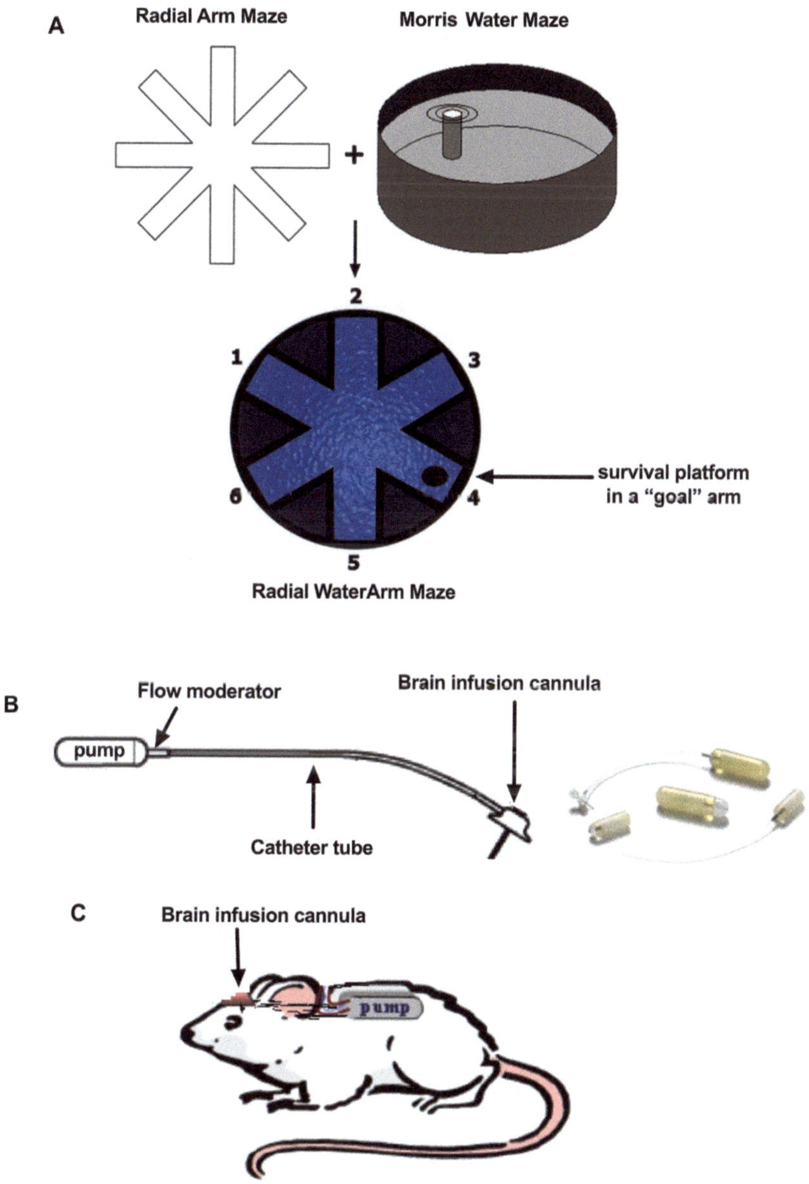

Fig. 1 Diagrams of (**a**) radial arm water maze (RAWM) in relation to Morris water maze and radial arm maze. In the RAWM, the light blue regions are the swimming central and arms field. The opaque oval in arm 4 is the submerged platform. (**b**) The mini-osmotic pump assembly. (**c**) Position of the mini-osmotic pump in animals

2.2 Animal Groups

To determine the influence of chronic psychosocial stress on the at-risk AD model, Wistar rats were assigned into four experimental groups: control, stress, subAβ, and stress/subAβ. The stress and stress/subAβ groups were exposed to psychosocial stress for 6 weeks. During the fifth and sixth weeks of stress, the subAβ and stress/subAβ groups were infused, icv, with $A\beta_{1-42}$ (sub-pathogenic concentration 160 pmol/day) using 2-week osmotic pumps. A similar concentration (160 pmol/day) of the inactive reverse peptide, $A\beta_{42-1}$ was infused in the control and stress groups [31, 32].

2.3 The Osmotic Mini Pump

The osmotic pump system, designed to deliver 160 pmol protein/day, consisted of an L-shape cannula, a flow moderator, a catheter, and a pump. The assembled pump system (Fig. 1b) was primed overnight in isotonic saline solution at 37 °C. The Aβ peptides were prepared as described [3]. Briefly, the Aβ1–42 peptide (AnaSpec Inc., San Jose, CA) was dissolved in a solution of 35% acetonitrile/0.1% trifluoroacetic acid (TFA) to prevent aggregation of Aβ peptides in the osmotic minipump. The day before pump implantation, the pumps were assembled and filled with the Aβ peptide using a syringe.

3 Methods

3.1 The "At-Risk" Aβ Rat Model of AD

The available rodent models of AD, both transgenic and those generated by exogenous Aβ administration have drawbacks as both models do not completely reproduce the complexity of the human AD pathology. Exogenously introduced Aβ peptide could cause neurodegeneration, inflammation and microglial activation [33–35].

The neurotoxic effects of Aβ appear only after it reaches a threshold concentration, because Amyloid β effect is dose-dependent both in vitro [36–39] and in vivo [40–42]. We wanted to determine a concentration of Aβ that does not affect cognitive function (i.e., sub-pathogenic concentration). To achieve that we constructed a "dose–response" relationship by icv infusion of various concentrations of Aβ in different groups of rats and measured the effects on performance in the RAWM as we have reported [31]. From that we determined the right concentration of Aβ peptides infused to produce an "at-risk" AD rat model in which various tests results indicated that this model of AD was not significantly different than control rats. This is the first non-transgenic rat AD model to reproduce a condition where there is a heighten vulnerability to AD with no detectable cognitive impairment. However, when this at-risk AD rat model was produced in chronically stressed rats they showed AD phenotype [31, 32] similar to that we previously reported in a full-fledged model of AD [3]. The at-risk model may represent individuals who seem normal but can develop AD dementia when exposed to stressful conditions.

3.2 Surgical Procedure: Implantation of Osmotic Pump

Animals were affixed with 14-day osmotic mini pumps (Alzet, Cupertino, CA) as reported [3, 31, 43]. Rats were anesthetized with i.p. injections of a mixture of ketamine (100 mg/kg), xylazine (2.5 mg/kg), and acepromazine (2.5 mg/kg) (Phoenix Pharmaceuticals, St. Joseph, MO) and the hair over the implantation site was shaved and skin disinfected with alcohol. The rat was then placed into a stereotaxic frame and the skull was exposed by a 2.5 cm midline sagittal incision of the skin starting slightly behind the eyes. A hole was drilled into the skull over the right cerebral lateral ventricle at (AP: −0.3, L: 1.2, V: 4.5), according to the rat brain atlas [44]. The cannula was inserted through the hole into the lateral ventricle and secured with dental cement. A subcutaneous pocket in the mid-scapular area of the back of the rat was made by lifting the skin and inserting and opening a hemostat, and the assembled osmotic pump was inserted into the subcutaneous pocket (Fig. 1c). The incision was closed with wound clips and treated with triple antibiotic ointment after disinfection with iodine solution. The rat was placed back into its cage and monitored until it emerged from anesthesia. Continuous icv infusion of nonpathogenic subAβ dose (160 pmol/day) was maintained for 2 weeks. The subAβ and stress/subAβ groups were infused with $A\beta_{1-42}$, while the control group was infused with a similar dose of the nontoxic, reverse peptide Aβ42-1.

3.3 Psychosocial Stress Procedure

For chronic stress, we used a variant of "intruder" psychosocial stress [45]. After arrival, rats were allowed 1 week of acclimatization to allow the rats to establish a social hierarchy in each cage. Then rats in the two stress group cages were chronically stressed by daily random switching of two rats from each cage to other cage for a period of 6 weeks. Signs of chronic psychosocial stress were indicated by a marked increase in blood corticosterone level [46] and elevated blood pressure [47].

3.4 Behavioral Tests: The Radial Arm Water Maze Task

All behavioral experiments were carried out at room temperature in a dimly lit room that has visual cues on the walls. Each rat was arbitrarily assigned an arm (the goal arm) that contains a hidden black platform at its far end. Each rat was released at an arm other than the goal arm where it would swim away to find the platform, which was submerged about 1 cm below the water surface. The rats were permitted no more than 1 min to locate the platform for each learning trial or memory test. An error was scored if the rat entered more than halfway into an arm other than the goal arm or if the rat entered the goal arm but failed to find the hidden platform. The rat was manually guided to the platform and scored with 7 errors if it

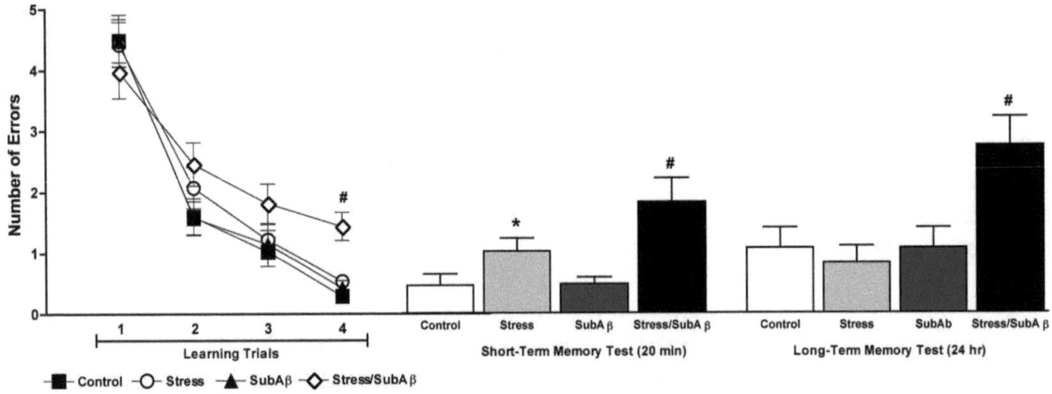

Fig. 2 Whereas our at-risk model (subAβ rats) showed normal learning and memory functions, chronic stress accelerated impairment of learning, short-term memory, and long-term memory of stress/subAβ rats measured in the RAWM. On days 6–8, there was a significant ($p < 0.001$) impairment in the ability of the stress/subAβ rats to locate the hidden platform. Short-term memory was significantly ($p < 0.05$) impaired in the stress rats compared to the control and subAβ rats. The stress/subAβ rats showed significant impairment of both short-term memory and long-term memory compared to control, subAβ rats ($p < 0.001$) and stress rats ($p < 0.05$). (*) Denotes significant difference from control and subAβ groups and (#) from control, stress, and subAβ groups ($n = 13$–14 rats/group). Values are expressed as mean ± S.E.M [32]

failed to locate the platform within 1 min. Once the rat mounted the platform, it was allowed 15 s sitting time on the platform before the next trial began. Rats were returned to the cage and 30 min later were similarly tested for short-term memory (STM), and 24 h later for long-term memory (LTM) [3–5, 31, 32, 48].

3.4.1 Results: Learning, Short-Term Memory, and Long-Term Memory

In all three cognitive tests, our model (subAβ rats) showed normal learning, short-term memory, and long-term memory (Fig. 2). The ability of rats in all groups to learn the location of the survival platform during the four acquisition phase trials was initially similar throughout the early days of testing as rats learned the location of the hidden platform. However, based on the number of errors in locating the hidden platform, the learning ability of the stress/subAβ rats was markedly impaired as days went by, particularly in the fourth trial, compared to the other three experimental groups (Fig. 2) [31, 32]. These findings were further confirmed by another parameter, the days to criterion (DTC), which showed that the stress/subAβ rats required significantly ($p < 0.05$) more days to reach a criterion (number of days in which the rat commits no more than one error in three consecutive days) than the other three experimental groups (Fig. 3a) [31, 32].

Rats were tested for short-term memory 20 min after the last learning trial. Again, the subAβ group performance was not significantly different than that of control rats. As expected, chronically stressed rats showed impaired short-term memory by making significantly higher number of errors compared to both control and

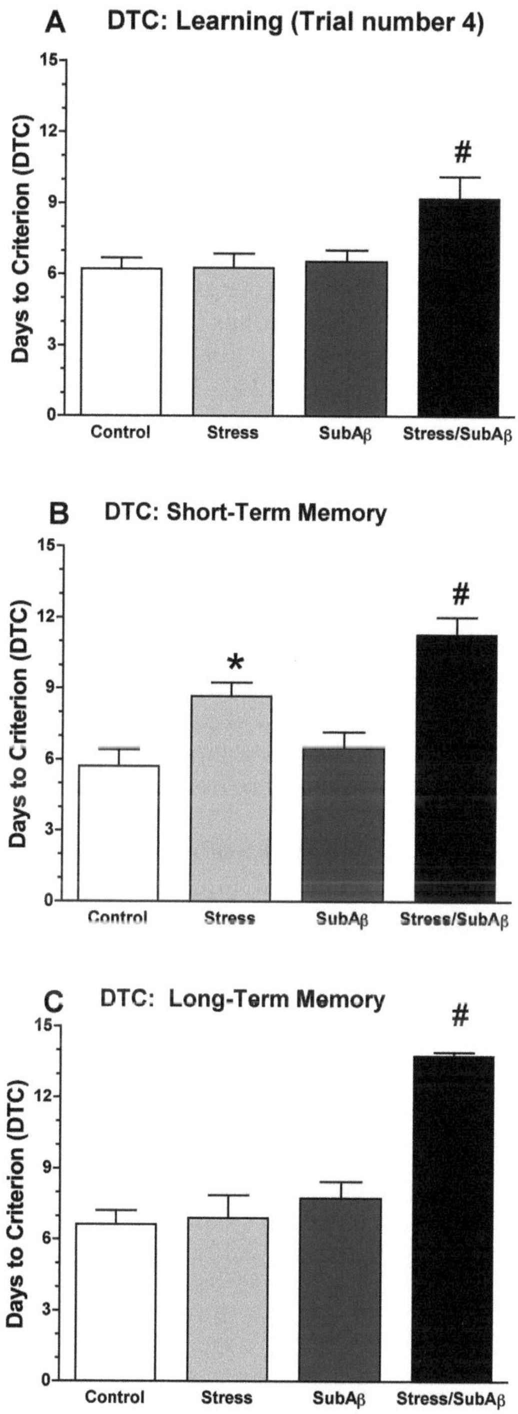

Fig. 3 Learning and memory functions measured as days required to reaching a criterion. The days to criterion (DTC) factor is reached when the animal does no more than one error in 3 consecutive days in a designated trial. Chronic stress significantly impaired (**a**) learning, (**b**) short-term memory, and (**c**) long-term memory in the subAβ rat model of AD. Stress/subAβ group required significantly

subAβ groups (Fig. 2). Dramatically, the stress/subAβ rats continued to commit significantly higher number of errors than all other groups, including the stress group. These findings were further confirmed by another parameter, the days to criterion (DTC), which showed that the stress/subAβ rats required significantly ($p < 0.05$; Fig. 3b) more days to reach the set criterion (see above) than the other three experimental groups [32].

In the long-term (24 h) memory test the stress/subAβ animals committed significantly more errors than the control, stress, and subAβ groups (Fig. 2). Note that our 6-week chronic psychosocial stress did not affect long-term memory confirmed our earlier reports [24]. In confirmation, the DTC test showed that the stress/subAβ rats required significantly more days to reach the criterion than the other three experimental groups (Fig. 3c) [32].

In conclusion, behavioral testing in the RAWM showed that the learning and memory abilities of the novel preclinical model of AD (subAβ rats) were normal and not different from those of the control rats (Figs. 2 and 3). However, when these rats were under chronic stress (Stress/subAβ group) they showed impaired learning, short-term and long-term memory similar to deficiencies seen in rats infused with a full pathogenic dose of Aβ peptides (300 pmol/day) [3–6]. Together these findings validate the subAβ as a model for seemingly normal (preclinical) individuals who may be at-risk for developing AD.

3.5 Electrophysiological Experiment: Synaptic Plasticity

As an index of synaptic transmission, we recorded extracellular field potentials from pyramidal layer of the Cornu Ammonis field 1 (CA1) region of the hippocampus in anesthetized rats before and after induction of E-LTP, L-LTP, and LTD. From the recorded field action potentials (Fig. 4), we measured the amplitue and slope of the field excitatory postsynaptic potential (fEPSP) and compared the resulting data across all groups (control, subAβ, stress, and stress/subAβ).

Anesthetized rats (i.p. urethane 1.2 g/kg; Sigma, USA) were positioned in stereotaxic frame (nose bar at 0.0). The skull was exposed and two holes were drilled for placement of electrodes. A bipolar concentric stimulating electrode was placed in left hippocampal CA3 region at a 5° angle toward the midline to stimulate the Schaffer collaterals/commissural pathway (according to the Rat Brain Atlas [44]; AP: −3, L: 3.5, V: 2.8). A glass capillary recording electrode, filled with 2 M NaCl, was placed in the right

Fig. 3 (continued) more days than the other three experimental groups to reach DTC in both learning and memory test. Stress group also required significantly more days than the control and subAβ groups to reach the DTC for short-term memory. (∗) Denotes significant difference from control and subAβ groups ($p < 0.05$) and (#) from control, stress, and subAβ groups ($p < 0.05$–0.001, $n = 10$–14 rats/group). Values are expressed as mean ± S.E.M. [31]

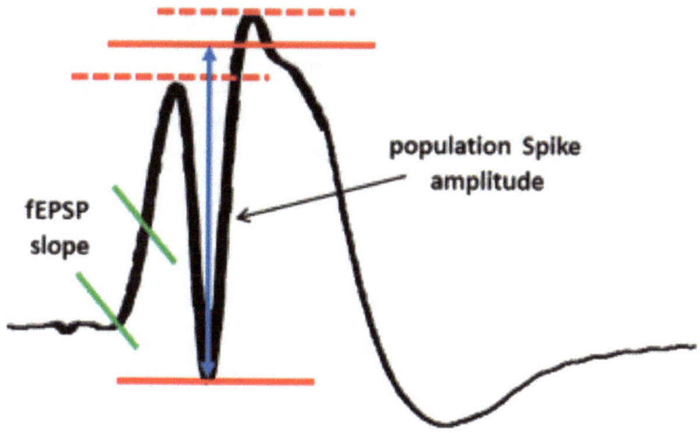

Fig. 4 Measurement of fEPSP slope and pSpike amplitude. The trace represents a typical population spike (pSpike) evoked by a single stimulus. The amplitude of the pSpike was measured from the mid-point between the red dashed lines to the lowest point of the spike. The slope of fEPSP is measured from upstroke between the two green lines

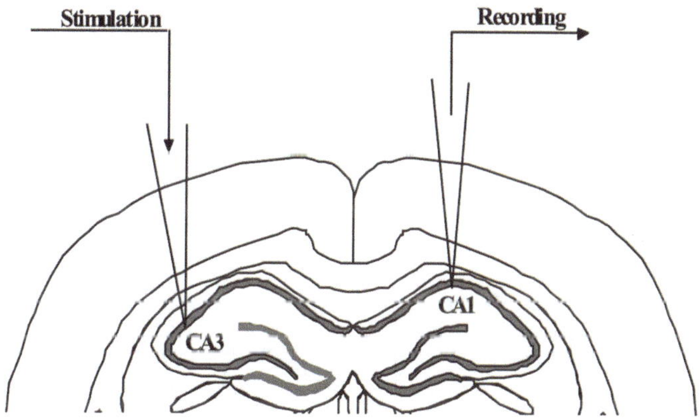

Fig. 5 Schematic illustration of the mid-sagittal section of the rat brain showing the hippocampal septotemporal structures. A stimulation electrode was positioned in the septal side of the left CA3 region of the hippocampus to evoke responses that are recorded with class microelectrode positioned in the septal side of the right CA1 region of the hippocampus

hippocampal area CA1 (AP: −3, L: 1.8, V: 2.0). The recording electrode was connected to an Axoclamp 2A amplifier (Axon Instruments, Inc.). Standard procedures for in vivo recording from the CA1 region of anesthetized rats were carried out as described [3, 49, 50]. A maximum extracellular population spike (pSpike, Fig. 4) was obtained by stimulating the left CA3 region and recording from the right area CA1 (Fig. 5). This was followed by a 30-min stabilization period after which a baseline was determined by applying a test stimulus every 30s for 20 min. Induction of early phase long-term potentiation (E-LTP) was achieved with high frequency stimulation (HFS) consisting of a train of eight pulses applied every 10 s for a period of 70 s (Fig. 6a) [51]. Late-

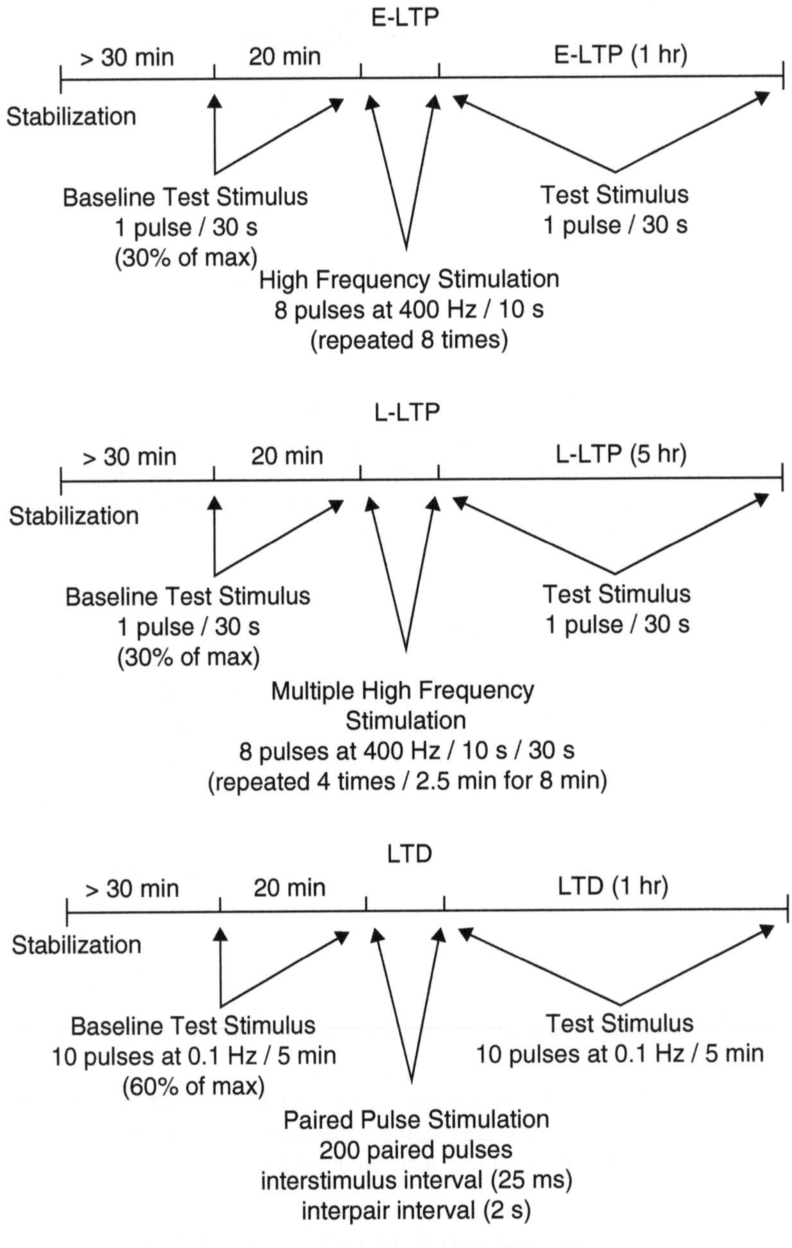

Fig. 6 Stimulation protocols for induction of E-LTP, L-LTP and LTD

phase LTP (L-LTP) was evoked by multiple high frequency stimulation (MHFS) consisting of 4 trains applied at 2.5-min intervals between trains. Each train consisted of 8 pulses (400 Hz) applied every 10 s for a period of 30 s (Fig. 6b). Long-term depression (LTD) was induced by paired pulse stimulation in the form of 3 trains, spaced by 15 min. Each train is composed of 200 paired pulses with an inter-pulse interval of 25 ms and an inter-pair interval of 2 s (Fig. 6c). For LTD, the stimulus intensity was adjusted to approximately 60% of the maximum response. The slope of the field excitatory postsynaptic potential (fEPSP), which represented the intensity of synaptic activity, was calculated from pSpikes (Fig. 4).

3.5.1 Results: Synaptic Plasticity Is Normal in the subAβ Rat Group

Electrophysiological testing showed all three forms of synaptic plasticity tested (LTD, E-LTP and L-LTP) were normal in the subAβ rat group. However, all three synaptic plasticity measures in the chronically stressed subAβ rat group (Stress/subAβ group) were markedly impaired compared to control, stress and subAβ rat groups (Fig. 7). These tests offered additional validation of our sub-clinical model and ascertained the negative influence of chronic stress on brain function.

3.6 Molecular Experiments

To test the effect of infusion of subAβ on the levels of various signaling molecules and AD-related molecules and to elucidate the molecular mechanisms by which stress accelerates impairment of cognitive ability in the at-risk (subAβ) rats we measured the basal levels of these molecules in this preclinical AD model using immunoblot analysis. These experiments allow us to correlate these finding with the results from the learning and memory and synaptic plasticity studies.

3.6.1 Hippocampal Dissection and Processing

Animals were sacrificed under urethane anesthesia and brains were rapidly removed and right hippocampi dissected to yield the dorsal (septal) and ventral (temporal) sides of the CA1 area. Once removed from the brain, the hippocampus was placed on a filter paper soaked in 0.2 M sucrose solution (to prevent the hippocampus from adhering to the filter paper) over a petri dish, filled with dry ice. The two ends of the hippocampus were trimmed. Then, the hippocampus was separated into three parts: septal, medial, and temporal parts. The medial portion was discarded. The septal and temporal parts of the hippocampus were placed in a vertical position to visualize the CA1, DG, and CA3 under a light microscope. The CA3 area was removed to facilitate the dissection of the CA1 area. The boundary between the CA1 area and the DG was marked by the hippocampal fissure, making it possible to separate and isolate the CA1 region. The dissected septal and temporal sides of the CA1 region were placed into prelabeled microcentrifuge tubes and stored at −80 °C until homogenization and processing of different cellular fractions [3, 32, 52, 53].

Stimulation of the CA3 area of the septal side of the left hippocampus evoked responses recorded in the CA1 area of the septal side of the right hippocampus. It has been reported that the Schaffer collaterals of the CA3 neurons project to as much as 75% of the CA1 neurons in the septal side of the right hippocampus while the CA1 neurons in the temporal side receive a much smaller input [54–57] (Fig. 8). Thus, the septal side of the contralateral (right) hippocampus, where LTP is generated, was considered as the

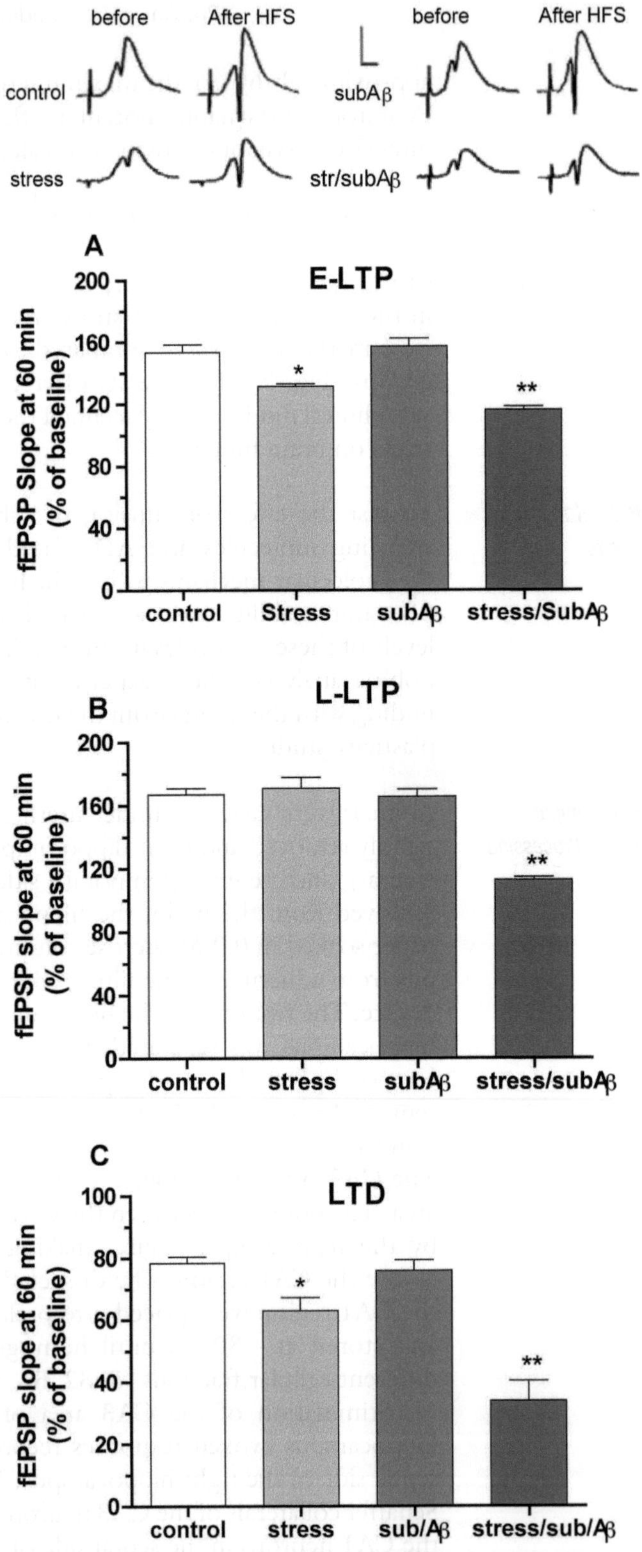

Fig. 7 Lack of effect of infusion of sub-pathological concentration (subAβ) on synaptic plasticity as revealed by electrophysiological recording from hippocampal area CA1 of anesthetized rats. (**a**) Early-phase long-term

"stimulated" side, whereas the temporal side, which receives relatively negligible input from the septal side of the left hippocampus, was considered as the unstimulated and used as "internal" control for the same animal. To reduce individual variation, the protein levels of the stimulated septal part were expressed as a percentage of the protein levels of the unstimulated temporal part of the CA1 as described [3, 14, 15, 32, 43, 48, 50, 52, 58].

3.6.2 Homogenization and Sample Preparation

Hippocampal area CA1 tissue was homogenized and sonicated in 200 μl isotonic lysis buffer (50 mM Tris–HCl (pH 7.4), 150 mM NaCl, 1 mM EDTA (pH 7.4), 1 mM EGTA (pH 7.4), 1% Nonidet P-40, 0.1% SDS, 5 mM $Na_4P_2O_7$, 100 μg/mL PMSF, 40 mM beta-glycerophosphate, 50 mM NaF, 1 mM PMSF) in the presence of protease inhibitor cocktail (Roche Applied Sciences, Indianapolis, IN). Protein concentration was assessed by micro bicinchoninic acid (BCA) assay (Pierce, Rockford, IL).

3.6.3 Immunoblotting

The following steps were completed for immunoblotting: (1) Sample preparation: CA1 tissue homogenate was diluted with 4X E-PAGE loading buffer 1 and 10X NuPAGE sample reducing agent (Invitrogen, Carlsbad, CA) and heated to 70 °C for 10 min; (2) Protein electrophoresis: 10 μL (containing 10 μg of protein) of the homogenate was loaded onto a 48-well protein electrophoresis gel (E-PAGE 48, Invitrogen) and resolved by

Fig. 7 (continued) potentiation (E-LTP) of hippocampal area CA1 induced by high frequency stimulation (HFS) of the Schaffer collaterals/commissural (SCC) pathway of anesthetized rats. E-LTP shown here was measured as increases in the slope of fEPSP at 60 min after HFS and expressed as percentage of the baseline values. Slope values in stress/subAβ rats are significantly lower than those of the three other groups at all time points after applying HFS. The stress group showed significantly decreased slope values compared to control, and subAβ rats. (∗) Denotes significant difference from control and subAβ groups and (∗∗) from control, stress, and subAβ groups ($p < 0.05$, $n = 4$–6 rats/group). Insets are pSpike traces from representative experiments. Calibrations (5 mV/5 ms) apply to all traces. (**b**) Late-phase long-term potentiation (L-LTP) in area CA1 of the hippocampus evoked by multiple high frequency stimulation (MHFS) of the SCC pathway of anesthetized rats measured at 5 h after MHFS and expressed as percentage of the baseline values. (∗∗) Indicates significant difference ($p < 0.05$, $n = 5$–7 rats/group). (**c**) Long-term depression (LTD) of area CA1 of the hippocampus measured at 60 min after induction. LTD was induced by stimulation of the SCC pathway in anesthetized rats by applying 3 trains, spaced by 15 min; each train is composed of 200 paired pulses with an inter-pulse interval of 25 ms and an inter-pair interval of 2 s. LTD was measured as decreases in fEPSP slope. Results are expressed as mean ± SEM. (∗) denotes significant difference from control group and (∗∗) from all other groups ($n = 4$–7 rats/group) [31, 32]

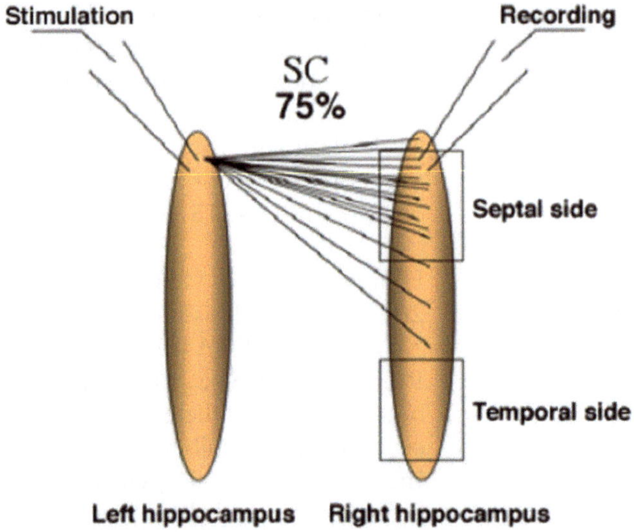

Fig. 8 Schematic representation of the hippocampal septotemporal structures. Stimulation of the septal side of the left CA3 region of the hippocampus evokes responses recorded in the septal side of the right CA1 region of the hippocampus. It is known that each CA3 neuron synapses to as much as 75% of the septal side of ipsilateral and contralateral CA1 field while the temporal side receives negligible input. Therefore, using the temporal side as an "internal" control, the protein level of the stimulated septal part was expressed as a percentage of that of the unstimulated temporal part of the CA1 area to minimize individual variations. SC is Schaffer-collateral nerve bundle

E-PAGE high-throughput protein electrophoresis system (E-BASE, Invitrogen); (3) Transfer: Proteins on the gel were tranfered onto polyvinylidene fluoride gel (iBlot gel transfer stacks, PVDF, Invitrogen) using Invitrogen dry blotting system (iBlot gel transfer device, Invitrogen). After the transfer, the PVDF was activated by submerging the membrane in methanol for approximately 1 min; (4) Blocking: To block any nonspecific binding, membranes were incubated overnight in 5% bovine serum albumin in Tris-buffered saline (TBS) with 0.1% Tween-20 (TBS-Tween); (5) Addition of the antibodies: Membranes were first incubated with unlabeled primary antibody (Table. 1) for 1 h at room temperature or overnight at 4 °C, then the membranes were washed three times with TBS-Tween to eliminate unbound antibody. Then the membranes were incubated in horseradish peroxidase conjugated secondary antibody (Table. 1) for 1 h at room temperature. After incubation of the secondary antibody, the blot was washed three times with TBS-Tween; (6) Detection: The antigen-antibody complex was visualized by enhanced chemiluminescence (ECL) western blotting substrate (Pierce, Rockford, IL) and picked up digitally using the Alpha Innotech imaging system. The intensity of

Table 1
Antibodies, dilutions and sources: summary of the primary and secondary antibodies that were used to detect the given molecules

	Primary antibody	Secondary antibody
p-CaMKII	Mouse anti-p-CaMKII mAb 1:1000 Santa Cruz Biotechnology, Inc., CA	Goat anti-mouse IgG, HRP 1:1000 Santa Cruz Biotechnology, Inc., CA
Calcineurin	Rabbit anti-calcineurin pAb 1:1000 Millipore, MA	Goat anti-rabbit IgG, HRP 1:2000 Santa Cruz Biotechnology, Inc., CA
BDNF	Rabbit anti-BDNF pAb 1:1000 Santa Cruz Biotechnology, Inc., CA	Goat anti-rabbit IgG, HRP 1:1000 Santa Cruz Biotechnology, Inc., CA
p-CREB	Goat anti-p-CREB-1 pAb 1:1000 Santa Cruz Biotechnology, Inc., CA	Rabbit anti-goat IgG, HRP 1:1000 Santa Cruz Biotechnology, Inc., CA
GAPDH	Mouse anti-GAPDH mAb 1:5000 Millipore, MA	Goat anti-mouse IgG, HRP 1:5000 Santa Cruz Biotechnology, Inc., CA
APP	Rabbit anti-APP pAb 1:1000 Abcam, MA	Goat anti-rabbit IgG, HRP 1:2000 Santa Cruz Biotechnology, Inc., CA
BACE	Rabbit anti-BACE1 pAb 1:1000 Abcam, MA	Goat anti-rabbit IgG, HRP 1:2000 Santa Cruz Biotechnology, Inc., CA

the immunoreactive bands was measured by densitometry using AlphaEase software and expressed as a ratio to that of the glyceraldehyde phosphodehydrogenase (GAPDH). GAPDH antibody served as a loading control to compensate for any discrepancies that may have occurred during protein concentration estimation.

Other different molecules in the same blot were detected in an additional stripping step to remove previously bound antibodies. Blots were washed for 5 min in distilled water and placed in the 0.2 M NaOH stripping buffer for 5 min. Then, the blots were washed for another 5 min with distilled water. After which, the membranes were blocked in 5% BSA in TBS-Tween for 30 min at room temperature. Then the steps were repeated starting from the addition of antibodies step.

3.7 Statistical Analysis

All statistics were carried out with one-way analysis of variance (ANOVA) followed by post-hoc test (Graphpad Prism). Minimal significance level accepted was set at $p < 0.05$. Results are expressed as means ± SEM.

3.8 Basal Levels of Memory- and AD-Related Signaling Protein Molecules

To study the involvement of the molecular mechanisms by which stress speeds up the impairment of cognitive ability in the at-risk (subAβ) rats and to correlate these findings with the results from the learning and memory and synaptic plasticity studies, we used immunoblot analysis (E-PAGE 48 and E-PAGE high-throughput protein electrophoresis system, Invitrogen, Carlsbad, CA) to assess the basal levels of AD-related peptides and cognition signaling molecules in this preclinical AD model and the effects of chronic psychosocial stress [32].

3.8.1 Results and discussion: Memory-Related Signaling Protein Molecules

Memory-related signaling molecules were not changed in the subAβ rat group. In contrast, the stress/subAβ showed marked change in the levels of all of these molecules (Fig. 9; Table 2) [59].

Calcium Calmodulin Kinase II (CaMKII)

Calcium-calmodulin dependent protein kinase II (CaMKII) is a crucial memory molecule that has been extensively studied in experiments to understand the process of learning and memory. Various forms of synaptic plasticity, particularly Long-term potentiation (LTP), are presently considered to be the cellular correlate of learning and memory and synaptic plasticity in which CaMKII plays as critical role. It has been suggested that activated CaMKII works as a molecular switch that changes temporary calcium signals into long-term biochemical changes necessary for synaptic plasticity and memory [60].

The putative molecular cascade of events leading to the formation of the active (phosphorylated) CaMKII (p-CaMKII) and eventually expression of LTP is experimentally initiated by high frequency stimulation of the synaptic pathway. This causes release from presynaptic terminals of the neurotransmitter glutamate, which activates glutamate receptors on the postsynaptic membrane, causing calcium influx through *N*-methyl-D-aspartate (NMDA) glutamate receptor [61–63]. This brief elevation of intracellular calcium leads to activation of protein kinase C gamma (PKCγ), which phosphorylates neurogranin causing dissociation of the neurogranin–calmodulin complex and freeing calmodulin [64–66]. The free calmodulin forms a complex with calcium, which binds to and activates CaMKII. Activation of CaMKII allows it to phosphorylate itself [67, 68] results in a constitutively active CaMKII (p-CaMKII), which phosphorylates and activates α-amino-3-hydroxy-5-methyl-4-isoxazole (AMPA) glutamate receptors and the synaptic vesicle-specific protein, synapsin, both of which are essential for expression of LTP and memory formation [63, 69–71]. Continuous activation of these substrates by p-CaMKII remains uninterrupted, even when calcium concentration returns to normal levels; it only stops when p-CaMKII is dephosphorylated by protein phosphatases including calcineurin [69, 72, 73].

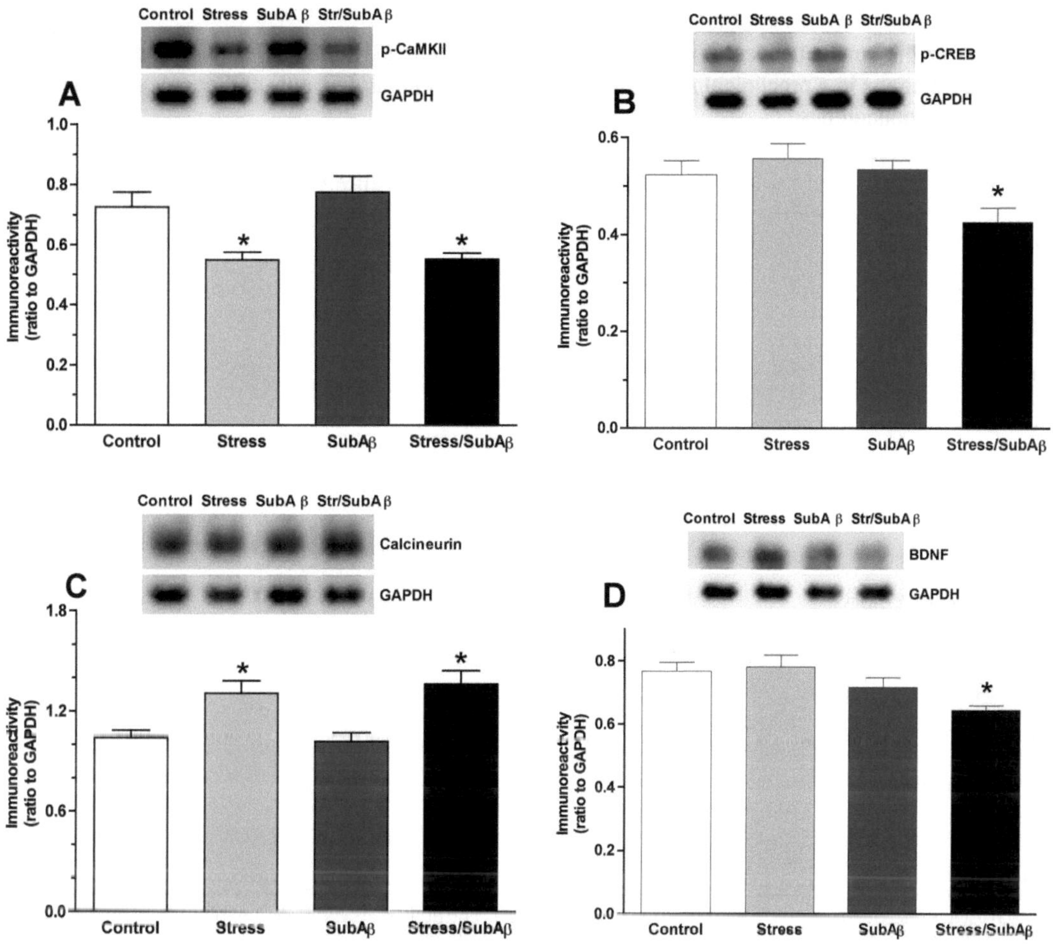

Fig. 9 Lack of effect of our at-risk model on memory-related signaling molecules. (**a**) Basal levels of phosphorylated calcium/calmodulin-dependent protein kinase II (p-CaMKII) in the hippocampal area CA1. Chronic stress (in stress and stress/subAβ groups) caused a significant ($p < 0.05$–0.01, $n = 6$–7 rats/group) decrease in the basal levels of p-CaMKII. (**b**) Basal levels of phosphorylated cyclic AMP response element binding protein (p-CREB) in the hippocampal area CA1. The combination of chronic stress and subAβ infusion caused a significant ($p < 0.05$, $n = 5$–6 rats/group) decrease in the basal levels of p-CREB. (**c**) Basal levels of calcineurin in the CA1 region of the hippocampus. There was a significant ($p < 0.05$–0.01, $n = 7$–8 rats/group) increase in the basal levels of calcineurin in animals exposed to chronic stress (stress and stress/subAβ groups). (**d**) Basal levels of brain-derived neurotrophic factor (BDNF) in the CA1 region of the hippocampus. There was a significant ($p < 0.05$, $n = 6$–7 rats/group) decrease in the basal levels of BDNF in stress/subAβ group. All results are expressed as mean ± S.E.M. (∗) Indicates significant difference from control and stress values. Insets are representative bands

In our model (subAβ rats), the basal levels of both p-CaMKII and total CaMKII were not significantly different from those in control rats [32]. However, in both the stress and stress/subAβ rat groups, the p-CaMKII levels were markedly reduced compared to control and subAβ groups (Fig. 9a, Table 2) suggesting impairment of the mechanism of CaMKII phosphorylation and/or enhanced

Table 2
Summary of the consequences of chronic psychosocial stress (STRESS) on cellular and molecular neuronal functions in a preclinical AD rat model (SUBAβ: 160 pmol/day; sub-pathogenic dose) in area CA1 of the hippocampus

	STRESS	SUBAβ	STR/SUBAβ
p-CaMKII	⇓	⇔	⇓
t-CaMKII	⇔	⇔	⇔
Calcineurin	⇑	⇔	⇑
BDNF	⇔	⇔	⇓
p-CREB	⇔	⇔	⇓
t-CREB	⇔	⇔	⇔
CaMKIV	⇔	⇔	⇓
Learning	⇔	⇔	⇓
Short-term memory	⇓	⇔	⇓⇓
Long-term memory	⇔	⇔	⇓
Early LTP magnitude	⇓	⇔	⇓
Late LTP magnitude	⇔	⇔	⇓
LTD magnitude	⇑	⇔	⇑⇑
BACE	⇔	⇔	⇑
APP	⇔	⇔	⇔

⇓ decreased, ⇑ increased, ⇔ no change compared to control rats. The prefixes p- and t- denote phosphorylated and total respectively

dephosphorylation, because total-CaMKII was not changed [32]. The finding that the basal levels of the dephosphorylating enzyme, calcineurin are markedly increased in the stress and stress/subAβ groups compared to control and subAβ groups (Fig. 9c, Table 2) [32] supports the notion of enhanced dephosphorylation.

Cyclic AMP Response Element Binding (CREB) Protein

CREB plays a crucial role in the generation of protein synthesis-dependent long-term changes necessary for long-term memory and synaptic plasticity in the brain [74, 75]. Active CREB signaling may play a pivotal role in the effects of chronic stress on neurogenesis, LTP and calcium currents in the DG area [76, 77]. Findings from this lab consistently showed that moderate chronic psychosocial stress in rats did not significantly affect the basal expression of p-CREB or total-CREB [3]. Levels of both CREB and BDNF mRNA were shown to decrease initially but recover within 24 h after stress [78].

Phosphorylated (active) CREB is vital for hippocampus-dependent long-term memory formation in mammals [74, 79].

A number of studies revealed the role of p-CREB in AD-associated impairment of memory and synaptic plasticity. For example, the translocation of pCREB from the cytoplasm into the nucleus (nuclear translocation) is blocked by Aβ resulting in decreased CRE-mediated responses [80]. Moreover, CREB phosphorylation is reduced by Aβ through decreasing glutamate NMDA receptor activation [81].

The basal level of p-CREB was normal in subAβ rat group as it was in control, stress groups. However, The basal level of p-CREB in brain area CA1 was significantly reduced in the stress/subAβ group (Fig. 9b). This suggests reduced synthesis of the CREB protein in area CA1 by chronic stress in the presence of subAβ concentrations. We have shown that infusion of full pathogenic dose of Aβ caused a marked decrease of p-CREB levels in area CA1 of rats [5, 78]. In light of the critical and highly conserved role of p-CREB in the generation of protein synthesis-dependent long-term changes required for long-term memory and synaptic plasticity in the brain [74, 75], the reduced basal levels of CREB may explain the compromised long-term memory in the stress/subAβ animals. These findings emphasize the validity of the AD at-risk model and further establish that chronic psychosocial stress can hasten the emergence of AD symptoms in at-risk individuals.

Calcineurin

Phosphorylated (activated) signaling molecules including p-CaMKII, p-CaMKIV, and p-CREB are typically dephosphorylated by a protein phosphatase, mainly protein phosphatase 2B (PP2B or calcineurin). Calcineurin dephosphorylates inhibitor 1 protein, the natural regulator of protein phosphatase 1 [82], which is effective in the dephosphorylation of CaMKII [83]. Calcineurin is thought to be the principal phosphatase involved in the regulation of synaptic plasticity in the hippocampus. Calcineurin is implicated in the induction and maintenance of long-term depression (LTD) [84] by diminishing postsynaptic activity in hippocampal neurons [85]. Furthermore, upregulation of calcineurin in the hippocampus inhibits LTP [86] and impairs hippocampus-dependent memory formation in mice [87].

Although chronic psychosocial stress results in over-expression of calcineurin in area CA1 of the hippocampus [3, 6, 14, 52], other types of stressors may have different effects on the levels of calcineurin. For example predator exposure stress produced no significant effect on the expression of calcineurin in hippocampal area CA1, amygdala or prefrontal cortex medial prefrontal cortex (mPFC) of rats [88]. Fascinatingly, calcineurin expression seems to be sex-specific in that it is decreased in male but not female animals in the hippocampus of previously maternally deprived rats [89]. Remarkably, within the hippocampal formation, the dentate gyrus area seems to be immune to the effect of moderately chronic stress in that chronic stress produced a significant decrease, rather than increase, in calcineurin levels. Thus, the dentate gyrus of chronically stressed rats seems to have a compensatory mechanism whereby calcineurin levels are reduced in order to maintain normal P-CaMKII levels, which may be responsible for the normal early LTP of the dentate gyrus in chronically stressed rats [90].

The involvement of calcineurin in AD pathology has been suggested by a number of reports [91–93]. For example, cognitive impairment in Tg2576 mice is reversed on treatment with tacrolimus, a calcineurin inhibitor [94]. Post-mortem examination of AD brain tissues showed that mRNA of calcineurin was upregulated in pyramidal neurons of the hippocampus, suggesting that calcineurin may play a key role in the pathogenesis of AD [95].

In area CA1 of subAβ at-risk rats, calcineurin levels were not significantly different from those in the control rat group. However, area CA1 of both the stress and the stress/subAβ rats showed significantly increased levels of calcineurin (Fig. 9c). Similar increases in calcineurin levels were seen in the CA1 area of unstressed rats treated with a full pathogenic dose of Aβ peptides [3].

Brain Derived Neurotropic Factor (BDNF)

Brain-derived neurotrophic factor (BDNF) is a member of a family of neurotrophic factors that includes nerve growth factor, neurotrophin-3, and neurotrophin-4/5, which activate various isoforms of tropomyosin kinase (Trk) receptors. BDNF supports neuronal growth, synaptic transmission and plasticity as well as pathogenesis of psychiatric disorders [96]. It exists at high levels in hippocampal neurons, and its expression is controlled by neural activity. BDNF promotes survival of existing and nascent neurons and supports synaptic plasticity in the central nervous system. Additionally, BDNF supports survival of cholinergic neurons of the basal forebrain [97], and other neurons in the hippocampus [98, 99] and cortex [100]. It may work as an activity-dependent modulator of neuronal structure, and its upregulation after exercise or tetanic stimulation modulates the induction and maintenance phases of LTP in the hippocampus [32, 101–104].

Ample evidence supports the role of BDNF in memory processes. For example, performance in various cognition tests was correlated with expression of BDNF mRNA [105–107]. Therefore, it may be assumed that BDNF loss may contribute to memory deficits associated with neurodegenerative diseases including AD. This was substantiated by showing marked decrease in BDNF mRNA and protein in the hippocampus and temporal cortex of autopsied AD brains [108]. Additionally, it has been reported that exposure to various stressors, both acute and chronic can significantly downregulate both BDNF mRNA expression and protein levels in the hippocampus [109–115]. However, in the brains of animal models of AD reports are conflicting concerning the levels of BDNF. This inconsistency in findings is unclear; it may be the result of various factors including different stages of the disease at which the levels of BDNF was determined, the type of AD model used, the dose level of Aβ administered etc.

In our experiments, the basal protein levels of BDNF were not significantly changed in stress and subAβ animals compared to control animals. However, in the chronically stressed subAβ (stress/subAβ) group, there was a significant ($p < 0.05$) decrease in the basal level of BDNF compared to those of control and stress groups (Fig. 9d) [116]. Comparable findings have been reported in cultured cortical neurons on the effect of sub-lethal dose of Aβ [117, 118]. We previously reported significantly ($p < 0.05$) higher levels of BDNF in rats infused with a pathogenic dose of Aβ [5]. This finding may be explained as an attempt by the brain to achieve repair at this stage of the disease.

3.8.2 Results: AD-Related Signaling Protein Molecules

The Aβ domain of the amyloid precursor protein (APP) is partly embedded in the plasma membrane. Splitting and processing of APP may occur through two major pathways: a non-amyloidogenic pathway, which produces harmless peptides, and an amyloidogenic pathway that produces the pathogenic Aβ peptides [119–122]. Three enzymes, α, β, and γ secretases, can cleave APP. The harmless non-amyloidogenic pathway, which precludes the formation of pathogenic Aβ fragments, is initiated by α-secretase splitting of APP within the Aβ domain [122]. In the amyloidogenic pathway, production of Aβ peptides starts with proteolytic cleavage of APP just above the Aβ domain by an integral membrane aspartyl protease called Aβ-site APP cleaving enzyme 1 (BACE1), which functions as the rate limiting step in the pathogenic production [123]. We used immunoblot analysis to determine the basal levels of APP in hippocampal area CA1 of the four experimental groups.

We analyzed the basal levels of BACE in hippocampal area CA1 and showed normal levels in subAβ rats. However, in the stress/subAβ animals, chronic stress significantly increased the levels of BACE compared with control, stress, and subAβ rats (Fig. 10, left panel, Table 2). Thus, stress in the stress/subAβ group produced an effect on BACE like that caused by that full pathogenic Aβ dose (300 pmol/day; [4]). There was no statistically significant change in the APP levels in stress/subAβ, which was not different from those of the control or the subAβ groups (right panel-Fig. 10; Table 2).

4 Notes

In summary, the at-risk model of preclinical AD can be used to represent the seemingly normal individuals predisposed to develop AD (at-risk for AD) and has been used successfully to study the effects of chronic stress. Our experimental findings using this

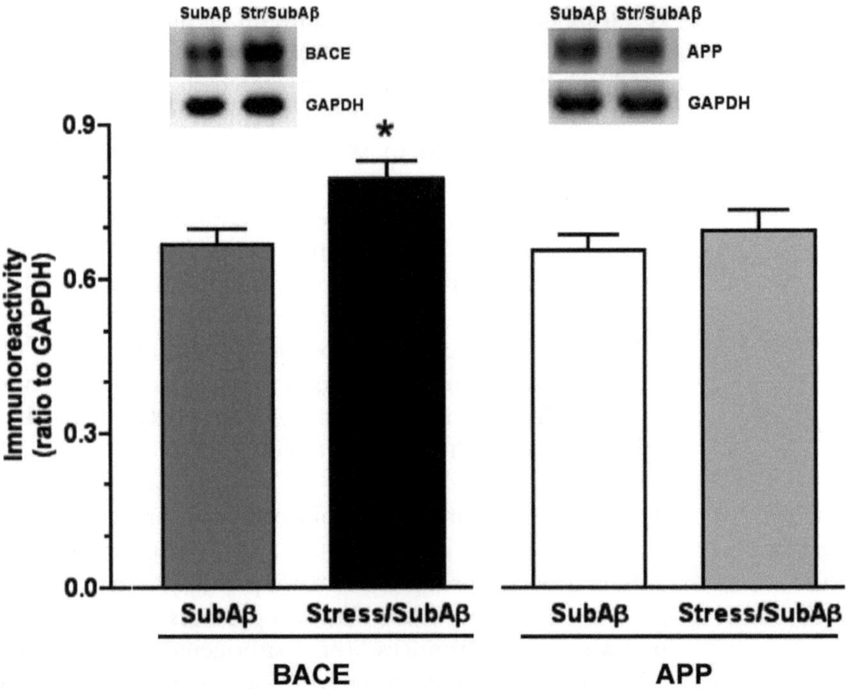

Fig. 10 Basal levels of AD-related major proteins. Left panel: Basal levels of beta-site amyloid precursor protein-cleaving enzyme (BACE) in hippocampal area CA1. Note that the preclinical AD model (SubAβ) is not significantly different from control or stress groups (not shown). Chronic stress of subAβ group (Stress/SubAβ) significantly increased the basal level of BACE in these rats. Right panel: Basal levels of amyloid precursor protein (APP) in the hippocampal area CA1. No significant changes in the basal levels of APP among the four experimental groups were observed. Results are expressed as mean ± SEM. *Significant difference ($p < 0.05$, $n = 6$–7 rats/group) from control, stress, and subAβ groups. Insets are representative blots

model of subclinical AD indicate that chronic stress can have a profound effect in accelerating the course of development of AD. The cellular and molecular interactions of stress and AD are summarized in Fig. 11.

We used this model to study the effects of chronic stress at the behavioral (learning and memory), electrophysiological (LTP) and molecular levels (western blotting). Although the behavioral and molecular studies are straightforward and unproblematic, the electrophysiological techniques can be challenging and may be time consuming to accomplish. For more detailed explanations of all three techniques the reader may want to review previous related publications from this laboratory some of which are listed in the reference section of this chapter.

Lately, we have used simplified RAWM protocol by cutting the days of testing from 6 to 10 days to a single day through increasing the number of trials for learning from 4 to 12 (Fig. 12). Although this method yielded reliable results like those obtained in the multiple-day method, the supportive data of the days to criterion

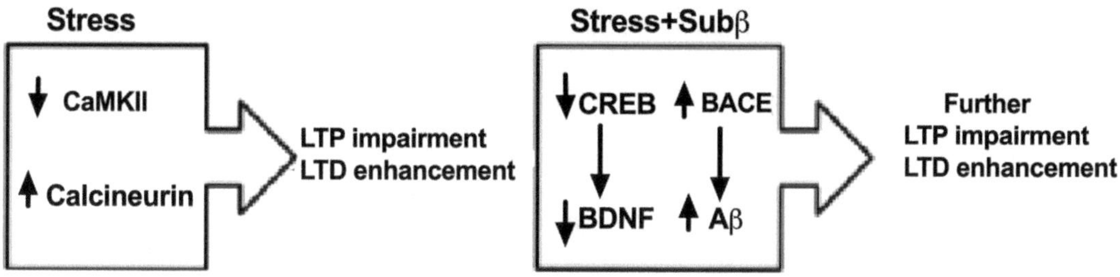

Fig. 11 Schematic of possible mechanism by which stress impairs synaptic plasticity in subAβ-infused animals. Stress causes a decline in kinase (CaMKII) levels and a rise in phosphatase (calcineurin) levels resulting in LTP impairment and LTD enhancement. Combining chronic stress with subAβ infusion in rats also causes a decrease in the levels of CREB and BDNF as well as an increase in the levels of BACE, the rate-limiting enzyme in Aβ production. These changes result in a greater impairment and enhancement of LTP and LTD, respectively, more than stress alone

(DTC) is lost. To test learning and short-term memory in the new methods, rats were first subjected to six learning trials (trials 1–6) followed by a 5 min rest period and then another six learning trials (trials 7–12). The animals were then tested 30 min for short-term memory and 24 h for long-term memory after the end of 12th trial. An example can be seen in (Fig. 12), in which the effect of regular exercise in rat model of full-fledged AD (use of pathological dose of Aβ) was tested [43, 104].

For accuracy, we used the temporal side of the hippocampus as an internal control in addition to GAPDH (loading control) as explained earlier in this chapter. However, in some of the representative blots, the temporal sides of the experimental groups and the temporal side of the control group were different. A possible reason for the effect may be inaccuracy in dissection or protein sample processing, in which the unstimulated temporal region may have been contaminated by stimulated septal region. Additional caution should be exercised in future experiments during dissection and sample processing to avoid sample contamination.

High levels of Aβ peptides disrupt phosphorylation of CaMKII and interfere with LTP induction [4, 124, 125]. We previously have shown that chronic stress negatively impacts LTP and downregulates basal levels of p-CaMKII in area CA1 of the hippocampus [3, 14]. Thus, it is likely that disruption of the vitally required CaMKII-dependent protein phosphorylation by stress could accentuate the initial Aβ-caused loss of p-CaMKII, thus augmenting the mechanism by which chronic stress impairs memory in this model of AD.

Normal brain levels of neurotrophic factors, including BDNF, are elevated in certain brain areas in reaction to several types of insults, including ischemia, seizure, traumatic brain injury, and neurotoxins [126, 127]. BDNF is known to plays a pivotal role in the health and survival of neurons [128, 129]. A protective

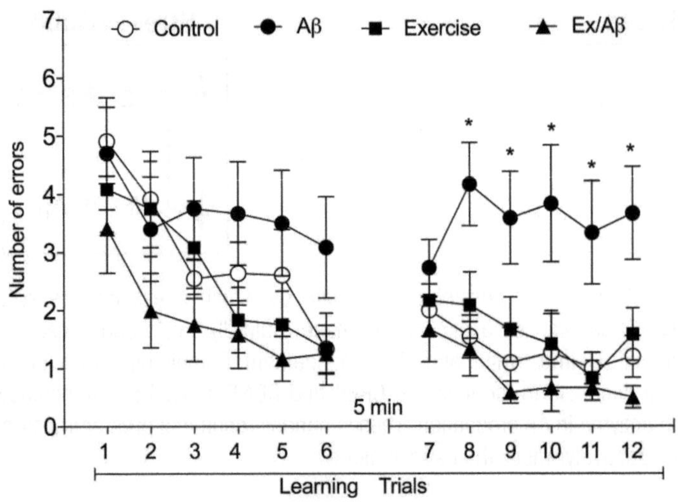

Fig. 12 A new time saving variation in the protocol of testing in the RAWM involving 1 day in which the learning "curve" is determined in 1 day by doing 12 trials. Short-term and long-term memory tests (not shown) were adequately determined in this protocol. (*) Denotes significant difference from all groups ($p < 0.05$, 10–12 rats/group)

mechanism seems to be activated to counter the Aβ-induced neurotoxicity in initial stages of AD. This was proposed by earlier reports that showed upregulation of protein levels of BDNF in the forebrain in APPsw mice model of AD [130] and upsurge in the mRNA of BDNF in the hippocampus [131] and in area CA1 in the full dose Aβ-treated rats [4]. In contrast, chronic stress has been reported to significantly downregulates BDNF levels in hippocampal area CA1 [45]. Therefore, by diminishing the availability of BDNF, stress impedes the repair process and consequently boosts the effect of Aβ.

Stress modifies the processing of various proteins involved in the expression of AD. For example, glucocorticoids or stress has been shown to increase the concentrations of APP and BACE proteins. This suggests that stress may be driving the processing of APP in the direction of the pathogenic pathway. Thus, we see enhanced levels of pathogenic Aβ protein [3, 4, 132] and increased formation of plaque [111] in the presence of stress.

References

1. Tanzi RE, Bertram L (2005) Twenty years of the Alzheimer's disease amyloid hypothesis: a genetic perspective. Cell 120:545–555
2. Castellani RJ, Lee HG, Zhu X, Perry G, Smith MA (2008) Alzheimer disease pathology as a host response. J Neuropathol Exp Neurol 67:523–531
3. Srivareerat M, Tran TT, Alzoubi KH, Alkadhi KA (2009) Chronic psychosocial stress exacerbates impairment of cognition and long-term potentiation in beta-amyloid rat

model of Alzheimer's disease. Biol Psychiatry 65:918–926

4. Srivareerat M, Tran TT, Salim S, Aleisa AM, Alkadhi KA (2011) Chronic nicotine restores normal Abeta levels and prevents short-term memory and E-LTP impairment in Abeta rat model of Alzheimer's disease. Neurobiol Aging 32:834–844
5. Alkadhi KA, Srivareerat M, Tran TT (2010a) Intensification of long-term memory deficit by chronic stress and prevention by nicotine in a rat model of Alzheimer's disease. Mol Cell Neurosci 45(3):289–296
6. Alkadhi KA, Srivareerat M, Tran TT (2011) Chronic psychosocial stress exacerbates impairment of synaptic plasticity in β-amyloid rat model of Alzheimer's disease: prevention by nicotine. Current Alz Res 8:718–731
7. Shankar GM, Li S, Mehta TH, Garcia-Munoz A, Shepardson NE, Smith I, Brett FM, Farrell MA, Rowan MJ, Lemere CA, Regan CM, Walsh DM, Sabatini BL, Selkoe DJ (2008) Amyloid-beta protein dimers isolated directly from Alzheimer's brains impair synaptic plasticity and memory. Nat Med 14:837–842
8. Mesulam MM (1999) Neuroplasticity failure in Alzheimer's disease: bridging the gap between plaques and tangles. Neuron 24:521–529
9. Rowan MJ, Klyubin I, Cullen WK, Anwyl R (2003) Synaptic plasticity in animal models of early Alzheimer's disease. Philos Trans R Soc Lond Ser B Biol Sci 358:821–828
10. Selkoe DJ (2004) Alzheimer disease: mechanistic understanding predicts novel therapies. Ann Intern Med 140:627–638
11. Sakono M, Zako T (2010) Amyloid oligomers: formation and toxicity of Abeta oligomers. FEBS J 277(6):1348–1358
12. Walker E, Mittal V, Tessner K (2008) Stress and the hypothalamic pituitary adrenal axis in the developmental course of schizophrenia. Annu Rev Clin Psychol 4:189–216
13. Whitworth JA, Mangos GJ, Kelly JJ (2000) Cushing, cortisol, and cardiovascular disease. Hypertension 36:912–916
14. Gerges NZ, Aleisa AM, Schwarz LA, Alkadhi KA (2004) Reduced basal CaMKII levels in hippocampal CA1 region: possible cause of stress-induced impairment of LTP in chronically stressed rats. Hippocampus 14:402–410
15. Gerges NZ, Alzoubi KH, Park CR, Diamond DM, Alkadhi KA (2004) Adverse effect of the combination of hypothyroidism and chronic psychosocial stress on hippocampus-dependent memory in rats. Behav Brain Res 155:77–84
16. Alzoubi KH, Aleisa AM, Alkadhi KA (2008) Effect of chronic stress or nicotine on hypothyroidism–induced enhancement of LTD: electrophysiological and molecular studies. Neurobiol Dis 32(1):81–87
17. Elgh E, Lindqvist Astot A, Fagerlund M, Eriksson S, Olsson T, Näsman B (2006) Cognitive dysfunction, hippocampal atrophy and glucocorticoid feedback in Alzheimer's disease. Biol Psychiatry 59(2):155–161
18. Hartmann A, Veldhuis JD, Deuschle M, Standhardt H, Heuser I (1997) Twenty-four hour cortisol release profiles in patients with Alzheimer's and Parkinson's disease compared to normal controls: ultradian secretory pulsatility and diurnal variation. Neurobiol Aging 18:285–289
19. Budas G, Coughlan CM, Seckl JR, Breen KC (1999) The effect of corticosteroids on amyloid beta precursor protein/amyloid precursor-like protein expression and processing in vivo. Neurosci Lett 276(1):61–64
20. Islam A, Kalaria RN, Winblad B, Adem A (1998) Enhanced localization of amyloid beta precursor protein in the rat hippocampus following long-term adrenalectomy. Brain Res 806:108–112
21. Wilson MA, McNaughton BL (1993) Dynamics of the hippocampal ensemble code for space. Science 261:1055–1058
22. Wilson RS, Evans DA, Bienias JL, Mendes de Leon CF, Schneider JA, Bennett DA (2003) Proneness to psychological distress is associated with risk of Alzheimer's disease. Neurology 61:1479–1485
23. Wilson RS, Schneider JA, Boyle PA, Arnold SE, Tang Y, Bennett DA (2007) Chronic distress and incidence of mild cognitive impairment. Neurology 68:2085–2092
24. Aleisa AM, Alzoubi KH, Alkadhi KA (2006b) Chronic but not acute nicotine treatment reverses stress-induced impairment of LTP in anesthetized rats. Brain Res 1097:78–84
25. McEwen BS (2008) Central effects of stress hormones in health and disease: understanding the protective and damaging effects of stress and stress mediators. Eur J Pharmacol 583:174–185
26. Alkadhi KA (2018) Exercise decreases BACE and APP levels in the hippocampus of a rat model of Alzheimer's disease. Mol Cell Neurosci 86:25–29
27. Buresova O, Bures J, Oitzl MS, Zahalka A (1985) Radial maze in the water tank: an

aversively motivated spatial working memory task. Physiol Behav 34:1003–1005
28. Hodges H (1996) Maze procedures: the radial-arm and water maze compared. Brain Res Cogn Brain Res 3:167–181
29. Diamond DM, Park CR, Heman KL, Rose GM (1999) Exposing rats to a predator impairs spatial working memory in the radial arm water maze. Hippocampus 9:542–552
30. Alamed J, Wilcock DM, Diamond DM, Gordon MN, Morgan D (2006) Two-day radial-arm water maze learning and memory task; robust resolution of amyloid-related memory deficits in transgenic mice. Nat Protoc 1:1671–1679
31. Tran TT, Srivareerat M, Alkadhi KA (2010) Chronic psychosocial stress triggers cognitive impairment in a novel at-risk model of Alzheimer's disease. Neurobiol Dis 37:756–763
32. Tran TT, Srivareerat M, Alkadhi KA (2011) Chronic psychosocial stress accelerates impairment of long-term memory and late-phase long-term potentiation in an at-risk model of Alzheimer's disease. Hippocampus 21:724–732
33. Pepeu G, Giovannelli L, Casamenti F, Scali C, Bartolini L (1996) Amyloid beta-peptides injection into the cholinergic nuclei: morphological, neurochemical and behavioral effects. Prog Brain Res 109:273–282
34. Nabeshima T, Itoh A (1998) Toxicity of beta-amyloid peptide. J Toxicol Sci 23(Suppl 2):177–180
35. Giovannini MG, Scali C, Prosperi C, Bellucci A, Vannucchi MG, Rosi S, Pepeu G, Casamenti F (2002) Beta-amyloid-induced inflammation and cholinergic hypofunction in the rat brain in vivo: involvement of the p38MAPK pathway. Neurobiol Dis 11 (2):257–274
36. Goodman Y, Mattson MP (1994) Secreted forms of beta-amyloid precursor protein protect hippocampal neurons against amyloid beta-peptide-induced oxidative injury. Exp Neurol 128:1–12
37. Green PS, Gridley KE, Simpkins JW (1996) Estradiol protects against beta-amyloid (25-35)-induced toxicity in SK-N-SH human neuroblastoma cells. Neurosci Lett 218:165–168
38. Lambert MP, Barlow AK, Chromy BA, Edwards C, Freed R, Liosatos M, Morgan TE, Rozovsky I, Trommer B, Viola KL, Wals P, Zhang C, Finch CE, Krafft GA, Klein WL (1998) Diffusible, nonfibrillar ligands derived from Abeta1-42 are potent central nervous system neurotoxins. Proc Natl Acad Sci U S A 95:6448–6453
39. Assis-Nascimento P, Jarvis KM, Montague JR, Mudd LM (2007) Beta-amyloid toxicity in embryonic rat astrocytes. Neurochem Res 32:1476–1482
40. Lue LF, Kuo YM, Roher AE, Brachova L, Shen Y, Sue L, Beach T, Kurth JH, Rydel RE, Rogers J (1999) Soluble amyloid beta peptide concentration as a predictor of synaptic change in Alzheimer's disease. Am J Pathol 155:853–862
41. Ingelsson M, Fukumoto H, Newell KL, Growdon JH, Hedley-Whyte ET, Frosch MP, Albert MS, Hyman BT, Irizarry MC (2004) Early Abeta accumulation and progressive synaptic loss, gliosis, and tangle formation in AD brain. Neurology 62:925–931
42. Carrotta R, Di Carlo M, Manno M, Montana G, Picone P, Romancino D, San Biagio PL (2006) Toxicity of recombinant beta-amyloid prefibrillar oligomers on the morphogenesis of the sea urchin Paracentrotus lividus. FASEB J 20:1916–1917
43. Dao AT, Zagaar MA, Levine AT, Salim S, Eriksen JL, Alkadhi KA (2013) Treadmill exercise prevents learning and memory impairment in Alzheimer's disease-like pathology. Curr Alzheimer Res 10 (5):507–515
44. Paxinos G, Watson C (1986) The rat brain in stereotaxic coordinates, 2nd edn. Academic Press Inc., New York
45. Aleisa AM, Alzoubi KH, Gerges NZ, Alkadhi KA (2006d) Chronic psychosocial stress-induced impairment of hippocampal LTP: possible role of BDNF. Neurobiol Dis 22:453–462
46. Gerges NZ, Stringer JL, Alkadhi KA (2001) Combination of hypothyroidism and stress abolishes early LTP in the CA1 but not dentate gyrus of hippocampus of adult rats. Brain Res 922(2):250–260
47. Alkadhi KA, Alzoubi KH, Aleisa AM, Tanner FL, Nimer AS (2005) Psychosocial stress-induced hypertension results from in vivo expression of long-term potentiation in rat sympathetic ganglia. Neurobiol Dis 20 (3):849–857
48. Zagaar M, Alhaider I, Dao A, Levine A, Alkarawi A, Alzubaidy M, Alkadhi K (2012) The beneficial effects of regular exercise on cognition in REM sleep deprivation: behavioral, electrophysiological and molecular evidence. Neurobiol Dis 45(3):1153–1162
49. Aleisa AM, Alzoubi KH, Alkadhi KA (2006a) Nicotine prevents stress-induced

enhancement of long-term depression in hippocampal area CA1: electrophysiological and molecular studies. J Neurosci Res 83:309–317
50. Alzoubi KH, Aleisa AM, Alkadhi KA (2006) Molecular studies on the protective effect of nicotine in adult-onset hypothyroidism-induced impairment of long-term potentiation. Hippocampus 16(10):861–874
51. Gerges NZ, Alkadhi KA (2004) Hypothyroidism impairs late LTP in CA1 region but not in dentate gyrus of the intact rat hippocampus: MAPK involvement. Hippocampus 14(1):40–45
52. Aleisa AM, Alzoubi KH, Gerges NZ, Alkadhi KA (2006c) Nicotine blocks stress-induced impairment of spatial memory and long-term potentiation of the hippocampal CA1 region. Int J Neuropsychopharmacol 9(4):417–426
53. Alzoubi KH, Aleisa AM, Alkadhi KA (2007) Nicotine prevents disruption of the late phase LTP-related molecular cascade in adult-onset hypothyroidism. Hippocampus 17 (8):654–664
54. Laurberg S (1979) Commissural and intrinsic connections of the rat hippocampus. J Comp Neurol 184:685–708
55. Tamamaki N, Abe K, Nojyo Y (1988) Three-dimensional analysis of the whole axonal arbors originating from single CA2 pyramidal neurons in the rat hippocampus with the aid of a computer graphic technique. Brain Res 452:255–272
56. Ishizuka N, Weber J, Amaral DG (1990) Organization of intrahippocampal projections originating from CA3 pyramidal cells in the rat. J Comp Neurol 295:580–623
57. Papatheodoropoulos C, Kostopoulos G (2000) Decreased ability of rat temporal hippocampal CA1 region to produce long-term potentiation. Neurosci Lett 279:177–180
58. Alhaider IA, Aleisa AM, Tran TT, Alzoubi KH, Alkadhi KA (2010) Chronic caffeine treatment prevents sleep deprivation-induced impairment of cognitive function and synaptic plasticity. Sleep 33(4):437–444
59. Alkadhi KA, Tran TT (2015) Chronic stress decreases basal levels of memory-related signaling molecules in area CA1 of at-risk (subclinical) model of Alzheimer's disease. Mol Neurobiol 52(1):93–100
60. Lisman J, Schulman H, Cline H (2002) The molecular basis of CaMKII function in synaptic and behavioural memory. Nat Rev Neurosci 3:175–190
61. Malenka RC, Kauer JA, Perkel DJ, Mauk MD, Kelly PT, Nicoll RA, Waxham MN (1989) An essential role for postsynaptic calmodulin and protein kinase activity in long-term potentiation. Nature 340(6234):554–557
62. Nicoll RA, Malenka RC, Kauer JA (1989) The role of calcium in long-term potentiation. Ann N Y Acad Sci 568:166–170
63. Fukunaga K, Stoppini L, Miyamoto E, Muller D (1993) Long-term potentiation is associated with an increased activity of Ca2+/calmodulin-dependent protein kinase II. J Biol Chem 268:7863–7867
64. Feng TP (1995) The involvement of PKC and multifunctional CaM kinase II of the postsynaptic neuron in induction and maintenance of long-term potentiation. Prog Brain Res 105:55–63
65. Kim JJ, Yoon KS (1998) Stress: metaplastic effects in the hippocampus. Trends Neurosci 21:505–509
66. Gerendasy DD, Sutcliffe JG (1997) RC3/neurogranin, a postsynaptic calpacitin for setting the response threshold to calcium influxes. Mol Neurobiol 15:131–163
67. Wang JH, Kelly PT (1995) Postsynaptic injection of CA2+/CaM induces synaptic potentiation requiring CaMKII and PKC activity. Neuron 15:443–452
68. Holmes WR (2000) Models of calmodulin trapping and CaM kinase II activation in a dendritic spine. J Comput Neurosci 8 (1):65–85
69. Fukunaga K, Muller D, Miyamoto E (1996) CaMkinase II in long-term potentiation. Neurochem Int 28:343–358
70. Nayak AS, Moore CI, Browning MD (1996) Ca2+/calmodulin dependent protein kinase II phosphorylation of the presynaptic protein synapsin I is persistently increased during long-term potentiation. Proc Natl Acad Sci U S A 93:15451–15456
71. Barria A, Derkach V, Soderling T (1997) Identification of the Ca2+/calmodulin-dependent protein kinase II regulatory phosphorylation site in the alpha-amino-3-hydroxyl-5-methyl-4-isoxazole-propionate-type glutamate receptor. J Biol Chem 272:32727–32730
72. Wang JH, Kelly PT (1996) Regulation of synaptic facilitation by postsynaptic Ca2+/CaM pathways in hippocampal CA1 neurons. J Neurophysiol 76(1):276–286
73. Fukunaga K, Miyamoto E (2000) A working model of CaM kinase II activity in hippocampal long-term potentiation and memory. Neurosci Res 38:3–17
74. Bourtchuladze R, Frenguelli B, Blendy J, Cioffi D, Schutz G, Silva AJ (1994) Deficient

long-term memory in mice with a targeted mutation of the cAMP-responsive element-binding protein. Cell 79:59–68
75. Lonze BE, Ginty DD (2002) Function and regulation of CREB family transcription factors in the nervous system. Neuron 35:605–623
76. Ecke LE, Cleck JN, White P, Schug J, Mifflin L, Blendy JA (2011) CREB-mediated alterations in the amygdala transcriptome: coordinated regulation of immune response genes following cocaine. Int J Neuropsychopharmacol 14(8):1111–1126
77. Datson NA, Speksnijder N, Mayer JL, Steenbergen PJ, Korobko O, Goeman J, de Kloet ER, Joëls M, Lucassen PJ (2012) The transcriptional response to chronic stress and glucocorticoid receptor blockade in the hippocampal dentate gyrus. Hippocampus 22(2):359–371
78. Li G, Wang Y, Yan M, Ma H, Gao Y, Li Z, Li C, Tian H, Zhuo C (2016) Time-dependent co-relation of BDNF and CREB mRNAs in adult rat brains following acute psychological stress in the communication box paradigm. Neurosci Lett 624:34–41
79. Alberini CM (2009) Transcription factors in long-term memory and synaptic plasticity. Physiol Rev 89(1):121–145
80. Arvanitis DN, Ducatenzeiler A, Ou JN, Grodstein E, Andrews SD, Tendulkar SR, Ribeiro-da-Silva A, Szyf M, Cuello AC (2007) High intracellular concentrations of amyloid-beta block nuclear translocation of phosphorylated CREB. J Neurochem 103:216–228
81. Snyder EM, Nong Y, Almeida CG, Paul S, Moran T, Choi EY, Nairn AC, Salter MW, Lombroso PJ, Gouras GK, Greengard P (2005) Regulation of NMDA receptor trafficking by amyloid-beta. Nat Neurosci 8:1051–1058
82. Mulkey RM, Endo S, Shenolikar S, Malenka RC (1994) Involvement of a calcineurin/inhibitor-1 phosphatase cascade in hippocampal long-term depression. Nature 369:486–488
83. Strack S, Barban MA, Wadzinski BE, Colbran RJ (1997) Differential inactivation of postsynaptic density-associated and soluble Ca2+/calmodulin-dependent protein kinase II by protein phosphatases 1 and 2A. J Neurochem 68(5):2119–2128
84. Thiels E, Kanterewicz BI, Knapp LT, Barrionuevo G, Klann E (2000) Protein phosphatase-mediated regulation of protein kinase C during long-term depression in the adult hippocampus in vivo. J Neurosci 20 (19):7199–7207
85. Wang JH, Kelly PT (1997) Postsynaptic calcineurin activity downregulates synaptic transmission by weakening intracellular Ca2+ signaling mechanisms in hippocampal CA1 neurons. J Neurosci 17(12):4600–4611
86. Winder DG, Mansuy IM, Osman M, Moallem TM, Kandel ER (1998) Genetic and pharmacological evidence for a novel, intermediate phase of long-term potentiation suppressed by calcineurin. Cell 92(1):25–37
87. Mansuy IM, Winder DG, Moallem TM, Osman M, Mayford M, Hawkins RD, Kandel ER (1998) Inducible and reversible gene expression with the rtTA system for the study of memory. Neuron 21:257–265
88. Zoladz PR, Park CR, Halonen JD, Salim S, Alzoubi KH, Srivareerat M, Fleshner M, Alkadhi KA, Diamond DM (2012) Differential expression of molecular markers of synaptic plasticity in the hippocampus, prefrontal cortex, and amygdala in response to spatial learning, predator exposure, and stress-induced amnesia. Hippocampus 22 (3):577–589
89. Takase K, Yamamoto Y, Yagami T (2012) Maternal deprivation in the middle of a stress hyporesponsive period decreases hippocampal calcineurin expression and causes abnormal social and cognitive behaviours in adult male Wistar rats: relevance to negative symptoms of schizophrenia. Behav Brain Res 232 (1):306–315
90. Gerges NZ, Aleisa AM, Schwarz LA, Alkadhi KA (2003) Chronic psychosocial stress decreases calcineurin in the dentate gyrus: a possible mechanism for preservation of early LTP. Neuroscience 117:869–874
91. Gong CX, Singh TJ, Grundke-Iqbal I, Iqbal K (1994) Alzheimer's disease abnormally phosphorylated tau is dephosphorylated by protein phosphatase-2B (calcineurin). J Neurochem 62(2):803–806
92. Billingsley ML, Ellis C, Kincaid RL, Martin J, Schmidt ML, Lee VM et al (1994) Calcineurin immunoreactivity in Alzheimer's disease. Exp Neurol 126:178–184
93. Berridge MJ (2010) Calcium hypothesis of Alzheimer's disease. Pflugers Arch 459:441–449
94. Taglialatela G, Hogan D, Zhang WR, Dineley KT (2009) Intermediate- and long-term recognition memory deficits in Tg2576 mice are reversed with acute calcineurin inhibition. Behav Brain Res 200(1):95–99

95. Hata R, Masumura M, Akatsu H, Li F, Fujita H, Nagai Y et al (2001) Up-regulation of calcineurin Abeta mRNA in the Alzheimer's disease brain: assessment by cDNA microarray. Biochem Biophys Res Commun 284:310–316
96. Hong CJ, Liou YJ, Tsai SJ (2011) Effects of BDNF polymorphisms on brain function and behavior in health and disease. Brain Res Bull 86(5–6):287–297
97. Knusel B, Beck KD, Winslow JW, Rosenthal A, Burton LE, Widmer HR, Nikolics K, Hefti F (1992) Brain-derived neurotrophic factor administration protects basal forebrain cholinergic but not nigral dopaminergic neurons from degenerative changes after axotomy in the adult rat brain. J Neurosci 12:4391–4402
98. Lindholm D, Carroll P, Tzimagiogis G, Thoenen H (1996) Autocrine-paracrine regulation of hippocampal neuron survival by IGF-1 and the neurotrophins BDNF, NT-3 and NT-4. Eur J Neurosci 8(7):1452–1460
99. Lowenstein DH, Arsenault L (1996) The effects of growth factors on the survival and differentiation of cultured dentate gyrus neurons. J Neurosci 16(5):1759–1769
100. Ghosh A, Carnahan J, Greenberg ME (1994) Requirement for BDNF in activity-dependent survival of cortical neurons. Science 263:1618–1623
101. Bramham CR, Messaoudi E (2005) BDNF function in adult synaptic plasticity: the synaptic consolidation hypothesis. Prog Neurobiol 76:99–125
102. Soule J, Messaoudi E, Bramham CR (2006) Brain-derived neurotrophic factor and control of synaptic consolidation in the adult brain. Biochem Soc Trans 34:600–604
103. Dao AT, Zagaar MA, Levine AT, Alkadhi KA (2016) Comparison of the effect of exercise on late-phase LTP of the dentate gyrus and CA1 of Alzheimer's disease model. Mol Neurobiol 53(10):6859–6868
104. Dao AT, Zagaar MA, Alkadhi KA (2015) Moderate treadmill exercise protects synaptic plasticity of the dentate gyrus and related signaling Cascade in a rat model of Alzheimer's disease. Mol Neurobiol 52(3):1067–1076
105. Thoenen H (2000) Neurotrophins and activity-dependent plasticity. Prog Brain Res 128:183–191
106. Huang EJ, Reichardt LF (2001) Neurotrophins: roles in neuronal development and function. Annu Rev Neurosci 24:677–736
107. Yamada K, Mizuno M, Nabeshima T (2002) Role for brain-derived neurotrophic factor in learning and memory. Life Sci 70:735–744
108. Connor B, Young D, Yan Q, Faull RL, Synek B, Dragunow M (1997) Brain-derived neurotrophic factor is reduced in Alzheimer's disease. Brain Res Mol Brain Res 49:71–81
109. Smith MA, Makino S, Kvetnansky R, Post RM (1995) Stress and glucocorticoids affect the expression of brain-derived neurotrophic factor and neurotrophin-3 mRNAs in the hippocampus. J Neurosci 15(3 Pt 1):1768–1777
110. Ueyama T, Kawai Y, Nemoto K, Sekimoto M, Toné S, Senba E (1997) Immobilization stress reduced the expression of neurotrophins and their receptors in the rat brain. Neurosci Res 28(2):103–110
111. Lee KW, Kim JB, Seo JS, Kim TK, Im JY, Baek IS, Kim KS, Lee JK, Han PL (2009) Behavioral stress accelerates plaque pathogenesis in the brain of Tg2576 mice via generation of metabolic oxidative stress. J Neurochem 108:165–175
112. Murakami S, Imbe H, Morikawa Y, Kubo C, Senba E (2005) Chronic stress, as well as acute stress, reduces BDNF mRNA expression in the rat hippocampus but less robustly. Neurosci Res 53(2):129–139
113. Tsankova NM, Berton O, Renthal W, Kumar A, Neve RL, Nestler EJ (2006) Sustained hippocampal chromatin regulation in a mouse model of depression and antidepressant action. Nat Neurosci 9(4):519–525
114. Nair A, Vadodaria KC, Banerjee SB, Benekareddy M, Dias BG, Duman RS, Vaidya VA (2007) Stressor-specific regulation of distinct brain-derived neurotrophic factor transcripts and cyclic AMP response element-binding protein expression in the postnatal and adult rat hippocampus. Neuropsychopharmacology 32(7):1504–1519
115. Rothman SM, Griffioen KJ, Wan R, Mattson MP (2012) Brain-derived neurotrophic factor as a regulator of systemic and brain energy metabolism and cardiovascular health. Ann N Y Acad Sci 1264:49–63
116. Alkadhi KA (2012) Chronic psychosocial stress exposes Alzheimer's disease phenotype in a novel at-risk model. Front Biosci (Elite Ed) 4:214–229
117. Tong L, Thornton PL, Balazs R, Cotman CW (2001) Beta -amyloid-(1-42) impairs activity-dependent cAMP-response element-binding protein signaling in neurons at concentrations in which cell survival is not compromised. J Biol Chem 276:17301–17306

118. Tong L, Balazs R, Thornton PL, Cotman CW (2004) Beta-amyloid peptide at sublethal concentrations downregulates brain-derived neurotrophic factor functions in cultured cortical neurons. J Neurosci 24(30):6799–6809
119. Liu F, Grundke-Iqbal I, Iqbal K, Oda Y, Tomizawa K, Gong CX (2005) Truncation and activation of calcineurin A by calpain I in Alzheimer disease brain. J Biol Chem 280:37755–37762
120. Thinakaran G, Koo EH (2008) Amyloid precursor protein trafficking, processing, and function. J Biol Chem 283 (44):29615–29619
121. LaFerla FM, Green KN, Oddo S (2007) Intracellular amyloid-beta in Alzheimer's disease. Nat Rev Neurosci 8(7):499–509
122. Querfurth HW, LaFerla FM (2010) Alzheimer's disease. N Engl J MedN Engl J Med 362(4):329–344
123. Vassar R, Bennett BD, Babu-Khan S, Kahn S, Mendiaz EA, Denis P, Teplow DB, Ross S, Amarante P, Loeloff R, Luo Y, Fisher S, Fuller J, Edenson S, Lile J, Jarosinski MA, Biere AL, Curran E, Burgess T, Louis JC, Collins F, Treanor J, Rogers G, Citron M (1999) Beta-secretase cleavage of Alzheimer's amyloid precursor protein by the transmembrane aspartic protease BACE. Science 286 (5440):735–741
124. Townsend M, Mehta T, Selkoe DJ (2007) Soluble Abeta inhibits specific signal transduction cascades common to the insulin receptor pathway. J Biol Chem 282:33305–33312
125. Zhao D, Watson JB, Xie CW (2004) Amyloid beta prevents activation of calcium/calmodulin-dependent protein kinase II and AMPA receptor phosphorylation during hippocampal long-term potentiation. J Neurophysiol 92:2853–2858
126. Durany N, Michel T, Kurt J, Cruz-Sanchez FF, Cervas-Navarro J, Riederer P (2000) Brain-derived neurotrophic factor and neurotrophin-3 levels in Alzheimer's disease brains. Int J Dev Neurosci 18:807–813
127. Lindvall O, Ernfors P, Bengzon J, Kokaia Z, Smith ML, Siesjo BK, Persson H (1992) Differential regulation of mRNAs for nerve growth factor, brain-derived neurotrophic factor, and neurotrophin 3 in the adult rat brain following cerebral ischemia and hypoglycemic coma. Proc Natl Acad Sci U S A 89:648–652
128. Barde YA (1989) Trophic factors and neuronal survival. Neuron 2:1525–1534
129. Zuccato C, Cattaneo E (2009) Brain-derived neurotrophic factor in neurodegenerative diseases. Nat Rev Neurol 5(6):311–322
130. Hellstrom-Lindahl E, Court J, Keverne J, Svedberg M, Lee M, Marutle A, Thomas A, Perry E, Bednar I, Nordberg A (2004) Nicotine reduces A beta in the brain and cerebral vessels of APPsw mice. Eur J Neurosci 19:2703–2710
131. Tang Y, Yamada K, Kanou Y, Miyazaki T, Xiong X, Kambe F, Murata Y, Seo H, Nabeshima T (2000) Spatiotemporal expression of BDNF in the hippocampus induced by the continuous intracerebroventricular infusion of beta-amyloid in rats. Brain Res Mol Brain Res 80:188–197
132. Catania C, Sotiropoulos I, Silva R, Onofri C, Breen KC, Sousa N, Almeida OF (2009) The amyloidogenic potential and behavioral correlates of stress. Mol Psychiatry 14:95–105

Chapter 5

Using Zebrafish to Study and Visualize Barriers of the CNS

Robyn Anne Umans

Abstract

Blood–central nervous system (CNS) barriers are structures critical for maintaining neural homeostasis. While these barriers preserve a healthy brain environment, their physical and chemical properties create obstacles for efficacious CNS drug delivery and are compromised in various CNS disease states. Even though the blood–brain barrier (BBB) was discovered over a century ago, molecular mechanisms and interacting cell types necessary for BBB function and integrity have only recently become uncovered. Furthermore, the field of BBB biology could benefit from utilizing an innovative model system to dissect elements crucial for barrier properties. Uncovered molecular pathways could bring to light novel CNS drug delivery targets and prevent CNS disease progression. There has been a surge of zebrafish use in biomedical research over the past few decades, which has provided researchers a valuable animal model system to genetically and chemically study these barrier structures deep within the brain parenchyma. Here, I describe genetic, chemical, and histology methods for utilizing zebrafish in studying blood–CNS barriers such as the BBB and how these zebrafish assays could be used to exploit the unanswered questions that remain in CNS barrier biology.

Key words Zebrafish, Blood–brain barrier, Central nervous system, Genetics, Microscopy, Genetic screen, Chemical screen

1 Introduction

1.1 History and Studies of CNS-Barriers

The brain barriers of the central nervous system (CNS) are crucial for tissue homeostasis and regulation of xenobiotic entry into the brain. Three physically distinct barriers exist within the CNS: the blood–arachnoid barrier, blood–cerebral spinal fluid (CSF) barrier, and blood–brain barrier (BBB). Each molecularly distinct interface acts together to survey the innate immune response, provide mechanical protection, and regulate molecular transport and ion metabolism within the brain [1]. While all of these structures contribute to a healthy CNS, methods to study the BBB will be the focus of this chapter.

Thousands of miles of blood vessels constitute the neurovascular unit (NVU) and are crucial for ionic balance and proper brain function. The brain is a highly metabolic organ, accounting for 20%

Nicholas J. D. Wright (ed.), *Basic Neurobiology Techniques*, Neuromethods, vol. 152,
https://doi.org/10.1007/978-1-4939-9944-6_5,

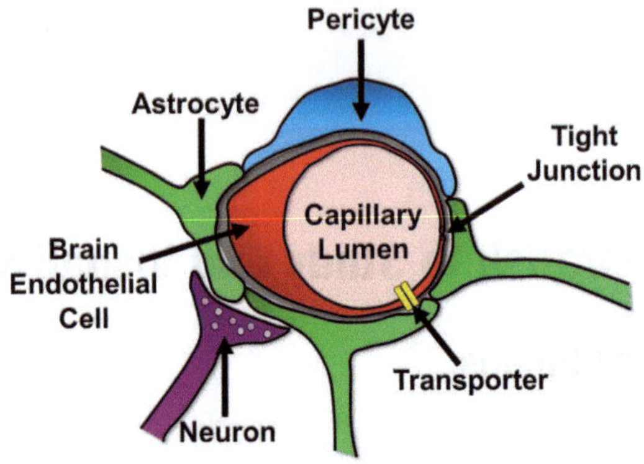

Fig. 1 The neurovascular unit: The NVU is composed of miles of cerebral blood vessels that establish a physiological arrangment with pericytes, astrocytes, and neurons to maintain CNS homeostasis. These intimate cellular relationships help foster cell autonomous and nonautonomous signaling pathways important for BBB development and maintenance. Additionally, a physical barrier is maintained by tight junction proteins and chemical regulation is mediated through transporter systems. While all of these cell types and proteins are required for a healthy brain, they inescapably limit the entry of most small molecule therapeutics into the CNS

of the body's energy consumption [2]. Oxygen demand plays a large role in the coupled process of functional hyperemia and accurate neuronal function at the tripartite synapse [3, 4]. In addition to its energetic requirements, the BBB physically limits the free passage of molecules in and out of the brain parenchyma. Tight junction proteins expressed between the endothelium regulate paracellular diffusion and transport systems along the vasculature modulating chemical flux supports these physiological processes and demands [1]. As it is deemed a unit, the NVU is composed of brain endothelial cells (BECs), astrocytes, neurons, and pericytes all establishing cell-autonomous and nonautonomous signals required for active BBB development and maintenance (Fig. 1) [5–8]. Consideration should be taken when choosing or analyzing various protein markers of the NVU. For example, astrocytes surrounding mammalian CNS blood vessels express distinctive markers from the fibrous astrocytes that interact with oligodendrocytes, although astrocytes are typically labeled with glial fibrillary associated protein (GFAP) [9, 10]. Additionally, GFAP does not label all astrocytes in the brain and not in the same pattern as certain ion channels or amino acid transporters [11]. Therefore, researchers should be tentative when interpreting glial–BEC interactions and when choosing a particular glial marker. Common NVU markers involved in these diverse signaling pathways as well as their functions are listed in Table 1.

While maintaining a healthy BBB is vital to protect the brain from xenobiotic damage, the physical and chemical properties of

Table 1
Common cellular markers of the neurovascular unit

Cell type	Protein marker	Marker function	Citations
Brain endothelial cells	CD31/PECAM	Cell adhesion and angiogenesis	[77]
	GLUT1	Glucose solute carrier transporter	[2, 78]
	MDR1/PGP/ABCB1	Multidrug resistance transporter	[79]
	BCRP/ABCG2	Multidrug resistance transporter	[80]
	CLAUDIN5	Tight junction component protein	[2, 20]
	ZO-1	Tight junction anchoring protein	[2, 81]
Neurons	NeuN	Neuronal nuclear antigen	[82]
Astrocytes	AQP4	Membrane protein and water channel	[8, 83]
	EAAT(1/2)	Sodium-dependent excitatory amino acid transporter	[2, 84]
	GFAP	Intermediate filament protein	[10]
	GLAST	Sodium-dependent glutamate transporter	[84]
	GLT1	Sodium-dependent glutamate transporter	[84]
Pericytes	PDGFRβ	Vascular development and stabilization	[5, 85–87]
	NG2	Structural chondroitin sulfate proteoglycan	[85, 88]
	DESMIN	Muscle intermediate filament	[85, 87]

the NVU prohibit the efficacious delivery of over 98% of small molecule therapeutics into the CNS [12]. Molecules that can diffuse across the BBB generally abide by Lipinski's rule of five, although they tend to be more lipid-soluble and lower in molecular weight [12, 13]. In addition, the BBB is compromised in CNS diseases such as glioma, multiple sclerosis, Parkinson's disease, and Alzheimer's disease, posing the question of whether a compromised BBB can be restored in disease states [2, 14]. Studies investigating mechanisms for healthy BBB development and maintenance could reveal signals important for improved CNS drug delivery and whether BBB impairment can be prevented in the progression of CNS disease.

The concept of the BBB was perceived over a century ago, yet experiments during the past few decades have only started to dissect the molecular and functional components of the BBB. For all of these biological processes, various model systems have been established. Therefore, researchers should consider the pros and cons to every BBB model to best answer scientific questions of interest. Table 2 highlights different model systems that have been used to study the BBB and considerations toward their use in experimental design.

Because of the array of systems used to study the BBB, researchers should use the most suitable model that fulfills their project goals. Groups across academia and industry have capitalized on the ease of culture techniques and generated in vitro models to study BBB biology and its potential use for drug discovery

Table 2
Model systems used in blood–brain barrier studies

Model system	Advantages	Disadvantages	Studies including system
In vitro cultures	• Cost efficiency • Easy genetic or chemical manipulation • Good tool for screens	• All BBB cell types complicated to co-culture • Simplified version of true physiology (i.e., blood flow, transendothelial resistance)	[16, 89]
Drosophila melanogaster	• Cost efficiency • Genetic tractability • Chemical and genetic screen capability	• Simplified hemolymph system	[23, 90]
Danio rerio	• Cost efficiency • Vertebrate • Genetic tractability • Chemical and genetic screen capability	• Models still being established • Not a mammalian system	[24, 26–28]
Avian	• Vertebrate • Developmental studies	• BBB is hard to access without surgical technique	[19]
Rodent	• Vertebrate • High homology with human • Genetic tractability	• BBB is hard to access without surgical technique	[5–7, 20, 22, 86, 87]

[15–17]. Such in vitro systems provide technical simplicity, but without co-culturing multiple cells types, it is hard to recapitulate complete aspects of blood flow, endothelial cell transcellular resistance, and the three-dimensional composition of the NVU. These in vitro models do have the benefits of straightforward manipulation and cost efficiency, but should ultimately complement findings in vivo.

The BBB is present in a spectrum of species, suggesting it was crucial during evolution to produce and retain this structure. Furthermore, in its simplest application in lower order species, the BBB may have evolved to maintain ionic homeostasis around synapses so that small changes in ionic flux would not disturb neuronal firing [18]. This evolutionary pressure to develop a barrier system has provided models in avian, fish, rodent, and even insect species for use in BBB biology studies. Seminal transplantation studies between quail and chick emphasized the importance of cell nonautonomous signaling during BBB differentiation and therefore reinforced BBB conservation among vertebrates [19]. Traditionally,

BBB studies have utilized rodent models to assess the histology and function of the BBB over development, in genetic knockouts, and in disease models [5, 20–22]. However, because the BBB is located deep within the brain, this feature has made it difficult to study the rodent BBB in vivo without complex surgical techniques and expensive equipment. Therefore, models that traditionally favor high-throughput approaches have emerged as tools to study BBB biology. *Drosophila melanogaster* are fabulous for genetic studies, model homologous BBB physiology, but their BBB is a more simplified hemolymph-based system [23]. Whether for the purpose of transplantation, genetic manipulation, or homology, all these species present limitations that can be overcome by using multiple complementary models. In the past decade, the fresh water zebrafish, *Danio rerio*, has emerged as another nontraditional vertebrate used to study the BBB [24–28]. In this chapter, I focus on methods to study the BBB using zebrafish as a model organism.

1.2 Zebrafish as a Model Organism in Genetics, Development, and Disease

Considered merely a household pet for years, it was not until the 1960's when Dr. George Streisinger introduced zebrafish into biomedical research by his groundbreaking work investigating mutational analysis during embryonic development [29]. Since these pioneering studies, zebrafish have been noted worldwide for their genetic tractability. In the late twentieth century, large-scale, forward genetic screens were performed in the United States and Germany, uncovering mutations in processes pertaining to retinal development, cardiac function, and human genetic diseases [30]. The zebrafish genome was fully sequenced by 2001, making additional genetic techniques such as whole-genome sequencing, positional cloning, insertional mutagenesis, germline transgenesis, and human disease model generation possible in the zebrafish [31]. Candidate approaches with reverse genetics are now widely established in zebrafish with methods ranging from transcription activator-like effector nucleases (TALENS) to the CRISPR/Cas9 system [32, 33]. Human mutations have also been introduced in the zebrafish that cause the same diseases derived in human. Mutations in *BRAF(V600E)* from human melanoma introduced into the fish cause the disease and identified molecular regulators to accelerate the malignancy, suggesting a conservation of genes between zebrafish and mammals [34]. Taken together, zebrafish are an ideal model organism for phenotypic or candidate genetic screens.

In addition to the ease of molecular manipulation, researchers are also drawn to zebrafish for physical and biomedical purposes. Zebrafish exhibit high fecundity, rapid maturation outside of the mother, a transparent appearance, and a conserved body plan with biological systems to that of mammals (Fig. 2). Widely used in developmental studies, zebrafish also are an ideal model for studying BBB development. For example, Daneman et al. (2009) demonstrated Wnt signaling, a crucial regulator of embryogenesis, is important for CNS angiogenesis [35]. Transgenic zebrafish lines

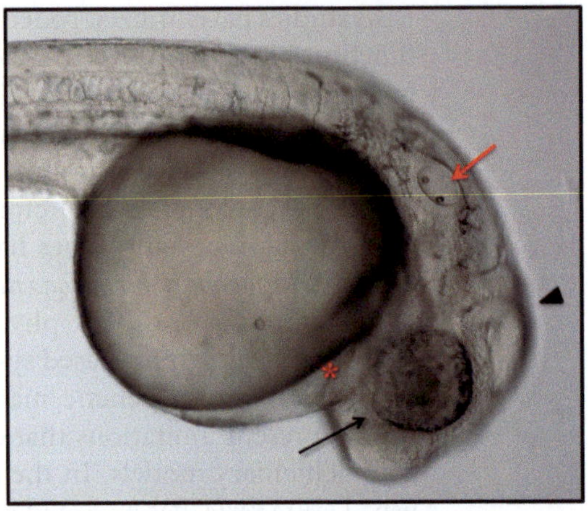

Fig. 2 Zebrafish are a suitable model organism for stdying developmental biological processes: A brightfield microscope image of an embryonic zebrafish illustrates advantages for this vertebrate model system. At merely 1 day post fertilization (dpf), the zebrafish embryo possesses a developing ear (**red arrow**), folds within the brain (**black arrowhead**), developing eyes (**black arrow**), and a beating heart with circulation (**red asterisk**). At young stages such as 1 dpf, young zebrafish can also be imaged without anesthesia by placing animals into a low melting point agarose solution or a methyl cellulose gel within a petri dish. While melanocyte pigment cells can already be seen at this stage, genetic mutants lacking pigmentation such as *casper* or chemicals added to the water such as PTU can prevent any pigment cells that may obstruct visualization of different body systems [65]

such as the inducible *Tg(hs:axin1)*, which over-expresses the negative Wnt signaling regulator, Axin1, under control of the heat shock promoter, would further dissect the importance of this signaling pathway at the BBB [36]. Such advantageous features have also deemed zebrafish a valuable tool in high-throughput drug screening studies. Being able to administer compound libraries to multiple vertebrates in a multiwell format opens the window for in vivo therapeutic drug discovery. For example, dorsomorphin, a potent and widely used bone morphogenetic protein (BMP) signaling antagonist, was discovered to disrupt dorsoventral axis formation in a zebrafish small molecule screen [37]. In addition to screens assaying developmental processes, zebrafish have been used to identify compounds that affect the cell cycle, cancer, and angiogenesis [38–41]. Pharmacokinetic testing such as absorption, distribution, metabolism, and excretion (ADME) studies have not been established in zebrafish yet, but hydrophobicity as measured by logP values can predict bioavailability in zebrafish treated with compounds with various molecular weights [42]. Zebrafish have also been a successful model to repurpose drugs and studies have validated zebrafish small molecule screen hits in mice [43]. Even with all of these physical and technical advantages, it was not until the past few decades that zebrafish were utilized to study the BBB.

1.3 Cerebral Vasculature Develops in Zebrafish Similarly to Mammals

Development of the zebrafish primordial hindbrain channels (PHBC) commences around 28 h post-fertilization (hpf) in the brain similarly as the formation of the perineural vascular plexus at E8.5 in the mouse and E11 in the rat [5, 44, 45]. Growth factors, including vascular endothelial cell growth factor (VEGF) signaling, have been widely established as important and specific mitogens in vessel formation [46]. Heterozygous lethality of one VEGF allele demonstrates the importance of this signaling pathway in blood vessel development [47]. Various zebrafish VEGF mutants also demonstrate genetically distinct VEGF profiles for particular blood vessel types [48].

After vasculogenesis, nascent blood vessels sprout into the developing brain parenchyma. This process of newly formed vessels budding off of preexisting ones is deemed angiogenesis [49]. CNS angiogenesis occurs similarly at 30hpf from zebrafish PHBCs as it does from the mouse perineural vascular plexus at E9.5–10 [35, 45, 50]. In the zebrafish, the main arterial route and vessel branching becomes hearty by 2 days post fertilization (dpf) [44]. In addition to CNS angiogenesis, markers of barriergenesis also differentiate similarly in the zebrafish. Glucose transporter 1 (GLUT1) has been identified as one of the initial markers of cerebral angiogenesis in mammals appearing at E12 in rat and E11 in mouse BECs [51, 52]. Zebrafish also specifically express Glut1 at the start of brain angiogenesis suggesting that brain endothelial cells express BBB properties during initial CNS angiogenesis [50, 53]. Recently, it has also been established that zebrafish express brain pericytes ensheathing endothelium as early as 48 hpf during brain development [54, 55]. The difference in timing of pericyte versus astrocyte development in the zebrafish brain may be an artifact of evolution, lack of specific transgenic lines to look at pericyte development in real time, as well as a limitation on established zebrafish astrocyte markers. Therefore, by using the zebrafish to study the BBB in real time, with genetic tools, chemical modulation, and developmental biology techniques, many unknowns about blood–CNS barriers could be answered.

1.4 Zebrafish Models and Characterization of Blood–CNS Barriers

In addition to the developmental conservation of CNS angiogenesis, zebrafish possess a BBB that is phenotypically and functionally comparable to that in mammals. Tight junction proteins, such as Claudin-5, anchored between BBB endothelium maintain a physical barrier between the circulation and the brain, as large molecular weight tracers are retained within the zebrafish brain vasculature when injected into circulation [24, 25]. When injections are performed earlier during zebrafish development, small molecules such as fluorescein display leakage only in the brain ventricles but not out of brain capillaries, suggesting the BBB develops before the blood–CSF barrier in zebrafish as it does in rodents [25, 56]. While conclusions have been based on a few studies, these functional assays have demonstrated functional BBB development early within 3 dpf in the zebrafish [25, 26]. Being able to visualize a functional BBB in a matter of a few days makes the zebrafish a prime model organism for studying this CNS-barrier.

Chemical transport systems such as the multidrug resistance transporter 1(Mdr1) and Glut1 are also expressed along the zebrafish brain endothelium, signifying there is a mediation of active efflux and xenobiotic entry at the zebrafish BBB [27, 28, 50, 57]. Fischer et al. (2013) demonstrated the influx of rhodamine after chemical and genetic inhibition of zebrafish *abcb4*, what they deem as the zebrafish ortholog to human *MDR1* [58]. Furthermore, zebrafish possess multidrug resistance proteins in the gut epithelium and can actively prohibit the absorption of doxorubicin, a common MDR1 substrate, into the body [27]. Because the gut lumen and BBB maintain a similar barrier phenotype, it is possible that Mdr1 expressed at the zebrafish BBB also excludes compounds similarly to that of MDR1 in humans [59]. Studies such as that by Tam et al. (2012) also validate the conservation and utility of the zebrafish model, as the same CNS-specific angiogenesis regulators were present in the zebrafish and mouse [28].

Traditionally, studies have focused on the molecular expression of the various aforementioned BBB related proteins and the functional retention of conjugated tracers to characterize blood–CNS barriers. These methods are reproducible among model organisms and have been fully implemented in zebrafish [24, 25, 27]. Transgenic animals modified to express nonnative genetic sequences have been studied to investigate developmental processes, tissue, and disease [31, 60, 61]. These models are genetically constructed by placing a specific promoter sequence upstream of a transgene that will express some bioluminescent or fluorescent protein. Many groups utilize endothelial cell promoters to study BEC development even though all blood vessels are labeled. For example, the *Tg (Tie2:GFP)* line is commonly used to study BBB development in the mouse [6, 35, 45]. The *Tie2* promoter, an endothelial specific receptor tyrosine kinase, was used to create and understand vascular endothelial cell growth and function [62]. Studies for zebrafish BBB development have also utilized a similar transgenic line, *Tg (fli1a:EGFP)*^{y1} which labels blood vessels throughout the whole organism with the *friend leukemia virus integration 1a* (*fli1a*) promoter driving expression of the enhanced green fluorescent protein (EGFP) reporter gene (as used in Figs. 3 and 4) [24, 26, 60]. These transgenics are sufficient to study vessel development, but their promoter sequences are not specific to brain endothelium. More recently, the *Tg(l-fabp:DBP-EGFP)* zebrafish reporter was generated to study the development and maintenance of brain barriers based on the expression and circulation of a 78 kDa Vitamin-D binding protein fused to EGFP [25]. The *Tg(l-fabp: DBP-EGFP)* line is useful for studying the functional development

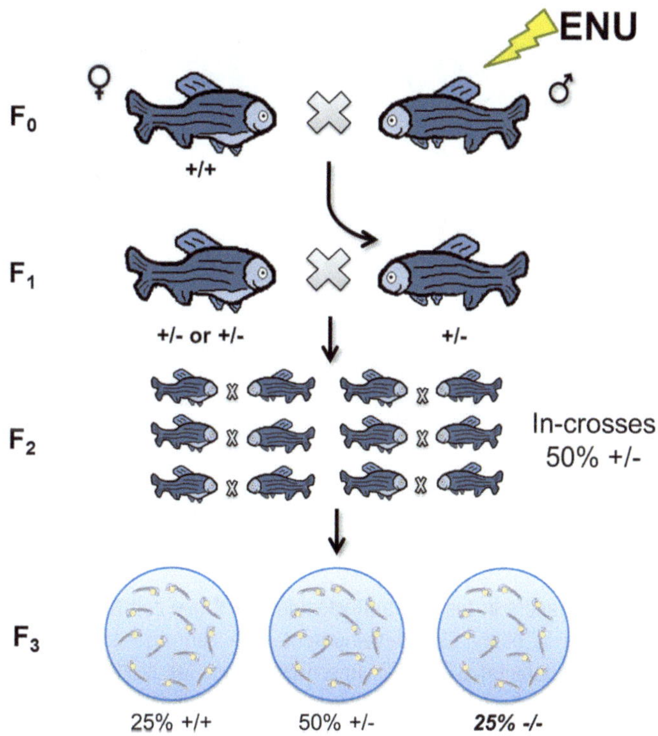

Fig. 3 Breeding scheme for a zebrafish F_3 forward genetic screen: The standard F_3 genetic screen identifies recessive mutations. It is important to identify or generate a specific line to treat with ENU based on the phenotype to be screened. Founder males are treated with the DNA alkylating agent ENU and then out-crossed to untreated females of the same genotype. The F_1 generation is produced and offspring are heterozygous for germline mutations. In-crosses in the F_1 or outcrossing F_1 heterozygotes to wild-type fish will create F_2 families. 50% of the fish within a F_2 family are heterozygous for mutations from their F_1 parents. F_2 fish are crossed to one another to generate F_3 progeny. These F_3 offspring will be screened for recessive mutations inherited in a Mendelian manner. Therefore, if a F_3 clutch carries a recessive mutation, around 25% of that clutch should present with a screenable phenotype

of the BBB endogenously without the need for tracer injection. However, this transgenic does not report on the barrier signals that are essential for maintaining BBB function. For example, as BEC-specific signals such as Gpr124 and Glut1 are discovered, transgenic tools could be generated to find the signals that regulate these BEC-specific properties during BBB development and maintenance [28, 45, 50]. Therefore, as more BEC-specific properties are identified, the field of BBB biology would benefit from the construction of conditional, transgenic BBB reporter lines, which could easily be accomplished with the zebrafish model system. As previously described, zebrafish are easily utilized in chemical and genetic screens. These assays may dissect novel pathways that affect

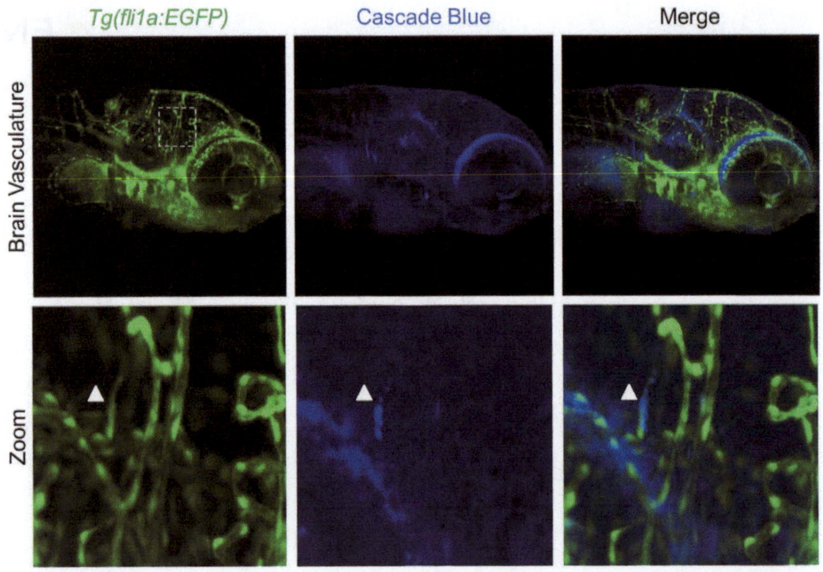

Fig. 4 Imaging tracers in the cerebral vasculature of live zebrafish larvae: Ubiquitous vascular reporters are commonly used to study BBB development. The zebrafish transgenic line *Tg(fli1a:EGFP)*^{y1}, whereby all endothelium are labeled with GFP, is a useful tool for studying vasculature development. Fluorescently labeled tracers can be injected intravenously into transgenics like this one and visualized to assess BBB integrity. Tracers are injected into the common cardinal vein and left in circulation for 30 min–1 h before imaging. At 4 dpf, the 3000 Da Cascade Blue tracer is excluded from the brain parenchyma and retained in the cerebral vasculature in a live larval zebrafish as visualized here with light sheet microscopy (**arrowhead**)

molecular BECs properties constituting the BBB versus signals that pertain to all endothelium. Once identified, these signals important to healthy BBB growth and preservation could lend themselves as useful drug targets to efficaciously deliver therapies into the CNS.

2 Materials

2.1 Forward Genetic and Chemical Screens to Identify BBB Modulators

2.1.1 Breeding and Husbandry

Specialized, aquatic, multilinked systems are offered from a variety of commercial manufacturers to house various sized zebrafish colonies. Green plastic floating plants can be used within home tanks or breeding boxes to provide environmental enrichment and help zebrafish separate from one another when desired. Embryos should be raised in egg water (refer to solution recipes in Table 3) until 4–5 dpf where they can be placed in larval water for rearing until adulthood. Under the counter incubators can be purchased to house embryos until they are ready to be raised. If not originally equipped, a light timer can be inserted into the incubator to match the appropriate light cycle that is in the main fish room. Breeding materials: system water; mating boxes; and fish nets. Egg water: 6 g Instant Ocean, 20 L Reverse Osmosis (RO) water and larval water: 18 g Instant Ocean, 20 L RO water.

Table 3
Solution recipes

Solution name	Component	Amount/ volume	pH	Notes
Egg water	Instant Ocean Reverse osmosis (RO) water	6 g 20 L	– –	–
Larval water	Instant Ocean RO water	18 g 20 L	– –	–
0.4% Tricaine stock	Tricaine powder (MS-222) 1 M Tris H_2O	400 mg 2.1 mL 9.79 mL	– pH 9 –	Store aliquots at −20 °C for long-term or 4 °C for weekly use; Final solution pH 7.0
1 mM $NaPO_4$ buffer	1 mM $NaPO_4$	100 mL	pH 6.5	For every week of ENU treatment
ENU treatment water	1 mM $NaPO_4$ (in fish system water)	1 L	pH 6.5	For every week of ENU treatment
ENU treatment water *with* tricaine	1 mM $NaPO_4$ (in fish system water) Tricaine	3 L 0.02%	– –	For every week of ENU treatment
10 mM $NaPO_4$	Made in egg water	1 mL	–	For every tank of fish to be treated; For spectrophotometric readings
ENU inactivating solution	$Na_2S_2O_3$ NaOH	20% 1%	– –	To inactivate any hazardous ENU spills and for cleanup; 5 L total
10X PTU stock	PTU powder Egg water	150 mg 500 mL	– –	Delays pigment development
1.2% low-melting point (LMP) agarose	LMP agarose powder Egg water	0.3 g 25 mL	– –	Agarose for imaging
PBST	Triton X-100 PBS	0.3% 1 L	– –	Used for IHC

2.1.2 ENU Treatment

N-ethyl-n-nitrosourea (ENU) is an alkylating agent that is commonly used in forward genetic screens. ENU confers single base pair mutations by transferring an ethyl group onto oxygen or nitrogen resulting in mispairing during DNA replication, and after two rounds, a single base pair substitution and mutation exists [63]. For each isopack of ENU, a particular amount of solutions must be prepared in a fume hood each week of treatment. The fume hood should also be covered in spill pads and the sash concealed in aluminum foil to maintain a dark environment for the treated fish. Tricaine, also known as MS-222 in the zebrafish community, is a common anesthetic used directly in the water. This compound is added during one of the treatments to help calm the zebrafish from any stress they may endure. 0.4% Tricaine stock (pH 7.0): 400 mg tricaine powder, 9.79 mL dH_2O, ~2.1 mL 1 M Tris (pH 9). Stocks should be stored long-term at −20 °C or kept at 4 °C for short-term (weekly) usage. For every week of treatment, the following solutions should be prepared; 10 mM $NaPO_4$ buffer: pH 6.5, 100 mL total; ENU treatment water *without* tricaine: 1 mM $NaPO_4$ (pH 6.5), made in fish system water, 1 L total; 1 mM $NaPO_4$ buffer: *with* tricaine, 3 L total. For every tank of fish to be treated: 10 mM $NaPO_4$ buffer (made in egg water) for spectrophotometric readings, 1 mL total; spectrophotometer; cuvettes; 5 L of inactivating solution: 20% $Na_2S_2O_3$, 1% NaOH (to inactivate any hazardous ENU spills and for cleanup). Four larger mating tanks are used for treatment on each set of fish; one tank with an insert and the three others without. Two Luer tip syringes and two 50 mL conical tubes are needed for preparation of the ENU isopack. For a highly detailed protocol solely discussing ENU screens in zebrafish, please also refer to the de Bruijn et al. (2009) protocol chapter [64].

2.1.3 Chemical Libraries and Materials

Chemical libraries can be sourced from academic screening core facilities or commercial vendors. For final dilutions, zebrafish can tolerate a solution with up to 1% dimethyl sulfoxide (DMSO). Phenylthiourea (PTU) delays pigment development in young zebrafish and can be used if not screening for phenotypes in a pigment-deficient background such as *casper* [65]. 10× PTU stock: 150 mg PTU powder, 500 mL egg water. Black bottomed 96 well plates are required for screening phenotypes with a fluorescent readout. To minimize evaporation from small volumes in a 96-well plate, a humidity chamber can be made that contains the screening plates. Simply moisten paper towels and line an appropriate sized sandwich container. If chemicals are to be deposited into wells without the aid of a liquid handling robot, multichannel pipettes can be used to save time and provide efficiency.

2.2 Live Imaging Materials

Zebrafish need to be immobilized for live imaging. To do so for imaging on an inverted lens microscope, zebrafish can be mounted in low melting point (LMP) agarose, using 35 mm dishes that contain a 14 mm microwell with glass coverslip insert. Refer to the "**Notes**" section for special guidelines when preparing LMP agarose. Dissection probes make good tools for orienting the zebrafish around until properly oriented. Tricaine solution will be needed to initially anesthetize the animal and cover the agarose once hardened. Working concentration of tricaine should be 0.02% for long-term (overnight) imaging and 0.04% may be used for short-term (a few hours) imaging if fish are to be recovered.

2.3 Microinjection Setup and Materials

For microinjections of DNA constructs into single-cell embryos or tracer dyes into the circulation of zebrafish larvae, a proper injection setup needs to be established. Refer to Fig. 5 to visualize the

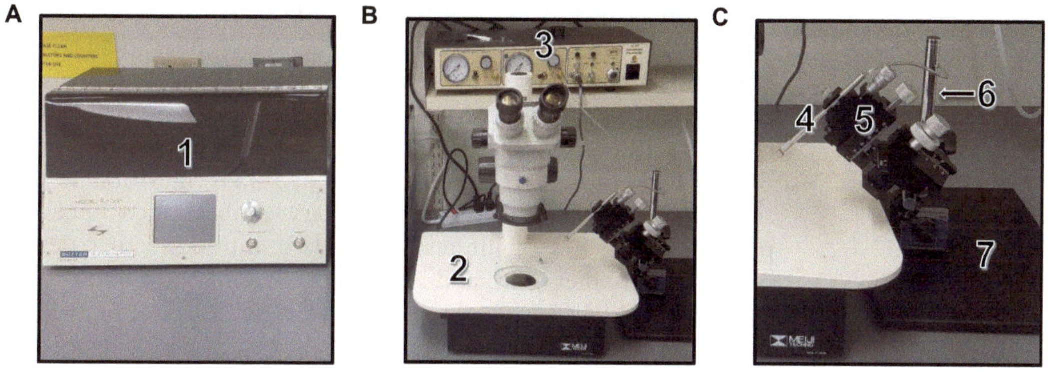

Fig. 5 Microinjection setup for assessing BBB integrity: Equipment for a microinjection setup is pictured throughout panels **a**, **b**, and **c**. In panel **a**, a horizontal pipettor puller (1) is stationed close to the microscope so that additional capillary needles may be pulled if needed during a round of injections. Panel **B** displays a brightfield dissection stereoscope with a lighted base (2) and a pneumatic picopump (3) attached to a footpedal (not shown) that controls the courser timing and length of the individual injection. Panel **c** zooms in on the various pieces that sit to the side of the microscope. These pieces include a pico nozzle (4) attached to a micromanipulator (5), a magnetic holder (6), and plate base (7). The air line can be set up with either a nitrogen or air supply, although air is more cost-effective

standard microinjection equipment and its arrangement. This setup includes a horizontal pipettor puller, capillaries (see the "Notes" section on size suggestions), capillary container, brightfield dissection stereoscope with a lighted base, a pneumatic picopump attached to a foot pedal, pico nozzle attached to a micromanipulator, and a magnetic base and a plate base. If possible, injection lines can be hooked up to an air line of choice to generate the appropriate pressures in the injection system. Compressed air can be used instead of a nitrogen line as a safer option and to reduce the necessity of air gas container purchases. An injection mold, glass Pasteur pipettes with bulbs, dissection probe, and tricaine is needed to position zebrafish for injection. To quantify the bolus that is injected into the animal beforehand, a 0.01 mm stage micrometer, halogen oil, and a 10% phenol red solution to mix the reagent into are required.

2.4 Conjugated Dyes

To assess size-selectivity of BBB integrity, various dextran conjugated dyes can be intravenously injected into the zebrafish common cardinal vein. Alexa Fluor-conjugated molecules, dextrans, albumin, or various tracers such as Cascade Blue or Rhodamine are commercially available for purchase. Some suggested concentrations are 2 nL of 1 mg/mL bovine serum albumin Alexa Fluor 647 Conjugate and 50 μM Fluorescein.

2.5 Immunohistochemistry (IHC) of BBB Markers in the Zebrafish

Unlike live imaging of transgenic animals, IHC is a common histological technique used to assay BBB markers at a moment in time. Standard BEC markers mentioned in Table 1 include transporters like Mdr1, Bcrp, Glut1, tight junction proteins Claudin5 and

ZO-1, and other NVU cell markers. Materials to perform IHC include: 1.5 mL tubes; a cryostat; charged glass slides; slide chamber holders; orbital shaker; mounting medium; cover glass; nail polish; phosphate buffered saline (PBS); 4% paraformaldehyde (PFA) in PBS; 30% sucrose in PBS; liquid blocker pen; PBST: PBS, 0.3% Triton X-100; 10% serum in PBST (it is best to use serum that is the same as the secondary antibody host species); primary antibodies of interest; secondary antibodies that detect primary antibodies used; and a humidity chamber. For adult brain dissections: 0.04% tricaine in system water; petri dishes; Kimwipes; dissection microscope; fine forceps; and razor blade or dissection scissors. For cryomold embedding: a dissection probe; cryomolds; Optimal Cutting Temperature (OCT) solution; a small container no greater than the width of the cryomold; 70% ethanol; and dry ice.

3 Methods

3.1 Breeding and Husbandry

The zebrafish housing facility should provide a 14-h light–10-h dark cycle and be maintained at 28 °C. Adults should be housed at a density of 5 animals per liter of system water, per home tank. Adults and larvae should be fed at least two times a day during the week and once on weekends. Breeding should be set up in the afternoon and clutches should be timed/collected the following morning.

1. To mate individual pairs, place a tank insert into a breeding box and fill it with system water. Try to minimize the amount of bubbles that persist in the water as these may startle the fish when they pop. Inserts may be purchased and put into mating boxes to separate pairs and stage the embryos they produce.
2. With a fish net, carefully transfer at least one male and female adult zebrafish into the breeding box (see "**Notes**" for comments on breeding).
3. After successful mating, embryos should be transferred into fresh embryo water. For raising embryos, clutches should be cleaned well when collected, and if possible, unfertilized eggs can be removed within the first 24 hpf. Because the first cell division occurs after 45 min, unfertilized eggs are easy to identify a few hours after clutches are generated. However, not all embryos will make it through the beginning stages of gastrulation and dead embryos should be removed early the next morning after clutches are collected.
4. When rearing larvae, zebrafish should be kept in embryo water until 4 dpf, when they may be put into static tanks of larval water.

5. Around 1–1.5 months of age, zebrafish juveniles (with the presence of a swim bladder) may be put on a slow drip on the main circulating system.

3.2 Tools and Procedures in Forward Genetic Screens

3.2.1 ENU Preparation and Handling

1. ENU is a highly potent mutagen and therefore extra care should be taken when handling this chemical. After the tank treatment solutions are prepared, the ENU preparation can take place. Place a Luer tip syringe in the ENU isopack to add 10 mM $NaPO_4$ buffer.
2. A second syringe (with tip attached) is used for pressure buildup in the isopack because of the displaced gas that is produced as the buffer is added to the container.
3. Once both tips are in place, first add 60 mL of 10 mM $NaPO_4$ buffer followed by another 25 mL, totaling 85 mL of 10 mM $NaPO_4$ buffer in an isopack.
4. Carefully remove 20 mL of air followed by 20 mL of ENU stock solution. This first 20 mL ENU aliquot is placed into one of the 50 mL conical tubes.
5. Repeat this step. The two conical tubes will have around 40 mL total between the two after aliquoting the ENU.
6. Carefully mix the ENU by pouring the tubes back and forth.

3.2.2 ENU Spectrophotometry Readings

Because each isopack of ENU varies between bottles, the final concentration of ENU needs to be calculated based on its optical density (OD) value with a spectrophotometer. Based on previously published screens, 3 mM ENU is a desirable concentration for each treatment [66]. ENU's molecular weight is 117.11 grams per mole and at a desired molarity of 3 mM that would constitute 0.351 mg/mL of ENU. For 1 mg/mL of ENU, at a spectrophotometer reading of 398 nm, the OD is 0.72 [67].

1. Dilute ENU 1:20 before being placed in a cuvette for readings on the spectrophotometer.
2. Calculate the amount of ENU based on the dilution factor in 1000 mL treatment water to total to 3 mM per ENU isopack. This is calculated as shown in the following equation:

$$\frac{[\text{O.D.}_{398}\ \text{value}] \times 20\ (\text{dilution factor})}{[0.72] \times [0.351\ \text{mg/mL}]}$$

3. 1000 mL is then divided by the dilution factor to decide what volume of ENU stock is needed to get a desired final concentration of 3 mM ENU.

3.2.3 ENU Treatment Paradigm and Fish Care

For a smaller-scale screen, ENU treatment can be performed on 25–30 adult males with positive spawning and fertility histories. The transgenic or genetic background of the males should be considered when establishing what the phenotypic readout will be for the screen. For example, genes that regulate BBB integrity, that is, hemorrhage formation or modulation of specific NVU structures as labeled by transgenic elements. Depending on the capacity of a fish facility, a genetic screen will mainly be limited by the initial amount of zebrafish that undergo mutagenesis and tank space availability for subsequent generations.

1. Separate ten to twelve males in two initial treatment tanks containing an egg collection insert and regular system water.
2. After the OD of the ENU is calculated and 3 mM ENU in system water is prepared, carefully transfer each set of fish with their tank insert into the 3 mM ENU treatment water.
3. Let fish bathe in this treatment water for 1 h in a quiet, dark fume hood.
4. After 1 h of 3 mM ENU treatment, carefully lift the insert and place into a subsequent water tank containing tricaine. Tricaine is added as a means to help calm the fish after ENU treatment.
5. Rinse the fish in this water for another hour in the quiet and dark.
6. Transfer fish once more to subsequent wash tanks for 1 h at a time, making two more rinses. All rinses will total to one ENU treatment bath and three rinses.
7. After rinsing, keep fish in the last rinse tank overnight in the fume hood to help reduce death that may be caused by immediately transferring them back into system water [68].
8. Assess the fish's health the following morning and then carefully transfer the animals back to the fish facility in a large tank with fresh tricaine wash water. A cardboard box can be placed over the tank during transport to maintain a dark environment.
9. Set up males for matings on Monday following ENU treatment and then perform the next round of ENU treatment on Thursday. This paradigm helps the fish get accustomed to mating for future breeding of the F_1 generation. ENU treatment should be done once weekly over the course of 3 weeks.

3.2.4 F_1, F_2, and F_3 Generations

Standard forward genetic screens identify recessive mutations within the F_3 generation. At least six successful paired matings within a F_2 family must be performed to optimize the chance of finding F_2 heterozygous pairs for a given mutation in each F_2 family.

1. Let the surviving ENU-treated males recover for 3 weeks in the fish facility.
2. Carefully label each founder male with a letter so that families can be tracked back to an individual.

3. Set up matings of ENU-treated males to untreated females of the same genotype. The offspring from these clutches are the F_1 generation.
4. Label F_1 clutches with the letter from their male parent and their spawn date. A few hundred adults (500+) F_1 should be generated for making the F_2 generation.
5. Generate F_2 families by either crossing F_1 zebrafish to one another or to untreated zebrafish of the same genotype.
6. Label the resulting F_2 families with their spawn date, a number (starting with family #1) and the letter of the original F_0 male that gave rise to their F_1 parents.
7. For F_3 generation and screening, set up random matings between F_2 family sibling pairs. F_2 pairs that spawn should be isolated and clearly labeled while their offspring are screened for the phenotype of interest. The individual F_2 pairs are labeled with an additional number after their F_2 family number as to track each pair. For example, the third pair from family #19 should be labeled as pair "19.3". Make sure to label the F_3 progeny with the same number as these isolated parent pairs as well as the F_3 spawn date.
8. Make sure to keep the individual F_3 clutches free of dead eggs and screen for the phenotype of interest. Based on Mendelian genetics, two F_2 heterozygous carrier parents will generally produce a clutch whereby 25% of the embryos will have the phenotype of interest (Fig. 3). However, not all mutants identified may be found at this ideal ratio as phenotypic penetrance is not always purely Mendelian.
9. Once a F_2 pair is identified to carry an unknown, recessive mutation for the phenotype being screened, the F_2 pair can also be individually crossed out to a polymorphic strain for future cloning. For example, if the mutation derives in the AB background, the F_2 parents should be crossed to a TL or WIK zebrafish line. Positional cloning can be achieved through other methods that are outside of the scope of this chapter. These methods include bulk segregant analysis or next generation sequencing.

3.3 Tools and Procedures in Chemical Screens

As with the forward genetic screen, a clear phenotypic readout in a certain developmental stage is needed for a chemical screen. Additionally, positive control compounds that cause the phenotype of interest should be identified before testing experimental compound libraries. 96-well plates can hold up to 300 μL of solution per well. PTU water can be used to prevent pigmentation in the animals and make fluorescence screening easier.

1. Set up matings with a divider so that embryos are staged for the screening time point of interest. See the "**Notes**" section in reference to a few tips on successful matings.
2. Collect embryos and be sure the plates are void of unfertilized embryos.
3. As most compound libraries are suspended in DMSO, embryos need to be acclimated to this solvent. Rinse embryos into a 1% DMSO solution made in egg water.
4. For the initial plating of the zebrafish, use a dissection microscope, plate up to three embryos per well (in the 1% DMSO water) onto a 96-well plate.
5. Keep the 96-well plate in a humidity chamber in the zebrafish incubator until drugging day. This will help any major evaporation and keep the eggs in sufficient solution.
6. On drugging day, check wells for any unfertilized eggs and remove them.
7. Remove the water in the wells, being careful to tilt the plate and not disturb any of the eggs. Based on how much final drug will be added to the individual wells, add the desired amount of 1% DMSO egg water. For example, if a total volume of 100 μL with 10 μL of compound is desired, add 90 L of 1% DMSO egg water back to the wells. An initial screening dose of 10 μM is suggested as a reasonable concentration in zebrafish small molecule screens [69].
8. Transfer compounds with a multichannel pipette from a drug library plate to the 96 well plate containing the zebrafish. Remember to keep a column or set of wells as a negative control, containing fish in 1% DMSO water only, as well as a column or set of wells as a positive control for the phenotype being screened.
9. After drugging, place the plates in a humidity chamber to prevent evaporation of the small volumes. This box containing the drugged embryos is then placed in the 28.5 °C incubator. Make sure the incubator is on a 14-h light/10-h dark cycle like the zebrafish facility.
10. Monitor wells for any abnormal development such as CNS necrosis or brain hemorrhage.
11. To score for changes in fluorescence, embryos should be anesthetized with 0.04% tricaine.
12. Carefully add 300 μL of cooled 1.2% low-melting point agarose to each well, let the agarose harden, and then flip the plates upside down so that larvae are now stationary at the bottom of the 96 well plate for screening. Images can be recorded on an appropriate plate-based imaging system.

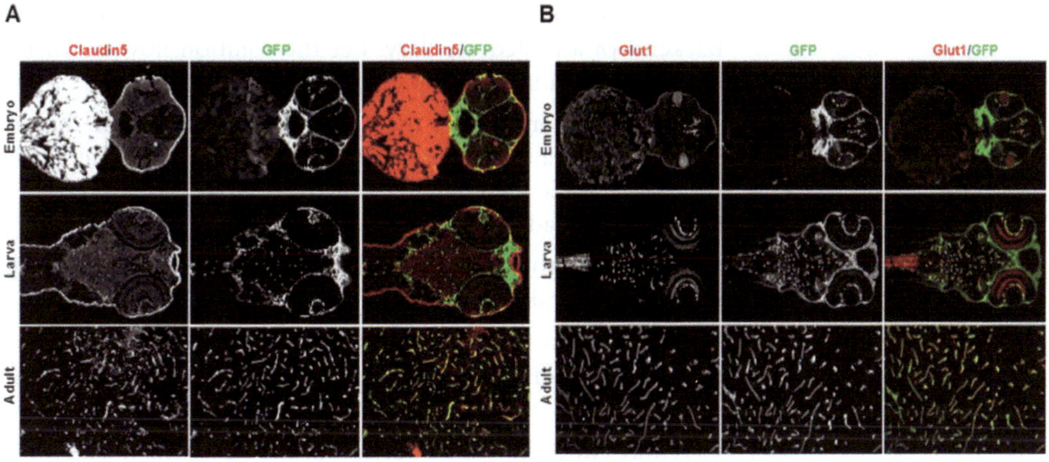

Fig. 6 Zebrafish brain endothelial cells express Claudin5 and Glut1 during development. *Tg(fli1a:EGFP)*^{y1} animals were sectioned and IHC was performed to look at the colocalization of BBB markers in GFP positive blood vessels. Just as in the mammalian BBB, embryonic, larval, and adult zebrafish express tight junction proteins like Claudin5 (**a**) and efflux transporters like Glut1 (**b**) in BECs. This standard IHC protocol is a useful tool to analyze a snaphsot of the zebrafish BBB. Such assays can be used to characterize mutants identified in genetic screens or the effects of potential small molecule modulators of the BBB

3.4 Procedures for Assessing the Integrity and Function of the Zebrafish BBB

The integrity of the blood–brain barrier is typically assessed through the vascular retention of molecules after tracer injection and histological location of tight junction and transporter protein expression. IHC is commonly used to survey protein expression within the NVU, using markers as those mentioned in Table 1. As two examples, Fig. 6 displays Claudin5 and Glut1 expression throughout the zebrafish lifespan in embryo, larval, and adult brain tissue. Refer to the "**Notes**" section pertaining to some important considerations when choosing antibodies for zebrafish.

3.4.1 Intravenous Microinjection of Tracers

Tracers of various sizes and colors are commercially available for microinjection in live zebrafish. These solutions are injected into the common cardinal vein near the heart and the retention of these dyes in circulation assess the structural character of the BBB. Unless in a pigment-deficient mutant, animals should be treated with PTU (*see* "**Notes**") to prevent melanocytes from interfering with imaging. Typically, around 2 nL of tracer is tolerated for microinjections. Each capillary needle needs to be calibrated before injection.

1. Mix around a few microliters of 10% phenol red solution with the solution to be injected. The phenol red colors the solution so that a volume per injection can be calculated when visualizing a bolus on the micrometer.
2. Place pulled capillaries into the capillary container so that they hang with the tip in the bottom and backend up.

3. Load 2–3 μL of the 10% phenol red/injection solution into the backend of a pulled capillary. Let the solution flow down via capillary action into the tip of the capillary.
4. Place the micrometer on the lighted base stage of the microscope and put a few drops of halogen oil onto the center. This should create a small circle of oil around the area where the scale will be read.
5. Carefully load the capillary needle into the pico nozzle, making sure not to break the tip end.
6. Slowly position the needle tip with the micromanipulator, so that the tip gradually meets the top of the oil. It is crucial that once a bolus is injected, it will settle over the scale.
7. Carefully cut back a small portion of the tip of the needle with fine forceps to create a small opening for solution to come through. Trim back smaller portions at first.
8. After an opening exists in the needle, adjust the hold and eject pressures so that there is positive flow of solution out of the capillary. This is crucial as any backward flow may damage the tissue inside of the animal.
9. A typical bolus should require only one to two foot presses per injection. To calculate a bolus for injection, use the following equation for a sphere:

$$V = \frac{4}{3}\pi r^3$$

where r is determined as the radius of the bolus injected based on the scale reading on the micrometer.
10. After the needle is calibrated, anesthetize the zebrafish with 0.02% tricaine so that it can be positioned into the injection mold for injection. It is crucial to position the animal so that the needle site meets down to the common cardinal vein, where blood is collected back to the heart. Injection into the pericardial pouch surrounding the heart will make it seem that the heart is filling but this will not be due to tracer in circulation.
11. Once injections are completed, allow tracers to get into circulation for around 3 min to an hour post-injection, remembering to limit light exposure to light-sensitive dyes.
12. Representative animals can be selected for embedding and subsequent imaging. Figure 4 is an example of a microinjection of Cascade blue in a larval *Tg(fli1a:EGFP)*^{y1} animal, expressing GFP in the vasculature.
13. For imaging of live animals, anesthetize the zebrafish with tricaine at a percentage based on the length of the desired imaging session. Keep in mind the type of microscope that

will be used to image the animal. For inverted microscopes, a glass coverslip-containing petri dish can be used.

14. To prepare LMP for embedding zebrafish, keep the LMP agarose at 50 °C during embedding so that it does not solidify. Additionally, a beaker containing warm water can also be used to keep the agarose nearby if embedding a large number of samples. The agarose can also be reheated if it solidifies. However, it is equally as crucial to not add the LMP agarose when it is still hot as this could harm the zebrafish. Once the conical tube with agarose does not feel hot to hold, this should be cool enough to add to the petri dish.
15. Using a glass Pasteur pipette, remove any tricaine water that is surrounding the zebrafish. Now, slowly drop the LMP agarose into the middle on the glass coverslip away from the fish. This will allow the agarose to cool and slowly surround the animal. As the agarose starts to pool, it will build up in a circle within the glass coverslip and the zebrafish will slowly float to the top.
16. Use a dissection probe to orient the zebrafish to the bottom, orienting the dorsal side of the head down to the coverslip.
17. Once the fish is embedded in the desired orientation, allow the agarose to harden.
18. After the agarose has hardened, cover the agarose with water containing tricaine to keep the animal anesthetized during imaging.

3.4.2 Collecting and Embedding Specimens for Cryosectioning

It is important to note that not all fixation strategies and formulations are compatible with every antibody and this information should be considered before starting an experiment.

1. To perform IHC in zebrafish with a NVU marker of interest, stage the animals for the time point (embryonic-larvae) desired for analysis.
2. Collect the animals in 1.5 mL tubes on ice and wait for the animals to sink to the bottom. At this point, the animals are anesthetized and the fish water can be changed out for 4% PFA for protein fixation. Allow fixation to proceed overnight at 4 °C.
3. For adult BBB IHC, adults must be deeply anesthetized in 0.04% tricaine for brain dissections. Dissections can be performed on a petri dish containing a Kimwipes soaked in 0.04% tricaine.
4. Tissues are dissected by using a razor blade or surgical scissors for decapitation followed by pulling the skin back with forceps and then opening of the skull to expose the brain.

5. Sink adult brains in 4% PFA at 4 °C overnight, the same as when younger animals are collected.
6. The next day, rinse specimens in PBS three times, 5 min for each wash, making sure to gently rock the tubes during the washes.
7. Exchange out the final wash with 30% sucrose made in PBS and place tubes back at 4 °C overnight to cryopreserve the tissue. Specimens will rise to the top of the meniscus once they are ready for embedding.
8. To embed the specimens for cryosectioning, label cryomolds and fill them with OCT, making sure not to introduce any bubbles.
9. Carefully place specimens at the top of the cryomold plastic front, and push the specimen gently into the OCT, trying to remove as much sucrose as possible. Any kind of excess sucrose may create a pocket around the tissue or form ice crystals that are deleterious during sectioning.
10. Orient the specimen as desired in the OCT with a dissection probe, remembering that they will eventually be adhered to a cryostat chuck for sectioning.
11. Flash freeze the cryomold on top of a small container with a dry ice/ethanol slurry. It is vital to make sure ethanol does not come in contact with the OCT as this will disrupt its integrity and consistency.
12. Cryomolds can be put on dry ice while others are being prepared, wrapped in Parafilm for longer-term storage, and stored at −80 °C.

3.4.3 Standard IHC Procedure for Zebrafish Cryosections

Tissue sections that are 10–50 μm thick can be prepared on a cryostat once all the specimens have been embedded and frozen in OCT molds. Cryosection slides can be stored in slide boxes for years at room temperature. After sections are collected, the following standard IHC protocol can be performed. All steps are designed to gradually introduce the tissue to the different IHC solutions throughout the procedure.

1. Carefully place slides in a chamber slide holder to wash off residual OCT with PBS. Put the chambers with slides on an orbital shaker to help the washing process. All washes should take place in the chamber slide holders.
2. Perform 3, 5 min washes in PBST to help permeabilize the tissue sections.
3. After these initialize washes, carefully dry the back of the slide as well as most of the front where the tissue remains. Make sure

to not remove the tissue, but lightly dab around it to remove excess liquid.

4. Once dry, use a liquid blocker pen to carefully circle around the tissue sections. Be extremely careful not to get the liquid blocker solution on the tissue as this will prevent exposure of the tissue to IHC solutions. At this point, it is good to determine which areas will be kept as a negative controls (either receiving primary antibody only and no secondary antibody solutions or no primary antibody and secondary antibody only solutions). As a general guideline, the top sections of the slide are best for negative control in case antibody solutions poor over down the rest of the slide.
5. Once the liquid blocker area is dry, place the slides into the humidity chamber.
6. Carefully drop 10% serum blocking buffer onto all of the outlined sections. Blocking buffer is added to offset nonspecific antibody binding.
7. Incubate the sections in blocking buffer for 1 h at room temperature.
8. Make the primary antibodies up at desired dilutions into the 10% blocking buffer close to the end of the 1 h incubation.
9. After the 1 h incubation, carefully remove the liquid from the outline sections. At this point, more liquid blocker can be applied if there is any compromise in the outline. Once adjusted, place the slides back into the humidity chamber.
10. Add the primary antibody solution to the experimental sections that will be assessed. Leave the control sections in the 10% blocking buffer and incubate the slides at 4 °C overnight.
11. The next day, perform a quick rinse of the slides in PBST followed by 6, 10 min washes in PBST. Thorough washing is required to remove any nonspecific primary antibody binding.
12. After washing, reassess the integrity of the liquid blocker outline and adjust as needed.
13. Make up secondary antibody dilutions in the 10% blocking buffer.
14. Place the slides back into the humidity chamber. At this point, the secondary antibody solution may be added to all circled sections if the negative control selected is for secondary antibody only.
15. Incubate the sections for 1 h at room temperature. If secondary antibodies are conjugated to fluorophores, this step and subsequent steps must be performed in the dark/with minimal light exposure.

16. Perform another round of 6, 10 min PBST washes on the orbital shaker.
17. After washes are finished, dab off the excess PBST on the slides, making sure to not disturb the tissue.
18. Carefully place a line of mounting medium (around 150 μL) along the edge of the slide by the tissue. Make sure that bubbles are not introduced into the medium as these bubbles will transfer over the tissue and impeded imaging.
19. Align the edge of a piece of cover glass to the mounting medium, creating a seal. Slowly lay the cover glass down, allowing the mounting medium to spread out over the tissue.
20. Let the mounting medium and cover glass settle for at least 15 min before sealing with nail polish.

3.5 Conclusions

As one of the three CNS barriers, the BBB is crucial for defense against circulating xenobiotics and endogenous materials that would otherwise pose a threat to the CNS [70]. The BBB is a complex, physiological arrangement of brain endothelium and interacting astrocytic endfeet, neurons, and pericytes [71]. Cell nonautonomous signals between these cells help the BBB maintain its physiological functions [6, 35, 45, 72–74]. However, the protection the BBB provides the CNS simultaneously confers a challenge for delivering therapeutics. In order to circumvent the difficulties associated with efficacious CNS drug delivery, models used to study the BBB should be able to dissect its molecular mechanisms, screen for new pathways and small molecule modulators, and assess BBB function. Zebrafish are trending in biomedical research usage, and in particular, have been established in the past decade as a model organism suitable for studying blood–CNS barriers. The assays described in this chapter highlight how zebrafish can be used to identify novel genetic and chemical BBB modulators as well as assay BBB integrity. Recognizing these processes in the zebrafish will prove critical for furthering the development of CNS therapies, discovering targets for better CNS drug penetration, and the prevention of CNS disease progression.

4 Notes

4.1 Matings

A few days before mating, adults can be separated into tanks of each gender to engage them to mate once they are placed in a mating box. For long-term housing, it is not recommended to store the animals in male or female only tanks as this can make females become egg-bound. Aside from the traditional pair-wise crosses, zebrafish can be separated into mating chambers with a ratio of two males per one female to help generate egg clutches. Separating the males from the female overnight with a divider may also aid in mating. Additionally, the level of water in the breeding box can be reduced the morning of mating so that the animals have more opportunities for contact with one another as well as the female having more frequent contact with the bottom of the tank. This contact can help stimulate her underside and spur mating behavior. Creating a "bank" within the mating box with a divider wedged underneath the tank insert may also create a distribution of shallow to high water.

4.2 Methylene Blue

An antifungal called methylene blue can be added to the egg water to prevent infestation of the embryo water. This powder can be added at a final less than 0.1% (w/v) of the embryo water. However, methylene blue fluoresces in the red channel and can be absorbed by the animal at early stages. Therefore, if proteins will be analyzed in the red channel, embryos should not make contact with methylene blue water or should be rinsed vigorously after 1 day of initial treatment.

4.3 PTU

PTU should be changed out every day to see the maximal effects on pigment development. Stock solutions of PTU should be kept at 4 °C and working stocks may be kept in the 37 °C incubator with the embryos. PTU should not be added until 24hpf as it can be toxic during beginning developmental stages [75].

4.4 Low Melting Point (LMP) Agarose Generation for Embedding Larval Zebrafish

As indicated by the name, LMP agarose boils more quickly than regular agarose and should be handled with caution. For short-term imaging (a few hours), a higher percentage of 1.2–1.5% can be generated. However, for overnight imaging, lower percentage (0.8%) should be generated so as not to delay the physical development of the larval zebrafish. To avoid over-boiling, around 25 mL of LMP agarose can be made within a 50 mL conical tube. After adding the appropriate amount of agarose powder to egg water, quickly vortex the agarose for around 10 s to help initially resuspend the powder into solution. Place the 50 mL tube in a small beaker of water while heating in a microwave. This surrounding water will help more evenly distribute heat to the LMP agarose. At the start, heat the solution for around 20–30 s. Remove the lid and let the solution cool for around 30 s. Close the conical tube and vortex the tube on and off for around 10 s. Vortexing during heating and cooling helps the undissolved gel melt into solution. Loosely unscrew the tube's lid and heat again for now 7–10 s. Repeat the heat, pause, vortex, heat pattern for about four times until the agarose is completely resuspended.

4.5 Embedding Live Zebrafish for Drug Screening

After zebrafish are anesthetized, around 150–200 μL of low melting point agarose should carefully be dispensed into the well with the animals. It is crucial to not introduce air bubbles into the wells as this could negatively affect visualizing zebrafish during screening. A repeater pipette (such as those from Eppendorf) are ideal for this application so that you can effectively distribute agarose among the plate. The low melting point agarose will solidify quickly when aliquoted from the larger amount that was originally generated, so choose a pipette tip that holds no more than a few milliliters at a time.

4.6 Capillaries and Settings for Microinjections

As highlighted in the Rembold et al. (2006) methods paper, horizontal pipette pullers, not vertical pullers, are ideal for pulling the finely tapered capillaries desired for zebrafish injections [76]. Capillaries with an inner filament should be used to pull injection needles so that solutions that are back-loaded into the needle will drain down into the fine end. A suggestion for capillary specifications include 4 inch length, thin-wall, single-barrel standard borosilicate with an OD of 1.5 mm and an ID of 1.12 mm. The heat settings may ultimately depend on the horizontal puller brand. Starting recommendations for heat settings such as with the Sutter P-1000 are, Heat = 510, Pull = 120, Vel = 70, Delay = 80, and Pressure = 200. After pulling one capillary, two injection needles will result. The longer the delay/cooling, the shorter the needle that will be pulled.

4.7 Tracer Injections

Evans blue is a common tracer used to assess BBB integrity via dye leakage. However, it should be noted that when injected I.V. into larval fish and imaged over time, Evans blue can cause phototoxicity in the animal (data unpublished). Therefore, knowing the possible toxicities of each dye during long-term imaging (at least 1 h) should be kept in mind when thinking about the health of the animals.

4.8 Finding Antibodies for IHC in Zebrafish

Caution should be taken when utilizing new antibodies in the zebrafish that have not already been validated for this system. While a reasonable conservation exists between zebrafish and mammals, the genome duplication event that occurred in teleosts may have created different paralogs of the homologous mammalian gene. To ensure an antibody will likely recognize the protein epitope in the zebrafish, a BLAST search and/or sequence alignment should be performed for the protein of interest. In general, a protein alignment can be performed for the general zebrafish protein against the species protein of which the antibody is generated. If an exact epitope is not given for an antibody, the N or C-terminus of the zebrafish protein can still be analyzed for sequence conservation.

References

1. Abbott NJ, Friedman A (2012) Overview and introduction: the blood–brain barrier in health and disease. Epilepsia 53(06):1–6. https://doi.org/10.1111/j.1528-1167.2012.03696.x
2. Zlokovic BV (2008) The blood-brain barrier in health and chronic neurodegenerative disorders. Neuron 57(2):178–201. https://doi.org/10.1016/j.neuron.2008.01.003
3. Attwell D, Buchan AM, Charpak S, Lauritzen M, MacVicar BA, Newman EA (2010) Glial and neuronal control of brain blood flow. Nature 468(7321):232–243. https://doi.org/10.1038/nature09613
4. Neuwelt EA, Bauer B, Fahlke C, Fricker G, Iadecola C, Janigro D, Leybaert L, Molnár Z, O'Donnell ME, Povlishock JT, Saunders NR, Sharp F, Stanimirovic D, Watts RJ, Drewes LR (2011) Engaging neuroscience to advance translational research in brain barrier biology. Nat Rev Neurosci 12(3):169–182. http://www.nature.com/nrn/journal/v12/n3/suppinfo/nrn2995_S1.html
5. Daneman R, Zhou L, Kebede AA, Barres BA (2010) Pericytes are required for blood–brain barrier integrity during embryogenesis. Nature 468(7323):562–566. https://doi.org/10.1038/nature09513
6. Ben-Zvi A, Lacoste B, Kur E, Andreone BJ, Mayshar Y, Yan H, Gu C (2014) Mfsd2a is critical for the formation and function of the blood-brain barrier. Nature 509(7501):507–511. http://www.nature.com/nature/journal/v509/n7501/abs/nature13324.html#supplementary-information. https://doi.org/10.1038/nature13324
7. Liebner S, Corada M, Bangsow T, Babbage J, Taddei A, Czupalla CJ, Reis M, Felici A, Wolburg H, Fruttiger M, Taketo MM, von Melchner H, Plate KH, Gerhardt H, Dejana E (2008) Wnt/β-catenin signaling controls development of the blood–brain barrier. J Cell Biol 183(3):409–417. https://doi.org/10.1083/jcb.200806024
8. Abbott NJ (2002) Astrocyte–endothelial interactions and blood–brain barrier permeability. J Anat 200(6):629–638. https://doi.org/10.1046/j.1469-7580.2002.00064.x
9. Molofsky AV, Krencik R, Ullian EM, Tsai HH, Deneen B, Richardson WD, Barres BA, Rowitch DH (2012) Astrocytes and disease: a neurodevelopmental perspective. Genes Dev 26(9):891–907. https://doi.org/10.1101/gad.188326.112
10. Lee S-W, Kim WJ, Choi YK, Song HS, Son MJ, Gelman IH, Kim Y-J, Kim K-W (2003) SSeCKS regulates angiogenesis and tight junction formation in blood-brain barrier. Nat Med 9(7):900–906
11. Lundgaard I, Osório MJ, Kress B, Sanggaard S, Nedergaard M (2014) White matter astrocytes in health and disease. Neuroscience 276:161–173. https://doi.org/10.1016/j.neuroscience.2013.10.050
12. Pardridge WM (2005) The blood-brain barrier: bottleneck in brain drug development. NeuroRx 2(1):3–14
13. Lipinski CA (2004) Lead- and drug-like compounds: the rule-of-five revolution. Drug Discov Today Technol 1(4):337–341. https://doi.org/10.1016/j.ddtec.2004.11.007
14. Watkins S, Robel S, Kimbrough IF, Robert SM, Ellis-Davies G, Sontheimer H (2014) Disruption of astrocyte–vascular coupling and the blood–brain barrier by invading glioma cells. Nat Commun 5:4196. http://www.nature.com/articles/ncomms5196#supplementary-information. https://doi.org/10.1038/ncomms5196
15. Naik P, Cucullo L (2012) In vitro blood–brain barrier models: current and perspective technologies. J Pharm Sci 101(4):1337–1354. https://doi.org/10.1002/jps.23022
16. Cho H, Seo JH, Wong KHK, Terasaki Y, Park J, Bong K, Arai K, Lo EH, Irimia D (2015) Three-dimensional blood-brain barrier model for in vitro studies of neurovascular pathology. Sci Rep 5:15222. http://www.nature.com/articles/srep15222#supplementary-information. https://doi.org/10.1038/srep15222
17. Helms HC, Abbott NJ, Burek M, Cecchelli R, Couraud PO, Deli MA, Forster C, Galla HJ, Romero IA, Shusta EV, Stebbins MJ, Vandenhaute E, Weksler B, Brodin B (2016) In vitro models of the blood-brain barrier: an overview of commonly used brain endothelial cell culture models and guidelines for their use. J Cereb Blood Flow Metab 36(5):862–890. https://doi.org/10.1177/0271678x16630991
18. Abbott NJ (2005) Dynamics of CNS barriers: evolution, differentiation, and modulation. Cell Mol Neurobiol 25(1):5–23. https://doi.org/10.1007/s10571-004-1374-y
19. Stewart PA, Wiley MJ (1981) Developing nervous tissue induces formation of blood-brain barrier characteristics in invading endothelial cells: a study using quail--chick transplantation chimeras. Dev Biol 84(1):183–192

20. Nitta T, Hata M, Gotoh S, Seo Y, Sasaki H, Hashimoto N, Furuse M, Tsukita S (2003) Size-selective loosening of the blood-brain barrier in claudin-5–deficient mice. J Cell Biol 161 (3):653–660. https://doi.org/10.1083/jcb.200302070
21. Bell RD, Winkler EA, Singh I, Sagare AP, Deane R, Wu Z, Holtzman DM, Betsholtz C, Armulik A, Sallstrom J, Berk BC, Zlokovic BV (2012) Apolipoprotein E controls cerebrovascular integrity via cyclophilin A. Nature 485 (7399):512–516. http://www.nature.com/nature/journal/v485/n7399/abs/nature11087.html#supplementary-information
22. Sohet F, Daneman R (2013) Genetic mouse models to study blood–brain barrier development and function. Fluids Barriers CNS 10:3. https://doi.org/10.1186/2045-8118-10-3
23. Stork T, Engelen D, Krudewig A, Silies M, Bainton RJ, Klämbt C (2008) Organization and function of the blood–brain barrier in Drosophila. J Neurosci 28(3):587–597. https://doi.org/10.1523/jneurosci.4367-07.2008
24. Jeong J-Y, Kwon H-B, Ahn J-C, Kang D, Kwon S-H, Park JA, Kim K-W (2008) Functional and developmental analysis of the blood–brain barrier in zebrafish. Brain Res Bull 75(5):619–628. https://doi.org/10.1016/j.brainresbull.2007.10.043
25. Xie J, Farage E, Sugimoto M, Anand-Apte B (2010) A novel transgenic zebrafish model for blood-brain and blood-retinal barrier development. BMC Dev Biol 10(1):76. https://doi.org/10.1186/1471-213x-10-76
26. Fleming A, Diekmann H, Goldsmith P (2013) Functional characterisation of the maturation of the blood-brain barrier in larval zebrafish. PLoS One 8(10):e77548. https://doi.org/10.1371/journal.pone.0077548
27. Umans RA, Taylor MR (2012) Zebrafish as a model to study drug transporters at the blood–brain barrier. Clin Pharmacol Ther 92 (5):567–570. https://doi.org/10.1038/clpt.2012.168
28. Tam Stephen J, Richmond David L, Kaminker Joshua S, Modrusan Z, Martin-McNulty B, Cao Tim C, Weimer Robby M, Carano Richard AD, van Bruggen N, Watts Ryan J (2012) Death receptors DR6 and TROY regulate brain vascular development. Dev Cell 22 (2):403–417. https://doi.org/10.1016/j.devcel.2011.11.018
29. Grunwald DJ (2002) Headwaters of the zebrafish — emergence of a new model vertebrate. Nat Rev Gen 3(9):717–724. https://doi.org/10.1038/nrg892
30. Patton EE, Zon LI (2001) The art and design of genetic screens: zebrafish. Nat Rev Genet 2 (12):956–966
31. Lieschke GJ, Currie PD (2007) Animal models of human disease: zebrafish swim into view. Nat Rev Genet 8(5):353–367. https://doi.org/10.1038/nrg2091
32. Huang P, Xiao A, Zhou M, Zhu Z, Lin S, Zhang B (2011) Heritable gene targeting in zebrafish using customized TALENs. Nat Biotech 29(8):699–700. http://www.nature.com/nbt/journal/v29/n8/abs/nbt.1939.html#supplementary-information
33. Hwang WY, Fu Y, Reyon D, Maeder ML, Tsai SQ, Sander JD, Peterson RT, Yeh JR, Joung JK (2013) Efficient genome editing in zebrafish using a CRISPR-Cas system. Nat Biotechnol 31(3):227–229. https://doi.org/10.1038/nbt.2501
34. Ceol CJ, Houvras Y, Jane-Valbuena J, Bilodeau S, Orlando DA, Battisti V, Fritsch L, Lin WM, Hollmann TJ, Ferre F, Bourque C, Burke CJ, Turner L, Uong A, Johnson LA, Beroukhim R, Mermel CH, Loda M, Ait-Si-Ali S, Garraway LA, Young RA, Zon LI (2011) The histone methyltransferase SETDB1 is recurrently amplified in melanoma and accelerates its onset. Nature 471 (7339):513–517. https://doi.org/10.1038/nature09806
35. Daneman R, Agalliu D, Zhou L, Kuhnert F, Kuo CJ, Barres BA (2009) Wnt/beta-catenin signaling is required for CNS, but not non-CNS, angiogenesis. Proc Natl Acad Sci U S A 106(2):641–646. https://doi.org/10.1073/pnas.0805165106
36. Wang X, Kopinke D, Lin J, McPherson Adam D, Duncan Robert N, Otsuna H, Moro E, Hoshijima K, Grunwald David J, Argenton F, Chien C-B, Murtaugh LC, Dorsky Richard I (2012) Wnt signaling regulates postembryonic hypothalamic progenitor differentiation. Dev Cell 23(3):624–636. https://doi.org/10.1016/j.devcel.2012.07.012
37. Yu PB, Hong CC, Sachidanandan C, Babitt JL, Deng DY, Hoyng SA, Lin HY, Bloch KD, Peterson RT (2008) Dorsomorphin inhibits BMP signals required for embryogenesis and iron metabolism. Nat Chem Biol 4(1):33–41. http://www.nature.com/nchembio/journal/v4/n1/suppinfo/nchembio.2007.54_S1.html
38. Peterson RT, Link BA, Dowling JE, Schreiber SL (2000) Small molecule developmental screens reveal the logic and timing of vertebrate development. Proc Natl Acad Sci U S A 97 (24):12965–12969

39. Stern HM, Murphey RD, Shepard JL, Amatruda JF, Straub CT, Pfaff KL, Weber G, Tallarico JA, King RW, Zon LI (2005) Small molecules that delay S phase suppress a zebrafish bmyb mutant. Nat Chem Biol 1 (7):366–370
40. Wang C, Tao W, Wang Y, Bikow J, Lu B, Keating A, Verma S, Parker TG, Han R, Wen XY (2010) Rosuvastatin, identified from a zebrafish chemical genetic screen for antiangiogenic compounds, suppresses the growth of prostate cancer. Eur Urol 58(3):418–426. https://doi.org/10.1016/j.eururo.2010.05.024
41. Ridges S, Heaton WL, Joshi D, Choi H, Eiring A, Batchelor L, Choudhry P, Manos EJ, Sofla H, Sanati A, Welborn S, Agarwal A, Spangrude GJ, Miles RR, Cox JE, Frazer JK, Deininger M, Balan K, Sigman M, Müschen M, Perova T, Johnson R, Montpellier B, Guidos CJ, Jones DA, Trede NS (2012) Zebrafish screen identifies novel compound with selective toxicity against leukemia. Blood 119(24):5621
42. Sachidanandan C, Yeh JR, Peterson QP, Peterson RT (2008) Identification of a novel retinoid by small molecule screening with zebrafish embryos. PLoS One 3(4):e1947. https://doi.org/10.1371/journal.pone.0001947
43. Ren B, Deng Y, Mukhopadhyay A, Lanahan AA, Zhuang ZW, Moodie KL, Mulligan-Kehoe MJ, Byzova TV, Peterson RT, Simons M (2010) ERK1/2-Akt1 crosstalk regulates arteriogenesis in mice and zebrafish. J Clin Invest 120(4):1217–1228. https://doi.org/10.1172/JCI39837
44. Isogai S, Horiguchi M, Weinstein BM (2001) The vascular anatomy of the developing zebrafish: an atlas of embryonic and early larval development. Dev Biol 230(2):278–301. https://doi.org/10.1006/dbio.2000.9995
45. Cullen M, Elzarrad MK, Seaman S, Zudaire E, Stevens J, Yang MY, Li X, Chaudhary A, Xu L, Hilton MB, Logsdon D, Hsiao E, Stein EV, Cuttitta F, Haines DC, Nagashima K, Tessarollo L, St Croix B (2011) GPR124, an orphan G protein-coupled receptor, is required for CNS-specific vascularization and establishment of the blood-brain barrier. Proc Natl Acad Sci U S A 108(14):5759–5764. https://doi.org/10.1073/pnas.1017192108
46. Gale NW, Yancopoulos GD (1999) Growth factors acting via endothelial cell-specific receptor tyrosine kinases: VEGFs, angiopoietins, and ephrins in vascular development. Genes Dev 13(9):1055–1066
47. Carmeliet P, Ferreira V, Breier G, Pollefeyt S, Kieckens L, Gertsenstein M, Fahrig M, Vandenhoeck A, Harpal K, Eberhardt C, Declercq C, Pawling J, Moons L, Collen D, Risau W, Nagy A (1996) Abnormal blood vessel development and lethality in embryos lacking a single VEGF allele. Nature 380 (6573):435–439. https://doi.org/10.1038/380435a0
48. Covassin LD, Villefranc JA, Kacergis MC, Weinstein BM, Lawson ND (2006) Distinct genetic interactions between multiple Vegf receptors are required for development of different blood vessel types in zebrafish. Proc Natl Acad Sci U S A 103(17):6554–6559. https://doi.org/10.1073/pnas.0506886103
49. Risau W (1997) Mechanisms of angiogenesis. Nature 386(6626):671–674. https://doi.org/10.1038/386671a0
50. Umans RA, Henson HE, Mu F, Parupalli C, Ju B, Peters JL, Lanham KA, Plavicki JS, Taylor MR (2017) CNS angiogenesis and barriergenesis occur simultaneously. Dev Biol 425:101. https://doi.org/10.1016/j.ydbio.2017.03.017
51. Dermietzel R, Krause D, Kremer M, Wang C, Stevenson B (1992) Pattern of glucose transporter (Glut 1) expression in embryonic brains is related to maturation of blood-brain barrier tightness. Dev Dyn 193(2):152–163. https://doi.org/10.1002/aja.1001930207
52. Bauer H, Sonnleitner U, Lametschwandtner A, Steiner M, Adam H, Bauer HC (1995) Ontogenic expression of the erythroid-type glucose transporter (Glut 1) in the telencephalon of the mouse: correlation to the tightening of the blood-brain barrier. Brain Res Dev Brain Res 86(1–2):317–325
53. Tam SJ, Richmond DL, Kaminker JS, Modrusan Z, Martin-McNulty B, Cao TC, Weimer RM, Carano RA, van Bruggen N, Watts RJ (2012) Death receptors DR6 and TROY regulate brain vascular development. Dev Cell 22(2):403–417. https://doi.org/10.1016/j.devcel.2011.11.018
54. Wang Y, Pan L, Moens CB, Appel B (2014) Notch3 establishes brain vascular integrity by regulating pericyte number. Development 141 (2):307–317. https://doi.org/10.1242/dev.096107
55. Bernardos RL, Raymond PA (2006) GFAP transgenic zebrafish. Gene Exp Patterns 6 (8):1007–1013. https://doi.org/10.1016/j.modgep.2006.04.006
56. Dziegielewska KM, Ek J, Habgood MD, Saunders NR (2001) Development of the choroid plexus. Microsc Res Tech 52(1):5–20. https://doi.org/10.1002/1097-0029(20010101)52:1<5::AID-JEMT3>3.0.CO;2-J

57. Zheng PP, Romme E, van der Spek PJ, Dirven CM, Willemsen R, Kros JM (2010) Glut1/SLC2A1 is crucial for the development of the blood-brain barrier in vivo. Ann Neurol 68 (6):835–844. https://doi.org/10.1002/ana.22318

58. Fischer S, Klüver N, Burkhardt-Medicke K, Pietsch M, Schmidt A-M, Wellner P, Schirmer K, Luckenbach T (2013) Abcb4 acts as multixenobiotic transporter and active barrier against chemical uptake in zebrafish (Danio rerio) embryos. BMC Biol 11(1):69. https://doi.org/10.1186/1741-7007-11-69

59. Daneman R, Rescigno M (2009) The gut immune barrier and the blood-brain barrier: are they so different? Immunity 31 (5):722–735. https://doi.org/10.1016/j.immuni.2009.09.012

60. Lawson ND, Weinstein BM (2002) In vivo imaging of embryonic vascular development using transgenic zebrafish. Dev Biol 248 (2):307–318

61. Hanahan D, Wagner EF, Palmiter RD (2007) The origins of oncomice: a history of the first transgenic mice genetically engineered to develop cancer. Genes Dev 21 (18):2258–2270. https://doi.org/10.1101/gad.1583307

62. Schlaeger TM, Qin Y, Fujiwara Y, Magram J, Sato TN (1995) Vascular endothelial cell lineage-specific promoter in transgenic mice. Development 121(4):1089–1098

63. Noveroske JK, Weber JS, Justice MJ (2000) The mutagenic action of N-ethyl-N-nitrosourea in the mouse. Mamm Genome 11 (7):478–483

64. de Bruijn E, Cuppen E, Feitsma H (2009) Highly efficient ENU mutagenesis in zebrafish. In: Lieschke GJ, Oates AC, Kawakami K (eds) Zebrafish: methods and protocols. Humana Press, Totowa, NJ, pp 3–12. https://doi.org/10.1007/978-1-60327-977-2_1

65. White RM, Sessa A, Burke C, Bowman T, LeBlanc J, Ceol C, Bourque C, Dovey M, Goessling W, Burns CE, Zon LI (2008) Transparent adult zebrafish as a tool for in vivo transplantation analysis. Cell Stem Cell 2 (2):183–189. https://doi.org/10.1016/j.stem.2007.11.002

66. Solnica-Krezel L, Schier AF, Driever W (1994) Efficient recovery of ENU-induced mutations from the zebrafish germline. Genetics 136 (4):1401–1420

67. Salinger AP, Justice MJ (2008) Mouse mutagenesis using N-ethyl-N-nitrosourea (ENU). CSH Protoc 2008:pdb-rot4985. https://doi.org/10.1101/pdb.prot4985

68. de Bruijn E, Cuppen E, Feitsma H (2009) Highly efficient ENU mutagenesis in zebrafish. In: Lieschke GJ, Oates AC, Kawakami K (eds) Zebrafish: methods and protocols, Methods in Molecular Biology, vol 546. Humana Press, New York, pp 1–12

69. Peterson RT, Fishman MC (2011) Designing zebrafish chemical screens. Methods Cell Biol 105:525–541. https://doi.org/10.1016/B978-0-12-381320-6.00023-0

70. Saunders NR, Ek CJ, Habgood MD, Dziegielewska KM (2008) Barriers in the brain: a renaissance? Trends Neurosci 31(6):279–286. https://doi.org/10.1016/j.tins.2008.03.003

71. Abbott NJ, Ronnback L, Hansson E (2006) Astrocyte-endothelial interactions at the blood-brain barrier. Nat Rev Neurosci 7 (1):41–53. https://doi.org/10.1038/nrn1824

72. Alvarez JI, Dodelet-Devillers A, Kebir H, Ifergan I, Fabre PJ, Terouz S, Sabbagh M, Wosik K, Bourbonniere L, Bernard M, van Horssen J, de Vries HE, Charron F, Prat A (2011) The hedgehog pathway promotes blood-brain barrier integrity and CNS immune quiescence. Science 334(6063):1727–1731. https://doi.org/10.1126/science.1206936

73. Armulik A, Genove G, Mae M, Nisancioglu MH, Wallgard E, Niaudet C, He L, Norlin J, Lindblom P, Strittmatter K, Johansson BR, Betsholtz C (2010) Pericytes regulate the blood-brain barrier. Nature 468 (7323):557–561. https://doi.org/10.1038/nature09522

74. Stenman JM, Rajagopal J, Carroll TJ, Ishibashi M, McMahon J, McMahon AP (2008) Canonical Wnt signaling regulates organ-specific assembly and differentiation of CNS vasculature. Science 322 (5905):1247–1250. https://doi.org/10.1126/science.1164594

75. Bohnsack BL, Gallina D, Kahana A (2011) Phenothiourea sensitizes zebrafish cranial neural crest and extraocular muscle development to changes in retinoic acid and IGF signaling. PLoS One 6(8):e22991. https://doi.org/10.1371/journal.pone.0022991

76. Rembold M, Lahiri K, Foulkes NS, Wittbrodt J (2006) Transgenesis in fish: efficient selection of transgenic fish by co-injection with a fluorescent reporter construct. Nat Protoc 1 (3):1133–1139. http://www.nature.com/nprot/journal/v1/n3/suppinfo/nprot.2006.165_S1.html

77. Woodfin A, Voisin M-B, Nourshargh S (2007) PECAM-1: a multi-functional molecule in inflammation and vascular biology. Arterioscler Thromb Vasc Biol 27(12):2514–2523.

https://doi.org/10.1161/atvbaha.107.151456

78. Cornford EM, Hyman S, Swartz BE (1994) The human brain GLUT1 glucose transporter: ultrastructural localization to the blood—brain barrier endothelia. J Cereb Blood Flow Metab 14(1):106–112. https://doi.org/10.1038/jcbfm.1994.15
79. Schinkel AH, Mayer U, Wagenaar E, Mol CAAM, van Deemter L, Smit JJM, van der Valk MA, Voordouw AC, Spits H, van Tellingen O, Zijlmans JMJM, Fibbe WE, Borst P (1997) Normal viability and altered pharmacokinetics in mice lacking mdr1-type (drug-transporting) P-glycoproteins. Proc Natl Acad Sci U S A 94(8):4028–4033
80. Eisenblätter T, Hüwel S, Galla H-J (2003) Characterisation of the brain multidrug resistance protein (BMDP/ABCG2/BCRP) expressed at the blood–brain barrier. Brain Res 971(2):221–231. https://doi.org/10.1016/S0006-8993(03)02401-6
81. Hawkins BT, Davis TP (2005) The blood-brain barrier/neurovascular unit in health and disease. Pharmacol Rev 57(2):173
82. Wolf HK, Buslei R, Schmidt-Kastner R, Schmidt-Kastner PK, Pietsch T, Wiestler OD, Blumcke I (1996) NeuN: a useful neuronal marker for diagnostic histopathology. J Histochem Cytochem 44(10):1167–1171
83. Nicchia GP, Nico B, Camassa LMA, Mola MG, Loh N, Dermietzel R, Spray DC, Svelto M, Frigeri A (2004) The role of aquaporin-4 in the blood–brain barrier development and integrity: studies in animal and cell culture models. Neuroscience 129(4):935–944. https://doi.org/10.1016/j.neuroscience.2004.07.055
84. Anderson CM, Swanson RA (2000) Astrocyte glutamate transport: review of properties, regulation, and physiological functions. Glia 32(1):1–14
85. Armulik A, Genové G, Betsholtz C (2011) Pericytes: developmental, physiological, and pathological perspectives, problems, and promises. Dev Cell 21(2):193–215. https://doi.org/10.1016/j.devcel.2011.07.001
86. Winkler EA, Bell RD, Zlokovic BV (2010) Pericyte-specific expression of PDGF beta receptor in mouse models with normal and deficient PDGF beta receptor signaling. Mol Neurodegener 5:32–32. https://doi.org/10.1186/1750-1326-5-32
87. Hellstrom M, Kalén M, Lindahl P, Abramsson A, Betsholtz C (1999) Role of PDGF-B and PDGFR-beta in recruitment of vascular smooth muscle cells and pericytes during embryonic blood vessel formation in the mouse. Development 126(14):3047
88. Ozerdem U, Grako KA, Dahlin-Huppe K, Monosov E, Stallcup WB (2001) NG2 proteoglycan is expressed exclusively by mural cells during vascular morphogenesis. Dev Dyn 222(2):218–227. https://doi.org/10.1002/dvdy.1200
89. Lippmann ES, Weidenfeller C, Svendsen CN, Shusta EV (2011) Blood-brain barrier modeling with co-cultured neural progenitor cell-derived astrocytes and neurons. J Neurochem 119(3):507–520. https://doi.org/10.1111/j.1471-4159.2011.07434.x
90. Mayer F, Mayer N, Chinn L, Pinsonneault RL, Kroetz D, Bainton RJ (2009) Evolutionary conservation of vertebrate blood-brain barrier chemoprotective mechanisms in Drosophila. J Neurosci 29(11):3538–3550. https://doi.org/10.1523/jneurosci.5564-08.2009

Chapter 6

Photometric Patch Electrode to Simultaneously Measure Neural Electrical Activity and Optical Signal in the Brain Tissue

Harunori Ohmori

Abstract

Photometric patch electrode (PME) measures optical signal of neurons simultaneously with electrical activity in deep brain tissues. As a light guide, PME transmits laser light to the tip of electrode to excite fluorophores within neurons. The emitted fluorescence from these neurons is captured by the PME simultaneously with the electrical activity. The optical signal is further transmitted through an optical fiber bundle to light detectors; either a photomultiplier tube or a spectrometer. The photomultiplier tube is used for a high-speed monitoring of fluorescence signal in a time range, while the spectrometer is used to analyze changes of a fluorescence signal-profile in a wavelength range. Captured electrical signal and fluorescence signal are highly correlated both in time course and amplitude. Furthermore, PME can apply chemicals locally in the brain tissue by pressure-control within the electrode. We will describe in detail the fabrication of PME and individual components of the PME recording system and demonstrate the application of PME in vitro in brain slices or in vivo in the brain tissue. As one of the limitations of using PME in in vivo experiments, difficulties of labeling neurons by calcium indicators in deep brain tissues are discussed.

Key words Photometry, Patch electrode, Calcium response, Spectrometer, Photomultiplier tube

1 Introduction

Neural information is transmitted by action potentials, which accompany changes of intracellular calcium concentration in most neurons. Calcium signal in neurons is known to affect neural electrical activity, such as to trigger synaptic transmission, and mediate synaptic plasticity. Intracellular calcium signal further influences gene expressions and eventually induces rearrangement of neural circuits [1–3]. Obviously, calcium signals contain more information than simply to inform neural electrical activity. It is, thus, pertinent to measure the electrical signal and calcium signal of neurons simultaneously in order to find clues to decipher the brain function. Neural electrical activity can be measured by an electrode, while the

Nicholas J. D. Wright (ed.), *Basic Neurobiology Techniques*, Neuromethods, vol. 152,
https://doi.org/10.1007/978-1-4939-9944-6_6,

intracellular calcium response can be measured as a fluorescence signal of compounds such as synthesized dyes or genetically encoded proteins [4, 5]. By labeling neurons with calcium-sensitive compounds, neurons can be imaged and calcium signal can be measured using the fluorescence microscope [6, 7]. However, the neural imaging is limited to a surface layer of the brain in most applications, a depth of less than 1 mm from the brain surface, because the imaging precision is greatly reduced by scattering of light in brain tissues [8, 9]. Simultaneous recording of electrical activity with neural imaging has a further difficulty of inserting an electrode to the brain region during fluorescence microscopy.

The depth limitation of fluorescence measurements has been challenged by invention of variety of imaging or photometric devices, such as the micro-endoscope [10, 11], and the fiber optics [2, 12]. A dual-core microprobe was invented to measure simultaneously the electrical activity and optical responses [13]. Each of these technologies has its own strengths and weaknesses.

We have developed a photometric patch electrode (PME) to measure the electrical activity and optical response of neurons by using a patch electrode as a light guide to both excite fluorophores and to capture the emitted fluorescence [14]. Because the neural electrical activity and fluorescence response captured by the PME are highly correlated in time course and size of responses, the PME can provide a useful measure of neural activity in deep brain tissues as corresponding electrical and calcium fluorescence signals.

1.1 Outline of the PME Recording System

PME recording system is composed of a part of laser transmission and a part of fluorescence detection (Fig.1a). The laser-transmitting part consists of a laser assembly (20–50 mW laser output is turned on and off through a shutter, reduced the intensity through 25–50% neutral density filter), one branch of a bifurcated optical fiber bundle (Fig. 4), and the electrode holder assembly (Fig. 3). Laser light is transmitted through the PME to excite the dye in neurons (Fig. 1b).

Emitted fluorescence is captured at the tip of PME and transmitted back to the other branch of bifurcated optic fiber bundle (Fig. 4), which leads the signal to one of the two fluorescence detectors. The fluorescence-detecting part has two measuring options (Fig. 1a). One option uses a grating-spectrometer coupled with a CCD camera to measure a spectrum of fluorescence signal (Fig. 1a-(i)). The other option uses a photomultiplier tube and a lock-in amplifier to measure the fluorescence signal at a fast speed (Fig. 1a-(ii)). The choice of detector depends on the type of experiment and the nature of fluorescence signal; namely the choice depends on the speed of phenomenon or whether or not the phenomenon includes spectrum change.

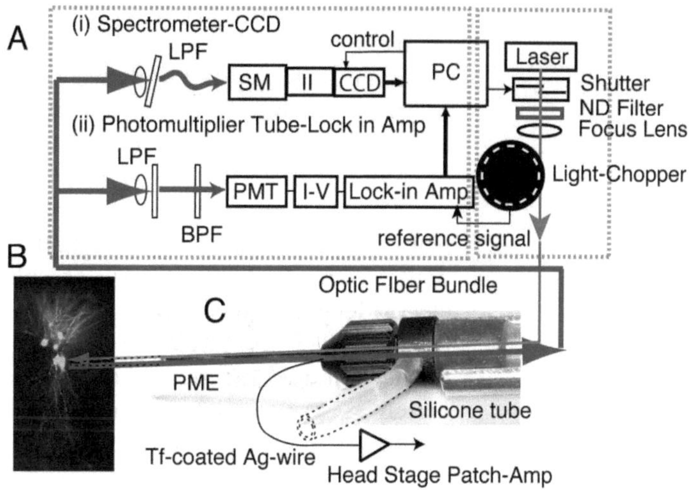

Fig. 1 Outline of photometric patch electrode (PME) recording system. (**a**) Laser transmission to the PME, and two modes of fluorescence detection. (i) uses a spectrometer and CCD camera for spectrometry, and (ii) uses a photomultiplier tube for high speed detection. Appropriate interference filters are used. Squared region on the right indicates a part of laser transmission, and on the left, indicates a part of detectors. (**b**) Images of cortical neurons to show the photometric measurement. (**c**) PME holder assembly. *LPF* long-pass filter, *SM* spectrometer, *II* image intensifier, *CCD* CCD camera, *PC* personal computer for control, *ND filter* neutral density filter, *BPF* band-pass filter, *PMT* photomultiplier tube, *Tf* Teflon, *Ag* silver

2 Materials

2.1 Animals

We treated all experimental animals in accordance with Kyoto University guidelines (approved number MedKyo 13344). Chicks 2–6 days after hatching (P2–6) were used for in vivo experiments and rat pups (P9) were used for slice experiments. In some experiments using genetic encoded calcium indicators (GECIs), the mouse hippocampus preparation was used in addition to the chick, after transfection of adeno-associated virus (AAV) that encodes GECIs. For the chick, avian strain of AAV (A3V, avian adeno-associated virus) was used to transduce GECIs. Details of viral transfection procedures will be described below.

2.2 Anesthesia and Craniotomy of the Chick

In contrast to mammals, the auditory cortex of chicken, field-L, is not located at the brain surface but is developed extending to the depth about 3–4 mm in the telencephalon. Surface layer of the telencephalon is occupied by a structure analogous to hippocampus of mammals [15]. The inferior colliculus (IC) is located more deep (7–8 mm) from the surface of the brain. In order to penetrate these brain regions with an electrode, craniotomy was made under anesthesia, which was either intra-peritoneal injection of urethane (1 mg/g body weight), or of a combination of anesthetics (medetomidine [0.15 mg/kg body weight], midazolam [3 mg/kg body weight], and butorphanol [0.5 mg/kg body weight] [16]. The anesthesia level was maintained during experiments by injection of chloral hydrate (0.2 mg/g body weight), or urethane.

The position of the head was fixed to a stereotaxic stage by a pair of ear bars and a metal rod that was fastened to the skull with dental cement. Then, a part of cranial bone was removed for electrode penetration. Electrode was inserted vertically to the field L at 1.5 mm lateral, and approximately 1.2–1.5 mm rostral of the lambdoid suture, and the inferior colliculus (IC) at approximately 3.7–4.0 mm lateral, and 1.7–2.0 mm rostral of the lambdoid suture as previously described [14, 17, 18].

2.3 Bolus Injection of OGB1-AM to the Acoustic Nuclei of Chicken In Vivo

Oregon Green BAPTA1-AM (1 mM in ACSF; OGB1-AM, Molecular Probes, Eugene, OR) was injected in these brain regions about 1 h ahead of PME recording. OGB1-AM was prepared by dissolving in 20% Pluronic F127 in DMSO, and diluted in ACSF (concentrations in mM: 125 NaCl, 2.5 KCl, 26 $NaHCO_3$, 1.25 NaH_2PO_4, 2 $CaCl_2$, 1 $MgCl_2$, and 17 glucose, pH 7.4) [14, 19]. The dye injection in brain tissues was made by pressure using a glass electrode. The glass electrode for injection was pulled from a hard glass capillary (1.5 mm OD, 1.0 mm ID), and had a tungsten wire (15 μm, Teflon coated) inside to monitor the sound evoked electrical field response. The tip of glass electrode was cut to make 20–30 μm diameter. The tungsten wire was also cut at the level of capillary tip and was made sensitive to neural electrical activity just at the exposed tip. The tungsten wire was fastened to the electrode at the proximal end by bending and thrusting through a silicone tube. Teflon coat was removed by cigarette lighter flame at the point where electrical contact was made with a lead to an amplifier. The electrode was filled with a small amount of dye solution at the tip (1–2 μL). The dye injection was made while monitoring acoustic response, at the depth where the acoustic response was the largest.

The ejection of dye solution from the tip in the brain tissue was confirmed as a depression of meniscus of filling medium within the electrode. A minimum level of positive pressure was applied manually to eject the dye, by using a subcutaneous injection syringe (1 mL). We maintained the positive pressure inside the capillary over 30–60 s. At the moment of dye ejection, the acoustic field response was reduced transiently but the response recovered immediately after the ejection (within 30 s in most cases). Therefore, we believe that the ejected volume should be small. The pressure was relieved and the electrode was maintained in that position for approximately 3 min after the dye ejection, and was then withdrawn slowly.

2.4 Acoustic Stimulation

Acoustic stimulus was made using a pair of small earphones (EF-1935; Knowles Electronics) that were attached to the hollow ear bars. The sound pressure level was calibrated using a standard microphone (type 4939; Bruel & Kjaer). The sound intensity was measured during experiments by a pair of small microphones (BT-1751; Knowles Electronics) equipped with a probe tube to the ear bar. The stimuli were either pure-tone or white-noise sound, with 5 ms rise and fall of a sine wave-shaped envelope. White noise stimulus was repeated at 1 s intervals to detect neurophonics while approaching the electrode to brain regions of interest as it was described previously [14, 17, 18]. The sound duration was 80 ms during electrode approach and 200 ms in most photometric experiments. Acoustic stimulation and data acquisition were controlled by customized software written in MATLAB [14]. All experiments were performed in an electrically shielded soundproof chamber, placed within a soundproof booth. Local field currents were low-pass filtered at 15 kHz (2-pole), and were recorded using a patch clamp amplifier in voltage clamp mode (Multiclamp-700B, Axon Instruments, Union City, CA).

2.5 Transfection of Brain Regions with Adeno-Associated-Virus Encoding mCherry or GCaMP6

Avian adeno-associated-virus (A3V, 10^{13} particles/mL) encoding GCaMP6s or mCherry was injected in auditory nuclei (IC or field L) of the chick (P2–3) by the same technique used for injection of OGB1-AM. A3V is the avian strain of adeno-associated virus (AAV), and is efficient to be transduced in chicken neurons [20]; although, the mammalian AAV vectors are not efficient to be transduced in avian cells. AAV is a nonpathogenic single stranded DNA virus that can be replicated only in the presence of a helper virus such as herpes virus or adeno virus [21]. The recombinant AAV vector permits nontoxic transduction of postmitotic cells and long-term gene expression in neurons but not in nonneuronal cells, which enables us to deliver genetically encoded tools for visualizing and manipulating neuronal activity [20, 22]. We applied A3V vectors that express mCherry or GCaMP6s under the control of Rous sarcoma virus promoter to the chick brain. These A3V vectors were kindly provided us by Drs. Matsui R and Watanabe D from Kyoto university. After the injection, the piece of craniotomied bone was returned to cover the brain surface and cranial skin was sutured. After 7–10 days from viral injection the chick was anesthetized, the craniotomied bone was removed, and the brain was penetrated with a PME. The site of virus transfection was searched as a response to white noise stimulus electrophysiologically, and was confirmed by the increase of fluorescence of mCherry (*see* Fig. 11).

For recording in the mouse hippocampus using AAV-GCaMP6f expression system, we injected AAV vector in the adult mouse hippocampus; the depth of about 1 mm at several adjacent locations. After 7–10 days, the PME recording was conducted in the hippocampus.

2.6 Brain Slices of Rat Pups

Slices (thickness 300 μm) of rat cerebral cortex or hippocampus were made using a microslicer (Pro1, Dosaka EM, Kyoto). Neurons there expressed either eGFP compound (Figs. 1b and 5a), or eCFP-eYFP FRET compound (Fig. 6) [23], or were labeled acutely by local perfusion of OGB1-AM on the slice (Figs. 7 and 8) [19]. Wistar rats of age P9 were deeply anesthetized with halothane (Fluothane; Takeda, Osaka, Japan) before decapitation, the brain was removed, and the block of brain was cooled in ice-cold 35 mM glucose saline (35GS; in mM: 130 NaCl, 4.5 KCl, 2 $CaCl_2$, 5 PIPES-Na, and 35 glucose, pH 7.4) saturated with 100% O_2 and then mounted in a 4% agarose gel (Low Gelling Temperature; Nacalai, Kyoto, Japan) prepared with the 35GS. Transverse brain slices were cut with the microslicer. Slices were incubated at least for 1 h at 35 °C, and then were maintained at room temperature (25–27 °C) in a 100 mM glucose ACSF (100-glucose ASCF; in mM: 75 NaCl, 2.5 KCl, 26 $NaHCO_3$, 1.25 NaH_2PO_4, 2 $CaCl_2$, 1 $MgCl_2$, and 100 glucose, pH 7.4) saturated with 95% O_2–5% CO_2. Experiments were conducted using an upright microscope (BX51WI, Olympus) in ACSF saturated with 95% O_2–5% CO_2 at room temperature [14].

3 Methods

3.1 Fabrication of PME

PME is pulled from a quartz glass capillary (1.5 mm OD, 1.0 mm ID) made of optic fiber grade quartz glass (Fig. 2), and the exterior is nickel (Ni) coated. Ni-coating is made before pulling the electrode by an autocatalytic chemical reaction (the electroless nickel-plating method, Melplate MB, Meltex, Tokyo). Center 10 mm and both ends of the capillary were prevented from coating (Fig. 2a). Both ends of capillary were sealed by caps of silicone tube during coating procedure, in order to protect Ni-coating media to enter into the capillary. The center was wrapped by a silicone tube and was prevented from coating, because the coating there was evaporated during electrode pulling by a laser puller (Fig. 2a). We expect the Ni coated surface to increase reflection of light in the electrode, protect the electrode surface from stains, and makes the electrode handling easier.

Both ends of the quartz capillary were polished by an optic fiber polishing machine to increase transmission of light at the interface between optic fiber bundle and PME (Fig. 2b, SpecPro polisher, Krell technology). After the end-polishing, capillaries were cleaned by aspiration of distilled water, dried, and pulled by a laser puller (P2000/G, Sutter Instrument). The electrode has a tip diameter of 2–3 μm and a resistance of 1–3 Mohm when filled with ACSF.

3.2 PME Holder Assembly

Proximal polished end of PME is apposed closely to the distal end of optic fiber bundle within the electrode holder assembly (Fig. 3a). The holder is made from black colored Delrin (polyoxymethylene,

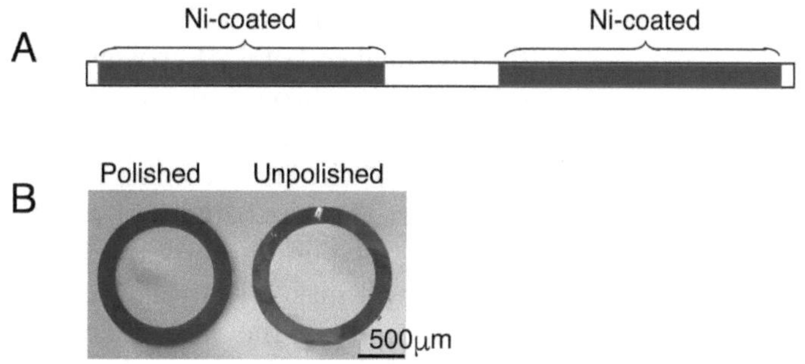

Fig. 2 Ni-coated quartz capillary. (**a**) PME is made of quartz capillary, and its surface is Ni-coated except for the center and two ends. (**b**) Both ends of the capillary are polished; left, after polishing, and right, before polishing. (**b** is reproduced with permission [14])

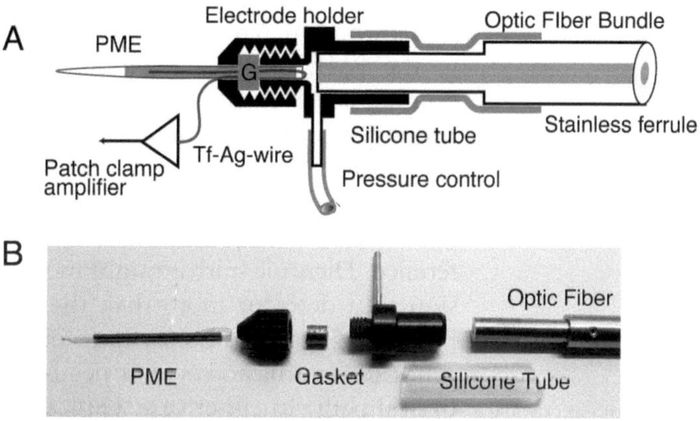

Fig. 3 PME holder assembly. (**a**) Illustration of the holder assembly. *G* gasket. (**b**) Individual parts of the holder assembly. (**a** is reproduced with permission [14])

Du Pont) and is made airtight by a gasket (G) of silicone tube. By another silicone tube the coupling between the holder and the stainless ferrule of optic fiber bundle is made airtight. The gasket surface is nickel-coated to reduce fluorescence emission from the silicone tube (Fig. 3b): some silicone tube emits fluorescence, when illuminated with a laser light. The gasket has a longitudinal slit to make the electrode thrust easier, but could still maintain the pressure inside the holder assembly by compression of the holder cap. Figure 3b shows from the left to the right individual parts of electrode holder assembly; PME, holder cap, gasket, main body of the holder that has a stainless tube connected to the side to control pressure inside the capillary, silicone tube to wrap the coupling between the holder and optic fiber bundle, and the stainless ferrule of optic fiber bundle. Electrical signal from the PME is led by a Teflon-coated silver (Ag)-wire directly to the head stage of a patch clamp amplifier (Figs. 1c, 3a, and 10c). The wire is folded at the proximal end of PME, passes through the gasket and comes out of the holder through the same hole with PME (Fig. 3a). Teflon coating of Ag-wire is necessary to insulate the wire, in order to protect the wire from contacting the nickel-coated surface of the electrode. The Ni-coated part of PME is likely bathed in the surrounding medium during experiments, and its contact with a bare Ag-wire will make electrically serious junction potential problem. At both ends of the wire Teflon coat is removed by flame of a cigarette lighter; one end is dipped into the pipette solution within the PME and the other end is formed like a ball to be pinched to lead electrical signal to the head stage patch clamp amplifier. The lead wire should be shielded to the ground at the head stage amplifier to reduce electrical noise (Fig. 10c).

3.3 Bifurcated Optic Fiber Bundle

Bifurcated optic fiber bundle is used to separate the fluorescence signal from excitation laser (Fig. 4). Excitation laser is transmitted to the PME through a thinner branch of bifurcated optic fiber bundle, which constitutes the center 10% of the fiber bundle. The rest 90% of optic fiber bundle captured the fluorescence from PME to transmit it to the light detectors. Alternative to the use of bifurcated optic fiber bundle is to use a dichroic mirror to separate the fluorescence light from the excitation laser by wavelength difference. Dichroic mirror might increase the level of light transmission to a detector more than the use of a bifurcated optic fiber bundle. This would be useful to increase the sensitivity of fluorescence detection; however, it needs more careful alignment of the optical path. In either case, appropriate interference filters should be inserted to the light paths. Laser light reflection from the proximal end of the PME is strong, and must be eliminated by filtering using a long-pass filter. The fluorescence signal of interest is transmitted to a photomultiplier tube detector through a sharp edge band-pass interference filter (Fig. 1a(ii)). In the case of grating-spectrometer detector, we only inserted a long-pass filter (Fig. 1a (i)), because the spectrometer measured a wide wavelength range signals (Fig. 6). In the excitation-light path, a band-pass interference filter is sometimes used to eliminate side bands of laser.

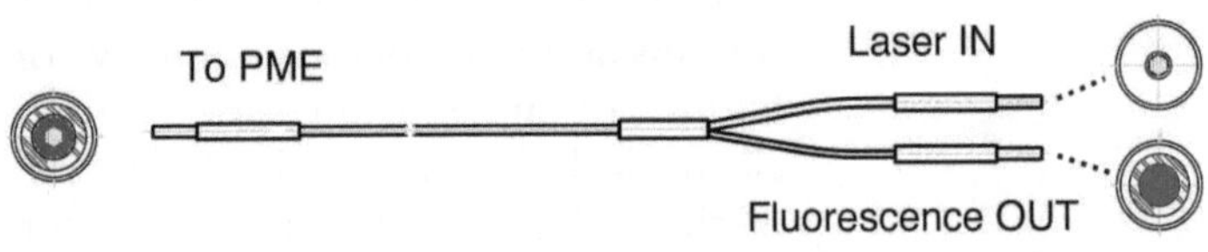

Fig. 4 Bifurcated optical fiber bundle. Fiber bundle is made of 100 of 100 μm diameter NA 0.22 optical fibers. Laser light was transmitted through 10% of optical fibers, and fluorescence was transmitted through 90% of fibers. The common end apposes to the PME

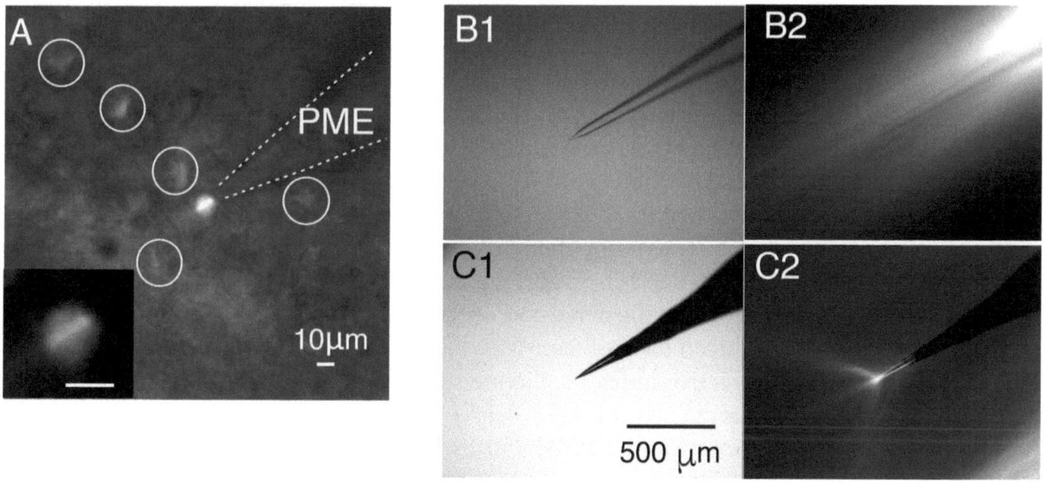

Fig. 5 Fluorescence excitation at the tip of PME and laser leak around PME. (**a**) Rat cortical neurons expressing eGFP in a slice preparation were excited and emitted fluorescence by laser light leaked and radiated from the PME. In the center of the panel a neuron was touched by a PME tip and most strongly emitted fluorescence. Edge of the PME is indicated by broken lines. Inset shows enlargement of the cell in the center. Other fluoresce emitting neurons are encircled. **b1**, **c1**; bright field image. **b2**, **c2**; radiated light from a PME was made visible in a fluorescence medium (Alexa Fluor 488). (**b**) A bare PME. (**c**) Sylgard coated PME near the tip. Sylgard was mixed with soot. **c2** shows that the laser leak was strong at the tip of PME and at the taper near the tip. (Reproduced with permission [14])

3.4 Fluorescence Excitation at the Tip of PME

Excitation of fluorophore was most effective at the tip of PME as is shown in Fig. 5a; a brain slice prepared from rat cortex, in which some neurons expressed eGFP. In this slice, there are several eGFP positive neurons (some are encircled). Because of slice experiments, the PME is in an ambient recording medium (ACSF). Accordingly, the laser light leaked out from PME illuminated the preparation diffusely. In Fig. 5a, some eGFP positive neurons are excited by laser light radiated from the PME and emit fluorescence. One that is located at the center of the figure is touched by the PME (indicated by a pair of broken lines) and emitted the strongest fluorescence (inset of Fig. 5a).

The condition of cellular illumination should be different in live brain tissues in vivo from this slice preparation. It is repeatedly discussed that laser light is scattered strongly by brain tissues in vivo, which limits the depth of high resolution imaging less than 1 mm from the brain surface by multiphoton microscopy [8, 9]. In case of PME in live brain tissues, the electrode is surrounded by cellular structures and closely contacted with them. Therefore, it is expected that the laser light leaked out from the PME reduced the intensity extensively by scattering in the surrounding tissue. By a model experiment, we may be able to observe a pattern of light scattering near the PME. Here, a PME was dipped in a fluorescence dye solution (Alexa Fluor 488). Near the PME there was a shower of laser light and the fluorescence emerged diffusely in the dye solution (Fig. 5b1, b2). Figure 5c2 illustrates the laser leak at the tip and from the tapered region near the tip of PME (Fig. 5c1) where the glass wall intersects with the laser light transmitted through the PME. The largest taper is formed at more proximal position to the photographed region of the electrode in Fig. 5b where a shower of laser leak occurs (Fig. 5b2). When the surface of PME was coated by Sylgard (Sylgard 184, Dow Corning) close to the tip, the surrounding fluorescence leak was extremely reduced and the fluorescence leak in the dye solution was observed only from the distal end of Sylgard coated region of PME [14]. This reduced level of laser leak should be attributed to a relatively lower refractive index of silicone resin (r.i. 1.40–1.43) that envelopes the quartz glass (r.i. 1.53–1.55). The laser leak was prevented efficiently by coating the PME with Sylgard that was mixed with microparticles like soot (Fig. 5c). We presume that the light scattering near the PME in live brain tissues in vivo might resemble the pattern after Sylgard coating of the PME [14] and the intensity of radiated laser light should be reduced. In contrast, the light transmitted inside the PME (most likely through the quartz glass wall) should reach the tip without much fading during transmission. Therefore, the emitted fluorescence is stronger at the tip (Fig. 5a), where also the fluorescence is most effectively captured. This idea is consistent with the observation of highly correlated electrical signal and fluorescence signal recorded by the PME in brain tissues in vivo as will be shown in the following section (Fig. 9).

3.5 Grating-Spectrometer

Two types of light detectors were switched manually (Fig. 1a). One detector is a grating-spectrometer and the other is a photomultiplier tube. Each detector has its own merits and demerits as will be discussed below.

Spectrum of fluorescence signal was obtained by the grating-spectrometer (C5094), image intensifier (C7979–81), and CCD camera (ORCA-R^2) in series (all from Hamamatsu Photonics, Hamamatsu, Japan) (Fig. 1a-(i)). The fluorescence light from the bifurcated optic fiber bundle was first focused through a lens on the detector probe of grating-spectrometer. The detector probe is made of total 15 fibers (NA = 0.22 and 200 μm core diameter). At the input of grating-spectrometer, each fiber channel is vertically aligned and the light beam is dispersed at 0.5-nm resolution in abscissa at the output, which forms in total a two-dimensional spectrum image (Fig. 6a). This spectrum image is projected, intensified by the image intensifier (II), on-chip integrated by the CCD camera and is captured by a personal computer (PC, Fig. 1a-(i)). Individual fiber channel corresponds vertically to 20–21 pixels on the CCD camera. Image data of the spectrum is frame-transferred to the personal computer, ensemble averaged at each wavelength to improve the signal to noise ratio, and the final spectrum data is computed as the superimposed line drawing on Fig. 6a. Image

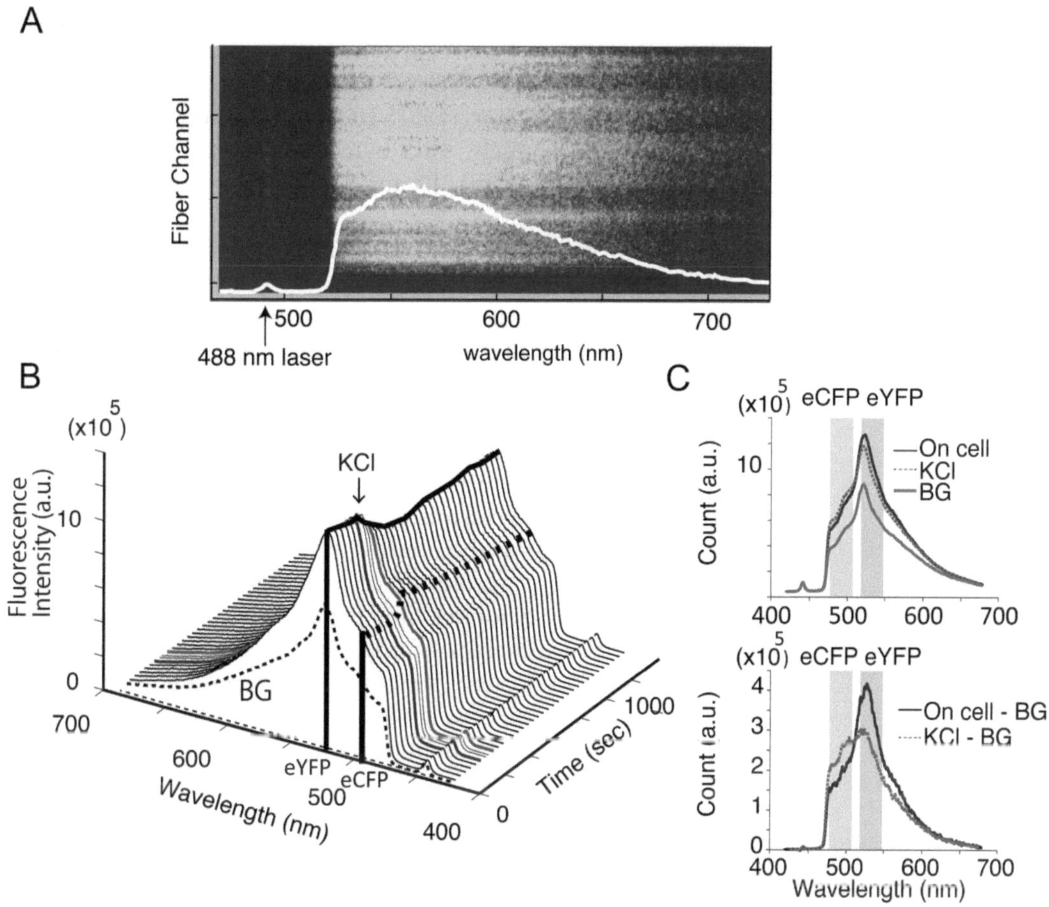

Fig. 6 Spectroscopy of a neuron expressing a FRET compound. (**a**) Image of spectrogram. Individual optical fibers in the detector probe of spectrometer are vertically aligned as fiber channels at the input of graded spectrometer. Optic signal of each fiber channel is dispersed along wavelength in abscissa and forms in total a two-dimensional spectrum image. The superimposed line is a vertical ensemble of individual spectrum. Upward arrow indicates the peak of 488 nm laser. (**b**) Time series of spectrum of FRET signal measured from a rat cortex pyramidal neuron in a slice that expresses a FRET compound [21]. A PME touched the neuron under visual manipulation, and KCl medium (30 mM) was applied to induce FRET response. Background (BG) spectrum, and signal levels for eCFP and eYFP are indicated. Spectrum was on chip integrated at the CCD camera for 0.3 s and measured every 20 s. (**c**) (Top), spectra before (on cell) and after KCl application (KCl), and the BG spectrum. (**c**) (Bottom), after subtraction of BG from two spectra before (On cell-BG) and after KCl application (KCl-BG). BG spectrum was obtained before PME approached the neuron. (**b**, **c** are reproduced with permission [14])

intensifier is gated by a trigger that is issued from the PC (Fig. 1a-(i)). On chip image integration time by CCD camera was set in most experiments at 0.3 s (with 2 × 2 binning), which limited the time resolution of recordings.

The PC issued trigger pulses, separately from the one that triggers the image intensifier, at programmed timings to synchronize external devices (not illustrated in Fig. 1a). Electrical signals such as the output of patch clamp amplifier, the acoustic stimuli and trigger pulses to synchronize external devices and the ensemble averaged spectrum data were recorded either by another PC or a multichannel data logger (EZ7510, 16-bit resolution, 20 kHz sampling rate, 8 channels, 40G byte mass storage HD recorder, NF corporation, Yokohama, Japan). Data logger is convenient as a backup recording device.

We intentionally tilted the laser-cut long-pass interference filter (BLP01-488 for a 488-nm laser, Semrock) inserted in front of the detector probe by 10–15 degrees from the normal angle in order to produce a small peak in the spectrum (Fig. 1a-(i), upward arrow in Fig. 6a). This spectrum peak was attributed to the reflected laser light from the proximal end of PME, and was not affected by experimental manipulation. It was used as the internal standard of spectrum profile in each experiment. The PME was replaced with a new one when the spectrum profile changed significantly after repeated penetrations in the brain.

Although the time resolution of optic signal is limited with the spectrometer (0.3 s or longer time intervals in most experiments), there are several advantages of using it; (1) a relatively high level signal to noise ratio because of ensemble averaging of multiple fiber channels data, and (2) availability of broad wavelengths range signal. The second point is important for detecting changes of fluorescence spectral profile, such as FRET (Fig. 6b, c), or BRET (bioluminescence resonance energy transfer) measurements [24, 25]. Furthermore, (3) a background spectrum or background fluorescence level was easy to be evaluated and subtracted from signal spectrum if necessary. Background subtraction sometimes improves the signal to noise ratio of the measurement (Fig. 6c).

We evaluated the fluorescence signal as a fraction of change ($\Delta F/F$) to the starting intensity level of each record; here F means the fluorescence level at the start of each recording and ΔF is the change of fluorescence from F at each moment. The background subtraction reduces F. In experiments presented in Fig. 6b, the background spectrum (BG) was evaluated before approaching the PME to the neuron (Fig. 6b, c top) and was subtracted from the response signal (Fig. 6c bottom). Here, measurements were conducted in a slice preparation from a single neuron by manipulating the PME to touch it under visual control. Individual spectrum was measured at 20 s intervals. Application of a high K^+ medium (30 mM, at the arrow, Fig. 6b) induced Ca^{2+} influx and evoked a FRET response [23]. FRET signals measured was the increase of eCFP signal (25%), the reduction of eYFP signal (27%), and the FRET ratio change of 35% [14].

3.6 Measurement of Temporal Calcium Signals in Slice Preparations

We measured calcium response in granule cells of dentate gyrus hippocampus of rat pup in a slice by manipulating the PME to touch an OGB1 fluorescence positive neuron (Fig. 7a). OGB1-AM was loaded in slices [19]. By triggering antidromic spikes by electrical stimulation (Fig. 7b), calcium ionic response was measured by the spectrometer (Fig. 7c) and further by the photomultiplier tube (Fig. 8), and features of measured signals were compared. The stimulus was a train of 1–100 bipolar pulses of 0.1 ms duration (50% duty cycle) at 100 Hz. The train of stimulus was applied at every 50 s to the mossy fiber projection and antidromic spikes were generated. The extracellularly recorded electrical signal showed a

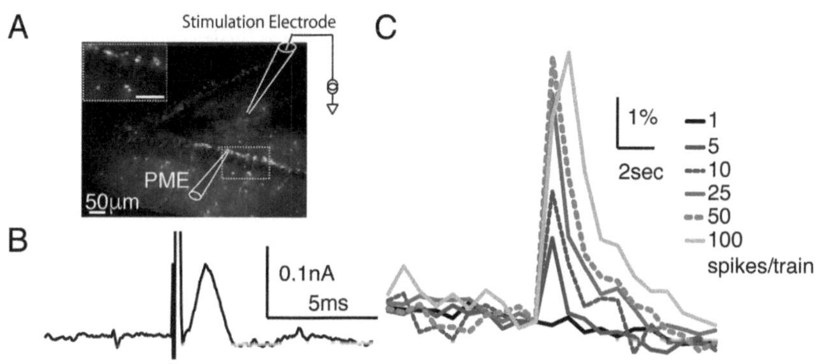

Fig. 7 Spectrometer measurements of OGB1 signal in a granule cell of dentate gyrus of rat hippocampus slice. (**a**) OGB1 loaded image of dentate gyrus granule cells. Locations of stimulating electrode and PME are indicated. Inset upper left shows the enlargement of squared area with a 50 μm calibration bar. (**b**) Cell attached recording by a PME from a granule cell that was activated by antidromic electrical stimulation. (**c**) PME fluorescence response to trains of electrical stimulation (100 Hz, 0.1 ms duration), sampled at 0.3 s intervals. Numbers indicate the number of stimulus in a train

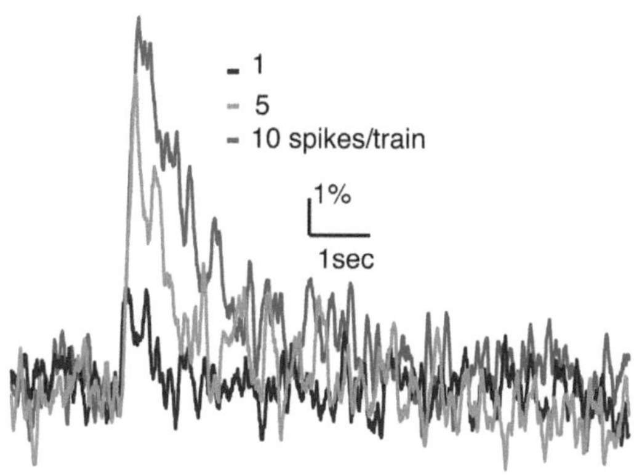

Fig. 8 Photomultiplier tube measurements of OGB1 signal in a granule cell of dentate gyrus of a rat hippocampus slice. Granule cells in a slice were loaded with OGB1, and cell attached recording was made with a PME. Response to electrical stimulation was generated by the same protocol as Fig. 7. The fluorescence was detected by a photomultiplier tube at 5 ms sampling intervals

transient current response after the stimulus artifact (Fig. 7b). The fluorescence signal size was depended on the number of stimuli in a train (Fig. 7c). When measured by a spectrometer (Fig. 1a-(i)), the peak response saturated at nearly 50 stimuli in a train. The peak position apparently delayed 1 s at 100 spikes in a train. This delay could be attributed to the long duration (1 s) of the stimulus train (100 stimulus of 100 Hz) that was longer than the time taken for charge integration on the CCD camera (0.3 s), namely the intervals of data sampling.

3.7 Photomultiplier Tube and Lock-in Amplifier to Measure Signals at Faster Speed

The slower time resolution of spectrometer measurement is overcome by using a photomultiplier tube (PMT) as a detector (H6780-02, Hamamatsu Photonics, Hamamatsu, Japan). The photo-current output was current-voltage converted (LI-76, NF corporation, Yokohama, Japan), low-pass filtered through a lock-in amplifier (LI-5640, NF corporation), and was AD sampled (Fig. 1a-(ii)). By using a light chopper (50% duty cycle, 5584A, NF corporation), reference signals (~2 kHz) were provided to the lock-in amplifier (Fig. 1a). Time constant of the lock-in amplifier was set at 1–5 ms in most experiments. This time constant determines the frequency of low pass filtering.

Time resolution of calcium signal was improved by the use of PMT as a detector. Figure 8 shows the fluorescence responses from granule cells of rat hippocampus slice (same as Fig. 7a). By comparison with the data obtained using the spectrometer (Fig. 7c), the fluorescence signal was fast rising and noisier, but the response feature to the stimulus number in a train was nearly the same (Fig. 8). Enhanced noise level of the signal may be attributed to the point by point sampling of photo-current by the photomultiplier-tube system, in contrast to the sampling by the spectrometer system, where signal was sampled after on-chip charge integration and ensemble averaging of signals from multiple fiber channels (*see* Fig. 6a).

In response to the trigger pulses issued by the PC, several data signals were recorded simultaneously, such as the fluorescence signal from the lock-in amplifier, the current signal from patch clamp amplifier, and the stimuli such as the left and right ear sound (~2 kHz sampling rate, Nidaq). Fluorescence detection was made after elimination of 488 nm laser light reflection by a laser-cut long-pass filter (520 nm and long wavelength, Di01-R514, Semrock), and the fluorescence signal of interest wavelength rage (520–570 nm) was extracted through an interference filter (FF01-542-27, Semrock).

3.8 In Vivo Recordings from Auditory Nuclei

We applied the PME recording system to auditory nuclei of the chick in vivo. Auditory responses are fast and synchronized to the sound stimulus. After 1 h of OGB1-AM injection, PME was inserted to the injection site, field L, the avian auditory cortex.

We applied acoustic stimulus periodically, while inserting the PME deep into the brain tissue. The dye injection site was confirmed by detecting the sound evoked electrical field response and OGB1 fluorescence signal. The fluorescence intensity of spectrum ranging from 500 to 600 nm increased up to 300 to 600% of the starting level measured at the brain surface when the penetrated PME hit the targeted region [14].

With a photomultiplier tube as a detector, we could observe a highly correlated local field current and optical signals (Fig. 9a). Here, acoustic responses were enhanced by local application of gabazine (0.1 –5 mM) through the PME. The local ejection of chemicals was made by applying a pressure of 2–10 psi inside the PME through the side tube attached to the PME holder assembly (Fig. 3a). This level of positive pressure did not affect the electrical field response or the optical signal. The small tip diameter (2–3 μm) of PME likely made the extrusion of fluid gentle from the tip.

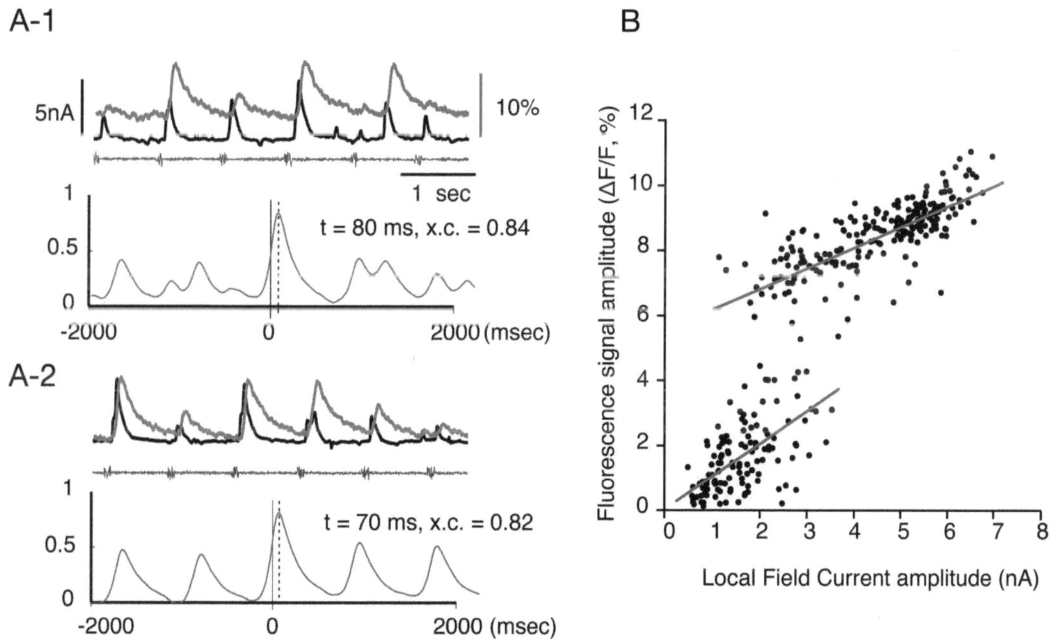

Fig. 9 Sound induced field responses of electric current and fluorescence in field L, the auditory cortex of chick. (**a-1**, **a-2**) Two examples of cross-correlation between the acoustic induced local field current (black) and fluorescence (gray) responses. Top three traces indicate field responses of calcium fluorescence and current, and the acoustic stimulation. Bottom trace indicates the cross-correlogram between current and fluorescence. *t* delay time, *x.c.* cross-correlation coefficient. (**b**) Correlation of amplitudes between the local field current and corresponding fluorescence. Amplitude of fluorescence signal was measured at the timing delayed 50–80 ms from the current response. The delay was calculated by corresponding cross-correlogram. Acoustic responses were enhanced by gabazine (0.5 mM), which was locally applied through the PME. Two best fit lines are indicated; $r = 0.58$ for a cluster of smaller amplitudes and $r = 0.76$ for larger amplitudes. (Reproduced with permission [14])

The OGB1 fluorescence calcium response and the local field current response are highly correlated (Fig. 9a, cross-correlation coefficient (x.c.) is larger than 0.8). The calcium response followed the sound driven local field current with a delay of 50–80 ms (Fig. 9a). This relatively long delay may be attributed to the fact that both local field current and calcium signals were the reflection of ensemble activity of group of neurons, and not the one to one correspondence of single neuron activity. Figure 9b shows two data clusters of sound evoked signals from consecutive recordings in field L, where optical response amplitude was plotted against corresponding local field current amplitude. The smaller amplitude cluster may indicate subthreshold responses (correlation coefficient, $r = 0.58$) and the larger amplitude cluster the suprathreshold responses ($r = 0.76$).

3.9 Use of Genetic Encoded Calcium Indicators

It is ideal to measure calcium response of neurons with genetic encoded calcium indicators (GECIs). GECIs could be expressed in neurons at a targeted region of the brain, and the response can be linked to a specific function of neurons [20, 26, 27]. Among GECIs, GCaMP6 is a highly recommended calcium indicator used in multiphoton microscopy because of the high responsibility to calcium ions. There are 3 variants in GCaMP6; 6 s, 6 m and 6 f. They are all excited by 488 nm laser light, but are different in response speed (faster in 6f > 6 m > 6 s), and response amplitude (larger in 6 s > 6 m > 6f). Calcium ion binding affinity is nearly the same between 6 s and 6 m (Kd = 144 nM (6 s) and 167 nM (6 m)) [28], but is slightly low in 6f (Kd = 375 nM). Hill's coefficient of calcium ion binding was close to 2 (6f) or 3 (6 s, 6 m).

3.10 Mouse Hippocampus Calcium Response Measured by GCaMP6f

Using adeno-associated virus (AAV) we transfected GCaMP6f to neurons of adult mouse hippocampus. Injection of AAV that encodes GCaMP6f was made at the depth of about 1 mm from the brain surface. After 1 week to 10 days of injection, we inserted a PME to the hippocampus region where AAV injection was made. The change of basal fluorescence level was very small (nearly 10%), when the PME approached the transfection site. However, at the depth of about 1 mm from the brain surface, we could measure Ca^{2+} signals of a significant size in response to a sensory stimulus applied to the tail, such as a tail lift stimulus. GCaMP6f fluorescence signal $\Delta F/F$ of 1–3% in amplitude was recorded simultaneously with local field potential or unit activity, which lasted several seconds after the stimulus. Since AAV injection was made at several adjacent locations in a single animal and covered a large area of the hippocampus, finding the transfected neurons and measurements of calcium signal by the PME were not difficult. We also tried measurements from a transgenic mouse that expressed GCaMP3. In this animal, all neurons expressed GCaMP3 and measurements of calcium signal together with electrical response was not a problem at all. From any locations in the brain we could detect fluorescence signals. However, because all neurons were responsible to the stimulus, we needed careful planning of experimental protocols to obtain physiologically meaningful results. More regional or conditional expression of GECIs were ideal to explore the specific function of neural circuits in details.

3.11 A3V-GCaMP6s Transfection in the Inferior Colliculus of the Chick

Expression of a proper reporter protein in a restricted brain region is ideal to investigate a specific function of neural circuits in vivo. However, finding a small transfection region by the PME was sometimes difficult, particularly in a deep brain region, and in animals under development in which the brain size grows substantially during the period required for viral transduction. We injected A3V encoding GCaMP6s into the inferior colliculus or field L of the chick of age P2 or P3. During 7–10 days of incubation, the brain size changed. Thus, the stereotaxic data of injection sites was not reliable at the time when a PME was inserted. Moreover, the quite low level of basal fluorescence of GCaMP6 increased the difficulty to find transfected neurons by a PME in deep brain regions. Accordingly, the following experiment is not a success story, but it might give a suggestion to design experiments.

We used A3V vector that encodes GCaMP6s together with another A3V vector that encodes mCherry. MCherry is a red fluorescence protein that has excitation maximum at 580 nm and emission peak at 610 nm, and was used to indicate the transfection site of GCaMP6s. The response amplitude of GCaMP6s is larger but the kinetics is slightly slower than the other two variants. Affinity to calcium ions of GCaMP6s is nearly the same as OGB1 (Kd = 170 nM) [29]. A3V-GCaMP6s and A3V-mCherry were mixed and injected together. It was expected that neurons expressed GCaMP6s, in a brain region where mCherry was expressed. In some experiments, we injected td-Tomato encoded in tandem with GCaMP6s in A3V. However, its expression in the chick brain was extremely small in number of transfected neurons and was spatially limited; therefore, we gave up its use.

3.12 Optical Platform for Excitation by Two Light Sources

In order to find expression region of mCherry in the brain and then to record optical signal from GCaMP6s by the same PME, two light sources must be switched for excitation; one is to excite mCherry (580 nm) and the other is to excite GCaMP6s (488 nm). These two light sources were mixed by a dichroic mirror (Semrock Di01-R514), and transmitted to the PME (Fig. 10a). Here, only one excitation light was allowed to be transmitted by a shutter control. The 580 nm light (from a high-pressure mercury lamp, filtered through Semrock FF01-580-23) was turned off when the fluorescence of GCaMP6s was measured, while 488 nm laser was turned off when mCherry was excited at 580 nm light. The spectrometer was used to detect the fluorescence from

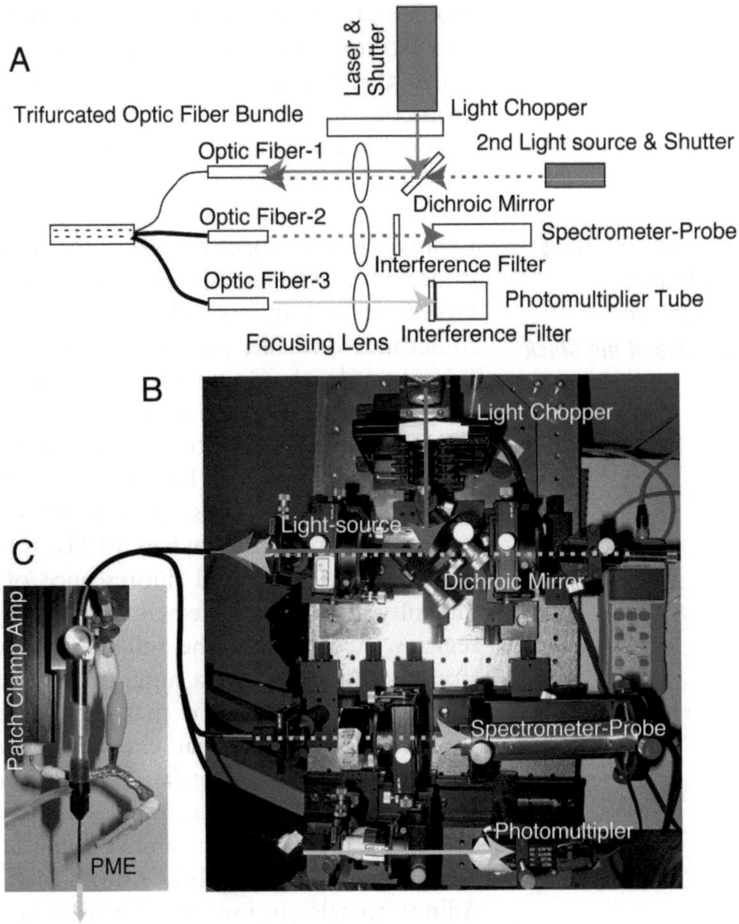

Fig. 10 Setup for measurement from two fluorescence compounds that have different excitation wavelengths. (**a**) Trifurcated optical fiber bundle and the arrangement of two light beams for excitation and fluorescence detection. Light beam was switched between 488 nm laser for the excitation of GCaMP6 and the second light of 580 nm for the excitation of mCherry. Second light was a projection through a 580 nm interference filter of a high-pressure mercury lamp. Detectors (spectrometer and photomultiplier tube) were also switched for different wavelength fluorescence measurement. (**b**) Set up of optical components of A. (**c**) Vertical setting of the PME, the holder assembly and optical fiber bundle for recordings in vivo

mCherry, with a long-pass filter (Semrock BLP01-594R). A trifurcated optic fiber bundle was adopted to excite two compounds and to detect two fluorescence signals. 10% of optic fibers of the bundle was used for the switched excitation, 40% of optic fibers led fluorescence to the spectrometer and 50% to the photomultiplier-tube. The design of the excitation light path and detector optical path is illustrated in Fig. 10a. Corresponding optical parts arrangements are shown in Fig. 10b. The distal end of trifurcated optical fiber bundle was connected to the PME holder assembly that was set vertically to a remote driven micromanipulator (DMA-1050, Narishige) within a sound-proof chamber where in vivo experiments were conducted (Fig. 10c).

We injected the mixture of A3Vs to the IC of chick. After 7–10 days of injection, the IC was penetrated with a PME and a small fluorescence response to the acoustic stimulation was recorded. When the PME approached the IC, mCherry fluorescence (emission 610 nm) was increased. But the increase of fluorescence level was very small and was only 30–100% of the baseline level measured at the brain surface (ΔF/F, Fig. 11a). Sound evoked local field current (LFC) responses were enhanced by local

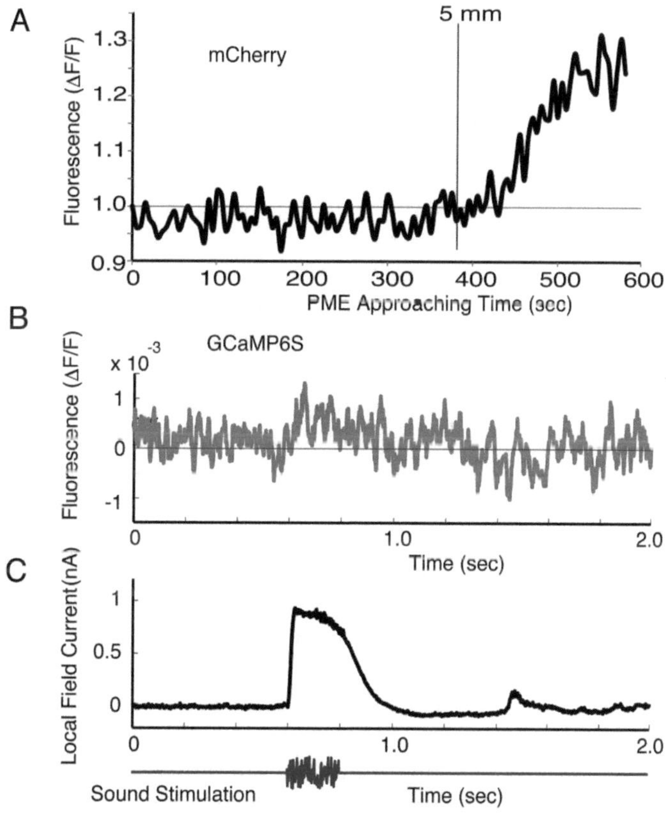

Fig. 11 mCherry fluorescence, and the acoustic response of GCaMP6S and local field current. (**a**) mCherry fluorescence was increased when the PME approached the A3V injection site in the chick IC. The vertical line indicates the depth of 5 mm from the brain surface, where mCherry fluorescence level started to increase. The fluorescence level saturated at the depth about 7 mm. Note that the number in abscissa indicates the approaching time for PME insertion. (**b**) GCaMP6s response to acoustic stimulation of 200 ms duration 70 dB sound pressure level white noise (22 traces were ensemble averaged). (**c**) Local field current response (top) to the sound stimulus (bottom) corresponding to the record in **b**

application of gabazine (0.5–5 mM) through the PME. The LFC (Fig. 11c) was large and the size (1 nA) was nearly equal to that we observed when OGB1was used as the calcium indicator. However, the GCaMP6s fluorescence signal to the sound stimulus was miserably small, about 0.1–0.2% of the baseline fluorescence ($\Delta F/F$) (Fig. 11b). With OGB1, application of gabazine enhanced both the sound evoked electrical response and optical response (Fig. 9) [14], but with GCaMP6s the sound evoked optical signal remained small. This result was discouraging. I will discuss several possibilities that reduced the sensitivity to the sound evoked optical signals.

4 Discussions and Conclusions

Activity of neurons is traditionally measured by an electrode. Here we report a new electrode technology to measure the optical signal of neurons, simultaneously with the electrical activity. PME is a patch electrode and at the same time the electrode can be used as a light guide to deliver excitation light to dye filled neurons and capture the optical signal that reflects the neural activity. Moreover, PME can locally apply chemicals by controlling the pressure inside the electrode. We succeeded in recording optical signal both from slice preparations by manipulating the PME under visual control to touch a fluorescing neuron, and from invisible neurons of deep brain tissues in vivo.

Loading indicator dyes to neurons in vivo was found most tricky and difficult step in the PME recording. Seeking a method of more efficient and reliable loading is mandatory for successful application of the PME technology in various regions of the brain. Accordingly, it is tempting to use variety of viruses (e.g., adeno-associated virus) [20] that encode indicator compounds, or transgenic animals that express the indicator compounds permanently or conditionally at specified region of the brain. However, viral expression of indicator compounds in brain regions was sometimes difficult, and the transfection efficacy might depend on animal species (Fig. 11). In the adult mouse hippocampus, we could record significant calcium fluorescence signals, but in the chick IC, the signal was miserably small.

Since the basal fluorescence level is low with GCaMP6, finding the transfected region is not easy. In the chick, particularly because the animal grew during the transduction period of 7–10 days after the virus injection at P2–P3, the brain size enlarged and the stereotactic readout at the time of injection was not reliable any more to find the region of viral transduction. Accordingly, we injected two virus constructs together in the chick IC; A3V that encodes mCherry and A3V that encodes GCaMP6s. We could find mCherry expression region in the IC as the increase of basal fluorescence. We expected that GCaMP6s was also expressed in that region.

However, the fluorescence response to the sound stimulus was extremely small there. This failure might be caused by several factors, such as (1) the expression of GCaMP6s was not efficient in the chick brain, or (2) the interference of viral transduction in neurons when a mixed injection of two viral constructs was made. We need further experiments to find the most appropriate method for the viral expression.

The small calcium response detected with GCaMP6s in the chick IC neurons might be due to some other factors that are associated specifically with GCaMP6 and with auditory neurons. Both OGB1 and GCaMP6s has a similar Kd (OGB1; 170 nM and GCaMP6s; 144 nM). However, the Hill's coefficient is different between two indicators (nearly 1 for OGB1 [29], about 3 for GCaMP6s [28]). The fluorescence signal is a sigmoidal function of calcium concentration, and the mid-point is determined by Kd and the steepness by Hill number. A dye that has a large Hill number will boost the fluorescence signal when a calcium concentration is higher than Kd, but the signal would remain minimum at lower concentrations than Kd. Considering the significant fluorescence responses detected by GCaMP6f in the adult mouse hippocampus, the acoustic evoked calcium influx in the chick IC neurons might have been too small to be boosted by the GECI (GCaMP6s). Most neurons have calcium influx during activity; however, calcium concentration must be maintained low even when the activity occurred at high frequency such as that occurred in auditory signaling. High level intracellular calcium concentration is toxic to any cells. Therefore, in order to detect small calcium signals from neurons by the PME, it seems critical to select fluorescence indicators that have a small Kd and a small Hill's coefficient for reporting calcium concentration as low as possible.

PME could be used with any other indicators than those for calcium ions, such as the indicator to report intracellular level of GPCRs (G protein coupled receptors), and furthermore in any application that uses light to measure or manipulate neuronal activity, such as optogenetic manipulation of neural activity, or detection of luminescence signal or BRET response [24, 25]. Luminescence emission does not need excitation light; thus, the background signal level seems small. However, the luminescence emission based on luciferase activity needs supply of substrates (luciferin, ATP), and these substrates can be applied locally through the PME. By using the PME to monitor electrical response and optical signal simultaneously, we may be able to find clues to the understanding neural activity in vivo that underlies and maintains neural functions in the brain.

Acknowledgments

We appreciate Dr. Eri Nishino who conceived the experiment together with H.O., and Y. Hirai who improved the software and conducted most experiments using PMT. We thank Drs. R. Matsui and D. Watanabe in Kyoto University for providing us A3V that encodes mCherry and GCaMP6. Experiments in the mouse hippocampus were conducted together with Dr. M. Ono in Kanazawa Medical University. This work was supported by Grants-in-Aid from Japan Society for the Promotion of Science to E. Nishino (23650205) and H. O. (20220008 and 26560464).

References

1. Berridge MJ, Lipp P, Bootman MD (2000) The versatility and universality of calcium signaling. Nat Rev 1:11–21
2. Adelsberger H, Garaschuk O, Konnerth A (2005) Cortical calcium waves in resting newborn mice. Nat Neurosci 8:988–990
3. Hagenston AM, Bading H (2011) Calcium signaling in synapse-to-nucleus communication. Cold Spring Harb Perspect Biol 3: a004564
4. Stosiek C, Garaschuk O, Holthoff K, Konnerth A (2003) In vivo two-photon calcium imaging of neuronal networks. Proc Nat Acad Sci U S A 100:7319–7324
5. Akerboom J, Chen T-W, Wardill TJ et al (2012) Optimization of a GCamp calcium indicato for neural activity imaging. J Neurosci 32:13819–13840
6. Ikegaya Y, Aaron G, Cossart R et al (2004) Synfire chains and cortical songs: temporal modules of cortical activity. Science 304:559–564
7. Ohki K, Chung S, Ch'ng YH, Kara P, Reid RC (2005) Functional imaging with cellular resolution reveals precise micro-architecture in visual cortex. Nature 433:597–603
8. Helmchen F, Denk W (2005) Deep tissue two-photon microscopy. Nat Methods 2:932–940
9. Svoboda K, Yasuda R (2006) Principles of two-photon excitation microscopy and its applications to neuroscience. Neuron 50:823–839
10. Jung JC, Mehta AD, Aksay E et al (2004) In vivo mammalian brain imaging using one- and two-photon fluorescence microscopy. J Neurophysiol 92:3121–3133
11. Hayashi Y, Tagawa Y, Yawata S et al (2012) Technical spotlight. Spatio-temporal control of neural activity in vivo using fluorescence microendoscopy. Eur J Neurosci 36:2722–2732
12. Vincent P, Makos U, Charvet I et al (2006) Live imaging of neural structure and function by fibered fluorescence microscopy. EMBO Rep 7:1154–1161
13. LeChasseur Y, Dufour S, Larvertu G et al (2011) A microbe for parallel optical and electrical recordings from single neurons in vivo. Nat Methods 8:319–325
14. Hirai Y, Nishino E, Ohmori H (2015) Simultaneous recording of fluorescence and electrical signals by photometric patch electrode in deep brain regions in vivo. J Neurophysiol 113:3930–3942
15. Jarvis E (2005) The Avian Brain Nomenclature Consortium. Avian brains and a new understanding of vertebrate brain evolution. Nat Rev Neurosci 6:151–159
16. Kawai S, Takagi Y, Kaneko S, Kurosawa T (2011) Effect of three types of mixed anesthetic agents alternate to ketamine in mice. Exp Anim 60:481–487
17. Fukui I, Sato T, Ohmori H (2006) Improvement of phase information at low sound frequency in nucleus magnocellularis of the chicken. J Neurophysiol 96:633–641
18. Nishino E, Yamada R, Kuba H et al (2008) Sound-intensity-dependent compensation for the small interaural time difference cue to sound source localization. J Neurosci 28:7153–7164
19. Murayama M, Miyazaki K, Kudo Y et al (2005) Optical monitoring of progressive synchronization in dentate granule cells during population burst activities. Eur J Neurosci 21:3349–3360
20. Matsui R, Tanabe Y, Watanabe D (2012) Avian adeno-associated virus vector efficiently transduces neurons in the embryonic and post-

embryonic chicken brain. PLoS One 7(11): e48730

21. Buller RM, Janik JE, Sebring ED, Rose JA (1981) Herpes simplex virus types 1 and 2 completely help adenovirus-associated virus replication. J Virol 40:241–247
22. McCown TJ, Xiao X, Li J, Breese GR, Samulski RJ (1996) Differential and persistent expression patterns of CNS gene transfer by an adeno-associated virus (AAV) vector. Brain Res 713:99–107
23. Takatsuka K, Ishii TM, Ohmori H (2005) A novel Ca^{2+} indicator protein using FRET and calpain-sensitive linker. BBRC 336:316–323
24. Hall MP, Unch J, Binkowski BF et al (2012) Engineered luciferase reporter from a deep sea shrimp utilizing a novel imidazopyrazinone substrate. Chem Biol 7:1848–1857
25. Couturier C, Deprez B (2012) Setting up a bioluminescence resonance energy transfer high throughput screening assay to search for protein/protein interaction inhibitors in mammalian cells. Front Endocrinol 3:100
26. Emery EC, Luiz AP, Sikandar S et al (2016) In vivo characterization of distinct modality-specific subsets of somatosensory neurons using GCaMP. Sci Adv 2(11):e1600990
27. Park JE, Zhang XF, Choi S-H et al (2016) Generation of transgenic marmosets expressing genetically encoded calcium indicators. Sci Rep 6:34931
28. Chen TW, Wardill TJ, Sun Y et al (2013) Ultrasensitive fluorescent proteins for imaging neuronal activity. Nature 499:295–300
29. Hendel T, Mank M, Schnell B et al (2008) Fluorescence changes of genetic calcium indicators and OGB-1 correlated with neural activity and calcium in vivo and in vitro. J Neurosci 28(29):7399–7411

Chapter 7

Patch-Clamp Fluorometry and Its Applications to the Study of Ion Channels

Esteban Suárez-Delgado and León D. Islas

Abstract

Patch-clamp remains the premier technique to study ion channel properties. Among the more useful extensions of patch-clamp, is the simultaneous use of fluorescence and spectroscopic techniques and electrophysiological recording known as patch-clamp fluorometry. This technique permits the simultaneous correlation of ionic current recordings with the activity of electrically silent protein conformational changes reported by the fluorescence measurement. Several recent and ongoing advances in fluorescent probes, genetically encoded fluorescent sensors based on fluorescent proteins or fluorescent noncanonical amino acids are making these methodologies more and more useful in the study of ion channel dynamics and regulation.

Key words Patch-clamp, Patch-clamp fluorometry, Electrophysiology, Membrane biophysics, Ion channels, Fluorescence, Spectroscopy, Live imaging, Fluorescent proteins

1 Introduction: Ion Channels

Transport and communication through cellular membranes have evolved a great variety of mechanisms that assure the availability of substrates and the coordination of different signals to maintain the fine balance in the organisms' metabolism.

Membrane-associated proteins that are responsible for transport can be categorized into two types: the carriers and the channels. The first ones are proteins that specifically bind the metabolite or solute that will be transported and then substrate is translocated across the bilayer through a conformational change that takes place during each transport cycle [1].

The second ones are ion channels, which are proteins that have a pore that traverses the cellular membrane and facilitates the transport of ions to both sides of the membrane. The source of energy for this transport is solely the electrochemical gradient for the specific ion.

Nicholas J. D. Wright (ed.), *Basic Neurobiology Techniques*, Neuromethods, vol. 152,
https://doi.org/10.1007/978-1-4939-9944-6_7, © Springer Science+Business Media, LLC, part of Springer Nature 2020

Ion channels are widely distributed in nature, practically every organism contains them; they respond to a great variety of stimuli, either chemical (hormones, second messengers, neurotransmitters) or physical (mechanical stimulation, temperature, voltage), by opening their pores [2, 3]. The responses to these diverse stimuli are directly linked to the extensive number of functions that the cell performs: regulation of acidity and osmolarity, liberation of hormones and neurotransmitters, and without doubt the most remarkable, the generation and propagation of action potentials in excitable tissues. All these functions allow the organism to perceive and process environmental information and to respond accordingly [4].

Despite exhibiting structural differences among them, ion channels are composed of several specific domains, the actions of which are allosterically coordinated to achieve the final result of opening the conduction pore that allows for the movement of ions across the membrane.

In most channels one can distinguish at least two domains. The first one related to the activation mechanism (the sensing domain) and another related to channel opening and ion conduction (the pore and selectivity filter). The activation domain can respond to one or several activation stimuli, generally producing a structural rearrangement or conformational change, which is transmitted to the activation gate. This coupling is known as gating. When the gate moves, the pore allows the flux of ions, which can be detected as an ionic current. The pore of the channel is a very efficient structure that permits transport of ions at a rate of approximately ten million per second when the channel is activated. Despite such high rate, the pore has a region called selectivity filter, which preferentially interacts with one kind of ion, allowing for the high selectivity generally seen in ion channels [5, 6].

1.1 Methods of Study of Ion Channels

In the middle of last century, the work done by Hodgkin and Huxley in the giant squid axon permitted the description of the cellular membrane as an equivalent electrical circuit, providing the basis for the electrophysiology techniques that are used nowadays [7].

In biophysical terms, the cellular membrane can be represented by an RC circuit where the lipid bilayer functions as a capacitor with capacitance C, and passive ion transport is represented as a resistance, R. Channels for each one of the permeating ions are represented as variable resistances, which can depend on voltage or other gating stimuli. Each one of these conductances is associated with an appropriate electromotive driving force given by the equilibrium potential of the ion.

To be able to study the properties of channels, techniques that can record the characteristics of the equivalent circuit were developed. One of the more widely used such techniques is the two-electrode voltage-clamp (TEVC) [8]. This technique introduces very thin glass pipettes into the cellular cytoplasm. The pipettes are electrically connected through a saline solution to electrodes, which go to the amplifier and allow both the injection and recording of current. Voltage control is achieved through feedback from the amplifier and currents can be recorded from medium to large size cells, being the *Xenopus laevis* oocyte one of the most commonly used. Another widely used technique is the cut-open oocyte voltage clamp (COVC), which allows for the recording of currents from a large area of the oocyte under better space-clamp conditions than TEVC [9, 10].

About 40 years ago electrical recording techniques had an important development. Neher and Sakmann created a method in which, instead of piercing the cellular membrane, a very strong seal is made between the cell membrane and the glass pipette [11]. This method, baptized as patch-clamp, has proven to be a spearhead in the study of the ion channels, so much so that, up to now, it is the main tool used for the fundamental evaluation of the biophysical properties of ion channels [12].

This technique offers several advantages, including the ability to record from only one channel, which permits direct estimates of the probability of the channel opening, as well as its unitary current behavior [13, 14]. Furthermore, it also allows recording of macroscopic currents arising from a large number of channels, as well as channel stimulation by both intracellular and extracellular modulators.

The patch-clamp method has multiple modalities that turn it into a useful and varied tool. Every seal begins with the approach and subsequent interaction of the glass pipette with the cellular membrane; this interaction is usually promoted by application of negative pressure in the pipette, which helps to introduce a portion of membrane into the pipette tip. The seal has been formed when an electrical resistance larger than 1 gigaohm (GΩ) is obtained and along with it, the first modality which is the *cell-attached* or *on-cell* configuration. This modality allows performing electrophysiological recordings with an intact intracellular medium. Other configurations are possible, which permit access to the cytoplasmic part of the membrane (*inside-out*), the extracellular face (*outside-out*), or the entire membrane of the cell (*whole-cell*) [15, 16].

1.2 Patch-Clamp Fluorometry

In the decades following its invention, the patch-clamp method was refined and found to have multiple applications in the study of ion channel function. At the same time, knowledge about structure–function relationships in ion channels increased though the implementation of other techniques, such as chemical modification [17], mutagenesis [18, 19], and more recently, cryo-electron microscopy [20, 21], X-ray crystallography [6, 22], nuclear magnetic resonance (NMR) [23, 24] and “in-silico” modeling [25, 26].

Despite the wealth of information that these methods provide, direct correlation between structure and functional states is difficult to establish. Thus, in the middle of the 1990-decade, fluorescence techniques, so helpful in other areas of neuroscience, were merged with electrophysiological recordings to investigate conformational changes in ion channels while activation was on course [27, 28].

One of the first published works that measured ion channel rearrangements with fluorescence methods and electrical recordings, was carried out by Mannuzzu et al. in *Shaker* potassium channels [29]. This experiment made use of *Xenopus laevis* oocytes, under TEVC and was named voltage-clamp fluorometry (VCF). VCF generally makes use of fluorescent probes attached to cysteine residues introduced by mutagenesis in the channel's sequence. The great usefulness of VCF stems from the fact that the fluorescence signals can provide information about conformational changes that do not directly produce an electrical signal, which cannot be recorded by conventional voltage-clamp methods [28, 30].

A few years after this initial report, Zheng and Zagotta went further to develop an approach combining fluorescence and electrophysiological recordings in the *inside-out* configuration, dubbed patch-clamp fluorometry or PCF [31, 32]. This method offers new information beside that inherent to patch-clamping (high time resolution of current recordings and access to the intracellular side of channels). First of all, and possibly the most important attribute, is the ability to detect conformational changes associated with electrically silent states (closed, inactivated, and desensitized states). And second, an increased quality of the fluorescence signals is attained by excluding the intracellular contents of a cell [33].

Because of these advantages, PCF has been applied to a large number of different ion channels using a great variety of combined spectroscopic and electrophysiological tools. Some of these tools rely on the introduction of fluorescent probes at cysteine residues (either endogenous to the channel or introduced by mutagenesis) [32, 34, 35], which can provide clues of local environmental changes during gating. By measuring changes in the photo physical properties of such fluorophores, the dynamics of ion channels during gating can be measured. Examples of these are the measurement of fluorescence intensity changes, the use of tryptophan residues as quenchers of introduced bimane derivatives [36], genetically encoded fluorescent proteins as FRET reporters [37–39], or as reporters of the local environment [38], the use of ionic quenchers of florescence to probe accessibility [32], fluorescently labeled ligands [40, 41] or introduced noncanonical fluorescent amino acids as probes within the protein sequence [42].

Every patch-clamp configuration already mentioned is useful for PCF [43]. Such versatility makes this technique a powerful tool for the study of not only channels, but also transporters [44] and receptors in general.

Given the enormous potential of patch-clamp fluorometry, we wish to provide a practical guide for the implementation of a PCF setup and an overview of the more recent possibilities of the technique.

This chapter describes the equipment and basic technical requirements to perform patch-clamp fluorometry, together with a series of tips that we hope can enhance the success rate of PCF experiments.

2 Materials

2.1 Microscopy Equipment

The preferred biological preparation for PCF is the oocyte of the frog *Xenopus laevis*. A microscope that permits easy access to these big cells is necessary. Generally, an inverted configuration with epifluorescence illumination allows for simultaneous measurements of fluorescent intensity while electrophysiological recording from membrane patches is performed. We use an inverted microscope (Nikon TE-2000 U). For illumination, the microscope is coupled via a custom optical system to one or more lasers (Fig. 1). Mercury or halogen lamps or even a power LED system can be used to excite the sample at different wavelengths of visible light. Optical components that allow control of the exposure time and the wavelength and light power that goes to the sample are placed after the light source. These are listed below:

- *Shutter.* A fast mechanical shutter (Uniblitz VMM-D1, Vincent Associates) that controls the laser exposure is used. This shutter is controlled by the imaging software.
- *Collecting lens.* A spherical lens (3 mm) that expands the laser beam diameter from 5 μm to 7 mm.
- *Collimation lens.* A lens that produces a collimated beam after expansion.
- *Neutral density filters.* Regulates the intensity of incoming light without disturbing the wavelength. These reduce the possible photo damage induced by the laser beam that can promote fluorophore photobleaching.
- *Focusing lens.* This lens focuses the collimated beam at the back focal plane of the objective, producing a homogeneous Köhler illumination at the sample plane.
- *Dichroic mirror.* A mirror with different transmission or reflection properties at two different wavelengths. In laser microscopy this is used for dividing the light that comes from the source of the light (laser) from the sample fluorescence.

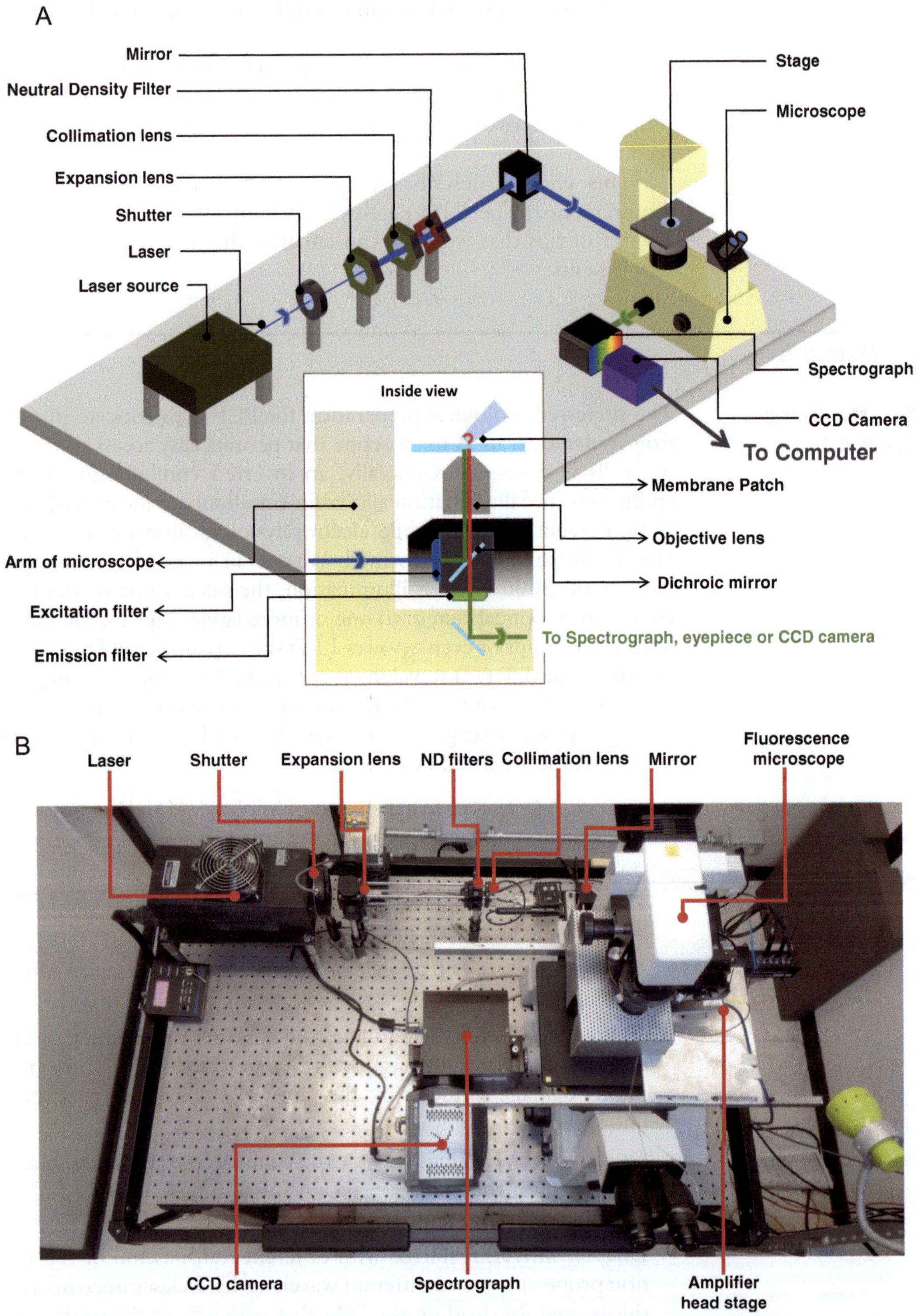

Fig. 1 Schematic of a patch-clamp fluorometry setup. (**a**) Schematic of the components for the illumination and detection paths. Laser is shown in blue and arrowheads mark its propagation direction. Note that the diameter of the laser beam becomes increases after going through a beam expander. The beam is attenuated

- *Excitation filter.* A filter that only passes certain wavelengths of light from the laser. Typically, it is a narrow band-pass filter and its bandwidth should fall completely inside the excitation range of fluorophore to stimulate.
- *Emission filter.* A filter that only passes certain wavelengths of light from the fluorophore attached to patch. Could be a long pass or bandpass, depending on the collection of light needed. To collect fluorophore spectra, a long pass filter is needed.

The three last elements are located in a preassembled filter cube that goes in the appropriate housing in the microscope and their properties are entirely dictated by the fluorophore to be used.

To collect emission light and serve as an illumination condenser, we use an oil immersion objective with a high numerical aperture (NA) of 1.4 (Nikon 60x) (*see* **Note 1**). For imaging purposes, the light collected by the objective is sent directly to a CCD camera (*see* **Note 2**). In experiments where the emission spectrum is needed, light is sent first to an imaging spectrograph (Spectra Pro 2150i, Acton Instruments).

Control of the camera, shutters and other devices can be carried out by the use of commercial software available from microscope manufacturers or other providers (MetaMorph, Molecular Devices). A versatile and more economical alternative is the use of open software such as Micromanager (version 1.4) (https://micro-manager.org) [45].

2.2 Electrophysiology Equipment

As it was mentioned before, the inverted microscope allows easy access to the sample from above. As in all electrophysiological setups, it is important to mount the microscope and additional optical equipment on a vibration isolation table (VW-3646, Newport) [46]. The amplifier head stage is mounted on a micromanipulator which can be of a fine mechanical type (Newport) or a motorized micromanipulator (Sutter Instruments, MP model). In our PCF experiments we employ an Axopatch 200B (Axon) amplifier but any other patch-clamp amplifier is suitable. Analog to digital conversion of recorded currents is carried out with an ITC-18 board (HEKA Electronik GMBH). The software we use for current recording and stimulus is Patchmaster (HEKA Electronik GMBH).

Patch-clamp amplifiers are extremely sensitive; this facilitates pick up of spurious noise signals. For noise reduction, it is

Fig. 1 (continued) and steered by neutral density filters and a mirror situated at the back illumination port of the inverted microscope. The inset shows the light pathway in the filter cube set. Changes of wavelength are represented with different colors in the spectrograph. The light from the sample is collected by the objective and guided to the eyepieces, CCD or spectrograph devices. (**b**) Photograph of the actual setup, showing the location of components as in **a**

important to carefully ground all the equipment, including the microscope, micromanipulator and the illumination and light detection accessories (*see* **Note 3**). Frequently, a Faraday cage that isolates the whole workstation or only the sample space is used.

To obtain giant membrane patches we use borosilicate glass capillaries. The patch pipettes are obtained from these capillaries (external diameter 1.5 mm, internal diameter 0.86 mm, length 10 cm, Sutter Instrument) using a micropipette puller (P97, Sutter Instruments). These pipettes are initially pulled to a large diameter (~50 μm) and then heavily fire-polished with a microforge (Narishige, MF-830) to a diameter of ~10 μm. The large size of the pipette tip is important for simultaneous recording of both fluorescence and patch currents. Note that the pipettes for the patch-clamp fluorometry method in oocytes are bigger than those used for mammalian cells (approximately 2–3 μm) (*see* **Note 4**).

2.3 Obtaining and Preparation of Oocytes

While the traditional patch-clamp method is used in different cell lineages of varied organisms, the mature oocyte of the frog *Xenopus laevis* is the most frequently employed in patch-clamp fluorometry (PCF) [47]. Obtaining and maintenance of these cells starts with the surgery of a mature female frog (bigger than 9 cm, Xenopus1, NASCO). In this procedure, a rigorous sterile technique is not necessary due to the antibiotic properties of the frog's skin; however, it is advisable to continually clean the surgical equipment to reduce even more the risks of infection. The frog is anesthetized in a solution of tricaine methanesulfonate (Sigma) at 0.22% (pH titrated to 7 with NaOH) for 5 min. The specimen is placed face up on wet gauze and a vertical incision is made in the abdominal region at 1 or 2 centimeters from the midline and approximately with 1 cm of length. When reaching the abdominal cavity, one can extract as many segments of ovary as needed by using dissection tweezers and scissors. The wound is closed by layers using a suture of absorbable material, usually chromic catgut, caliber 3-0 (Atramat) for the muscle and a nonabsorbable nylon suture of 3–0 caliber for the skin.

The ovary segments are placed in a petri dish with OR2 solution (in mM): 82.5 NaCl, 2 KCl, 1 $MgCl_2$, and 5 HEPES, titrated to a 7.5 pH with NaOH (*see* also Table 1) and they are manually disaggregated in small cumuli of oocytes, being careful not to damage them. Oocytes are covered by a layer of follicular cells that should be removed. To achieve this, we use a solution of collagenase (from *Clostridium histolyticum*, type IA, Sigma) in OR2 at a concentration of 1.5–2 mg/mL. The ovary segments are placed in this solution and are dispersed in an oscillating shaker at 60 rpm for about 30–40 min (*see* **Note 5**). After this enzymatic treatment, the solution is decanted and the oocytes are washed 3 or 4 times with OR2, to later be shaken again for 30–50 min in OR2 (*see* **Note 6**). After this treatment the defoliated oocytes are

Table 1
Components of solutions used in the extraction and maintenance of *Xenopus laevis* oocytes

OR 2 solution	ND 96 solution	ND 96 without calcium solution
• 82.5 mM NaCl • 2 mM KCl • 1 mM $MgCl_2$ • 5 mM HEPES	• 96 mM NaCl • 2 mM KCl • 1.8 mM $CaCl_2$ • 1 mM $MgCl_2$ • 5 mM HEPES • pH = 7.5 with NaOH	• 96 mM NaCl • 2 mM KCl • 1 mM $MgCl_2$ • 5 mM HEPES • pH = 7.5 with NaOH

transferred to an incubation solution. In our case, we use the ND96 solution that contains (in mM): 96 NaCl, 2 KCl, 1.8 $CaCl_2$, 1 $MgCl_2$, 2.5 of pyruvic acid, 5 HEPES, and 20 μg/mL of gentamicin, titrated to 7.5 pH with NaOH (*see* also Table 1) (*see* **Note 7**).

Oocyte selection for mRNA injection and subsequent use in PCF experiments is an important step. Usually oocytes in stages V or VI are preferred. These are characterized by having a diameter of 1000–1300 μm and clearly separated animal and vegetal poles.

The chosen oocytes must be incubated at 18 °C for at least 3 h before injecting, so that they recover from the collagenase treatment and to uncover those that have been damaged by the process. For mRNA injection, we use a micro-injector device (Nanostepper, Drummond Scientific) and borosilicate glass pipettes with a tip of 40–60 μm of internal diameter. The pipettes are completely filled with light mineral oil (Sigma) before mounting on the microinjector. When the pipette is already mounted, the solution of mRNA can be aspirated. About 30–50 nL of mRNA will be injected to each oocyte (*see* **Note 8**). After the injection, the oocytes in ND96 solution are returned to the incubator for a few days until the required level of channel expression is achieved [48].

2.4 mRNA for Injection

As it has been already stated, mRNA is used for the expression of the channel that will be studied. The mRNA that is injected in the oocyte is usually obtained through the in vitro transcription technique from a vector that contains the channel sequence, along with a tail of polyadenines in the 3′ end and a cap in the 5′ end to avoid mRNA degradation inside the cytoplasm. The mMesage mMachine in vitro transcription kit from Ambion is suitable for this purpose.

2.5 Solutions

The composition of each solution employed in the experiments will vary accordingly to the type of channel to be studied, the possible ligands that activate it and the permeant ion; nonetheless, some general considerations that may facilitate patch formation are appropriate [46].

The first aspect to consider is the osmolarity. In the *Inside-out* configuration, it is advisable that the bath solution (intracellular) and the one inside the pipette (extracellular) have the same osmotic potential.

The second aspect is pH control. A buffer system with a pKa close to the desired pH should be used. For physiological pH solutions, the most common buffer is HEPES (4-(2-hydroxyethyl)-1-piperazineethanesulfonic acid, Sigma) that has a pKa of 7.5. Finally, for the intracellular solution the calcium concentration should be kept to a minimum by using chelating agents, such as EGTA (Sigma) at an appropriate concentration (10 mM). On the other hand, the extracellular solution should contain divalent cations (Ca^{2+}, Mg^{2+}) 1–2 mM, to facilitate patch formation.

3 Method

3.1 Obtaining the Patch

Experience with traditional patch-clamp techniques is useful for PCF experiments [49]. As stated in the introduction, the patch-clamp method consists in forming a patch of cellular membrane at the tip of a polished glass pipette and measuring the currents that flow through the channels present in this patch during activation.

Now, we will detail the steps to follow in order to achieve PCF in the *inside-out* modality. The steps described here apply to channels that are already fluorescently labeled, perhaps with genetically encoded fluorescent proteins or noncanonical fluorescent amino acids (Fig. 2). In a latter section we will describe methods to perform in-situ labeling of channels present in the patch.

3.1.1 Election of the Oocyte

Oocytes to be used for PCF should be as healthy as possible. This generally means that no discoloration of the animal pole or swelling of the cell has occurred during the incubation period. This period will dictate the day in which the experiments will be performed and it depends on how long the channel takes to be expressed in the membrane of the oocyte. Usually, this waiting time is between 2–5 days after injection of mRNA. The chosen oocyte is transferred to a Petri dish where the vitelline membrane is removed with the help of a pair of fine-tipped tweezers (*see* **Note 9**).

3.1.2 Setting up the Experiment

The recording chamber should have a small volume (less than 1 mL) and the bottom should be made of a thin glass coverslip. A perforation on the body of the chamber is useful to apply immersion oil to the objective without having to remove the chamber each time. Before transferring the oocyte to the recording chamber, all equipment and software should be running, thus reducing the time the oocyte is exposed before the experiment can commence.

The oocyte should be placed near the center of the recording chamber and its edge focused initially with a 10× objective. The gigaseal will be obtained under this magnification. The pipette is moved into the solution. Just before making contact with the

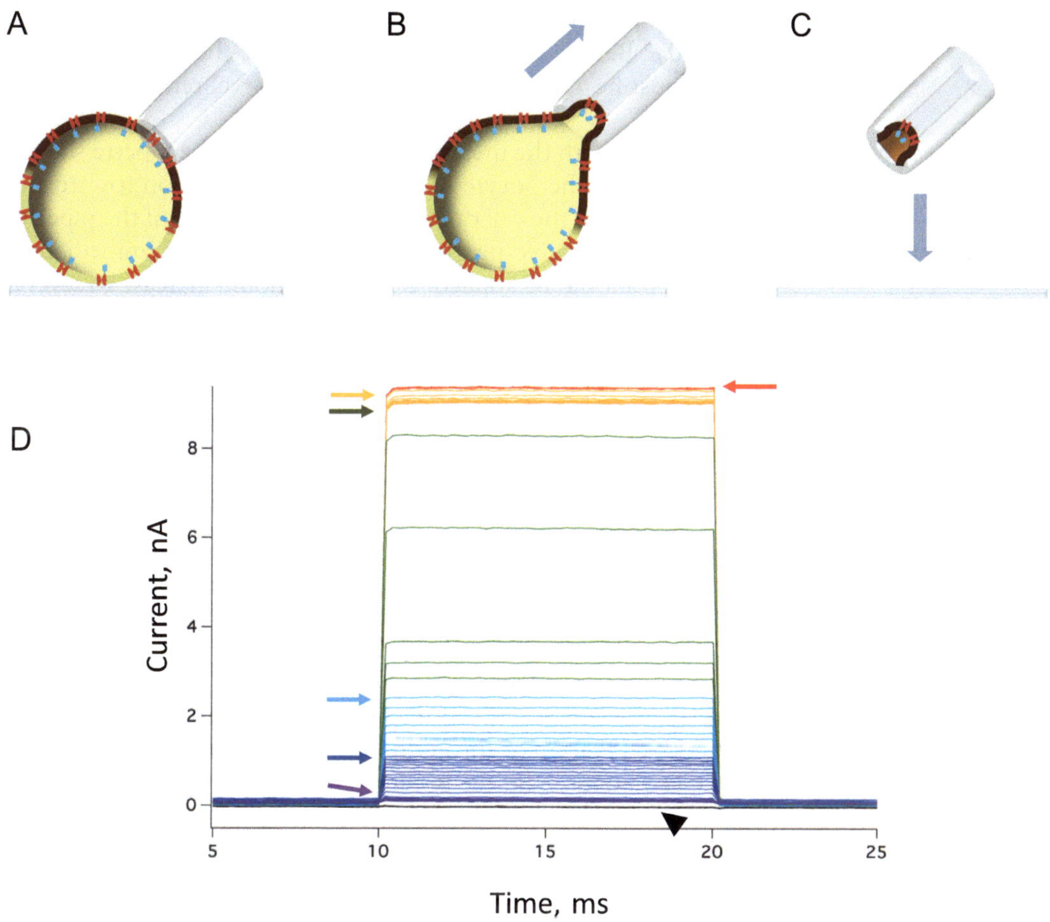

Fig. 2 Seal formation in *Xenopus laevis* oocytes. (**a**) Microinjected oocytes with the channel (red) fussed to a fluorescent protein (blue) correctly expressed in the cell membrane. The patch pipette is sealed to the membrane, helped by a little suction. (**b and c**) The inside-out configuration is obtained by pulling the glass pipette away from the cell, the pipette tip moves toward the bottom of experimental chamber until sharply focused. (**d**) A family of pipette currents in response to pulses of 10 mV and 10 ms. Traces at 400 ms intervals are shown. In red is the pipette current before touching the membrane at the initial configuration (Red arrow). When the pipette makes contact with the oocyte membrane (yellow arrow), the current slightly decreases (yellow traces). Positive pressure in the pipette is released (green arrow), promoting entrance of the membrane into the pipette tip and the current quickly diminishes (green traces), indicating a sudden increase of resistance. Then soft suction is applied (light blue arrow) to improve the contact of the membrane with the glass, further increasing resistance (light blue traces). If the increase in resistance slows down or stops (dark blue arrow) the intensity of suction has to be gently increased to continue seal formation (dark blue traces). When the resistance is in the order of 700–900 MΩ (purple traces), suction can be increased (purple arrow) to finally achieve the gigaseal, which is also observed as small current of a few pA amplitude (black traces and black arrowhead)

surface of the bath solution, positive pressure is applied to the pipette to maintain the tip as clean as possible. This positive pressure will be kept up to the moment of contact with the oocyte membrane. As is customary, pipette resistance is monitored via the application of a small voltage step (10 mV for 20 ms), (*see* **Note 10**). Before touching the membrane, the current of the pipette must be zeroed to provide a current level reference.

3.1.3 Obtaining the Seal

Upon contact with the membrane of the oocyte, the pipette should be pushed until the oocyte shows a slight deformation toward the internal region of the cell due to the horizontal push of the pipette. At this point, positive pressure will be released and immediately, soft suction should be applied. Resistance should increase to several tens of megaohms and suction will be maintained until resistance reaches a value larger than 1 GΩ (Fig. 2) (*see* **Note 11**). When the gigaseal is obtained, the tip of the pipette is removed in one even movement (*see* **Note 12**). Immediately after, the voltage is set to a value at which the channels are closed. This voltage is termed "holding potential."

3.2 Recording Fluorescence

Before fluorescence imaging can begin, a good image of the patch at the tip of the pipette should be achieved. The pipette is moved as close as possible to the bottom of the chamber and the high NA objective is placed in the light path. Care should be taken that the pipette tip does not come in contact with the chamber bottom, to minimize movement artifacts during imaging. This can be achieved by direct observation or through observation of the imaging field in the CCD camera. The tip is centered in the imaging field, away from the rest of the oocyte. At this point the microscope lamp is switched off and the patch can be illuminated with the appropriate excitation light to record fluorescence. The focus can be adjusted to form a good fluorescence image of the patch (*see* **Note 13**) (Fig. 3).

3.3 Simultaneous Measurement of Current and Fluorescence

For ligand-activated channels, fluorescence images can be acquired in the steady-state, before perfusion of the bath with a solution containing the agonist and also in the presence of it. For voltage-gated channels the situation is a bit more complicated, since image capture should be precisely synchronized to the voltage stimulus. Some camera control software provides TTL pulses that can be used to trigger acquisition by the electrophysiology software. We use an approach provided by micromanager software, where a trigger pulse is delivered by the Patchmaster software to both open the shutter and take a picture, with precise timing. This permits acquisition of fluorescence when the channel is closed (negative potential) or open (positive potentials). Since photobleaching of the fluorophore can be an issue, especially when using high intensity illumination, it is advisable to take paired pictures at a fixed negative potential and at varying positive potentials, thus reducing the influence of photobleaching.

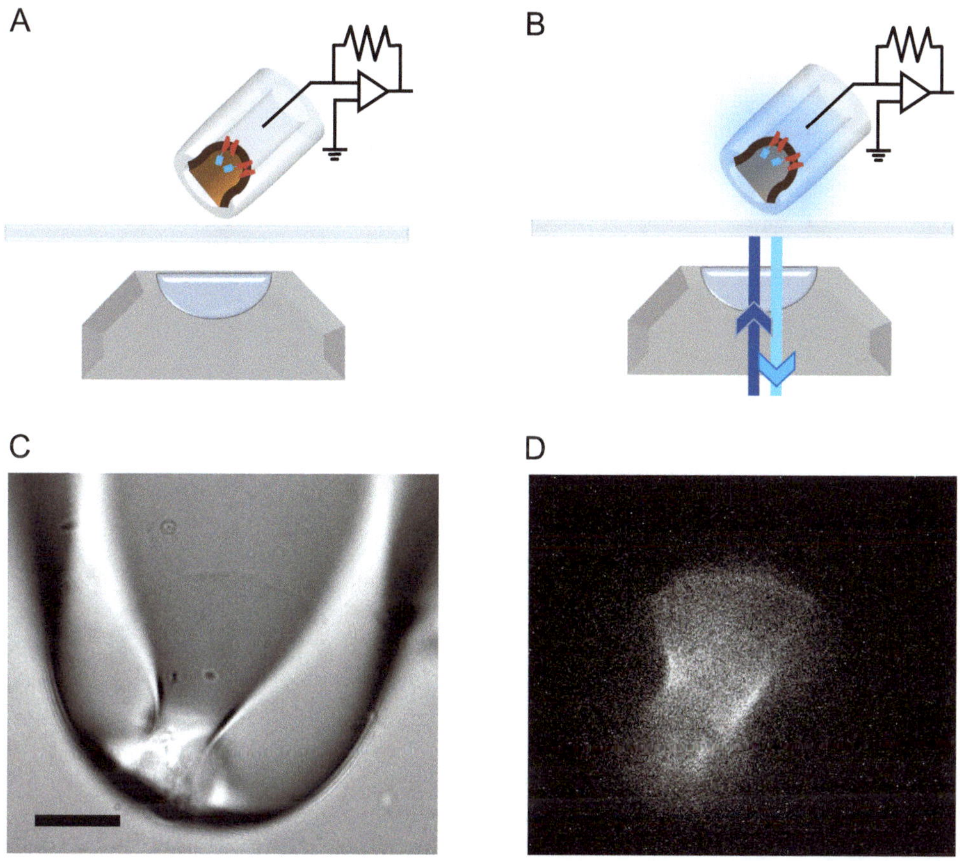

Fig. 3 Scheme of steps to achieve a patch-clamp fluorometry experiment. (**a**) The patch in the pipette is lowered toward the bottom of the chamber and focused. (**b**) The excitation source is turned on and fluorescence can be recorded. Images of a membrane patch from an oocyte which expresses the proton channel H_V1 fused to the Venus fluorescent protein. (**c**) Bright-field image, black bar represents 10 μm. (**d**) Fluorescence image of the patch in **c**. The scale bar is the same as in C. Fluorescence was excited at 488 nm and imaged with a band pass filter centered at 525 nm

3.4 Quantification of Fluorescence Intensity

After the acquisition of images, an initial step toward the analysis of fluorescence is to remove the background fluorescence. A region of interest (ROI) is selected in a part of the image where patch fluorescence is not expected. Then another ROI of the same size is placed around the patch. The value of counts of grey scale of the background is subtracted from the value of the patch ROI to acquire the absolute fluorescence and use it for more complex analysis (e.g., FRET) [28, 31]. Fluorescence can be quantified and plotted as relative fluorescence counts to allow comparison between patches with different channel expression levels.

3.5 In-Situ Labeling for Patch-Clamp Fluorometry

Another technique frequently used to monitor conformational changes in ion channels is the site-specific labeling with fluorescent dyes attached with thiol-reactive chemistry [27, 28, 50, 51]. This

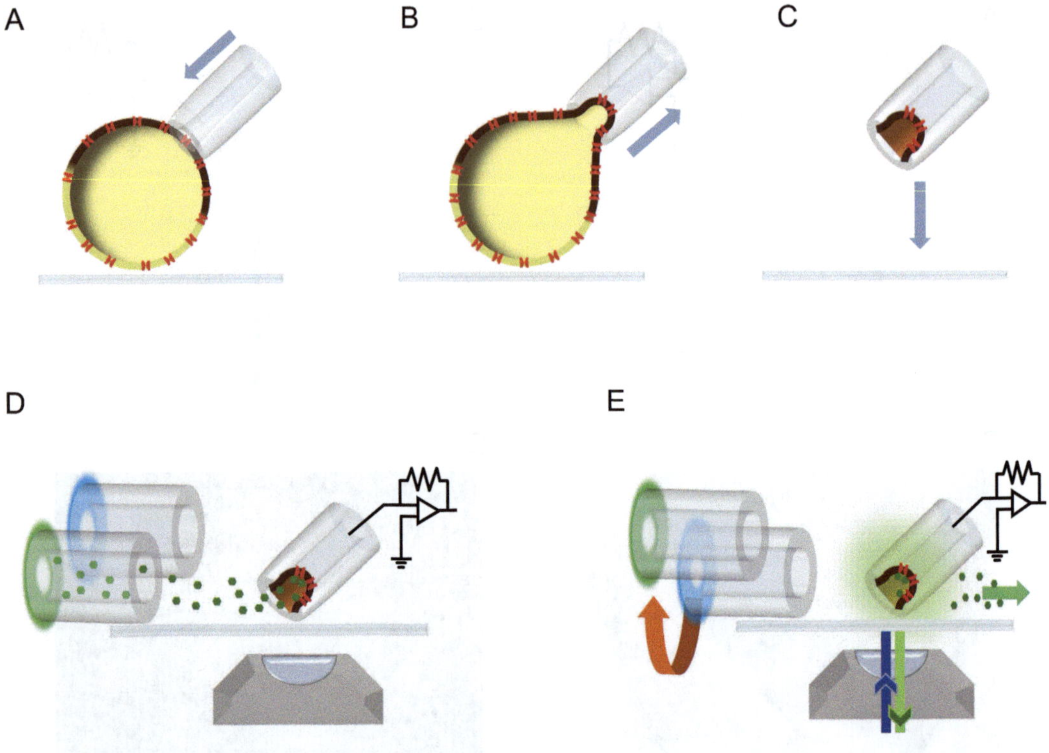

Fig. 4 Labeling of channels in the inside-out configuration. (**a–c**). Following the same steps as in Fig. 2, an inside-out patch is obtained. (**d**) The patch is perfused with the cysteine-reactive fluorescent dye that will be attached to a cysteine residue in the channel (green hexagons). (**e**) Later, dye excess is washed out to reduce the background signal and allow electrophysiology and fluorescence recording

method creates a covalent bond between a reactive group in the fluorescent probe and a sulfhydryl group of a cysteine in the amino acid sequence of the channel. The cysteine-reactive groups usually are methanethiosulfonate (MTS), iodoacetamide (IAA) or maleimide (Mal). Since very small volumes of dye are used to label the patch, a special perfusion system should be used. This is accomplished by using a small tip diameter (approximately 10 μm) glass pipette containing 10–20 μL of reactive dye. Light mineral oil is then used to fill up the rest of the pipette and this in turn is connected to plastic tubing and a 10–20 cc syringe with a two-way valve. The dye pipette is placed in front of the patch pipette and the dye is then fed by gravity. The labeling time will be dependent on the reactivity of the cysteine being labeled [31]. Before recording starts, the residual nonreacted dye is washed, reducing the intensity of background signal and improving signal to noise ratio (Fig. 4).

One important consideration is that the channel to be used should be a cysteine-less protein in which the specific cysteine to be labeled is introduced by mutagenesis. Most channels can be made cysteine-less, although generally at the cost of reduced expression levels. The relative low frequency of cysteines in proteins helps improve specificity of labeling and introduction of cysteine residues at different sites in intracellular and extracellular domains is well tolerated. Cysteine introduction and subsequent labeling must be done being careful as not to significantly perturb the overall protein structure and activity. An advantage is that organic fluorophores with a relatively small size such as fluorescein (ThermoFisher Scientific), Alexa fluorophores (ThermoFisher Scientific), bimane (Sigma), and Badan (ThermoFisher Scientific) can be linked to cysteines without any gross modification in channel structure and function. This allows for the study of conformational changes with better resolution than that obtained with fluorescent proteins [52, 53].

Despite its advantages, dye labeling at cysteine residues cannot be totally devoid of background fluorescence largely due to labeling of other native transmembrane proteins that contain cysteine residues. To reduce this interfering factor, a residue with state-dependent accessibility is the best target. A nonfluorescent reactive molecule, like tetraglycine maleimide, can be used to react with exposed cysteines before treatment with the fluorescent probe [36]. Another method that can be applied is blocking of cysteines with maleimide at low temperature followed by incubation at higher temperature, allowing for new channels to be inserted in the membrane to more specifically label them [54]. Another way to protect the cysteines of the ion channel is to bind them temporarily with Cd^{2+} of Zn^{2+} while nonfluorescent modification happens [55].

The use of a wide variety of fluorophores provides diverse experimental configurations for the study of structure and conformational changes of ion channels. Organic fluorophores can be employed as a pair with FP (or other organic fluorophores) to measure FRET [32, 34, 40, 41, 56] or LRET changes [57] or as donors to a transition metal ion like Ni^{2+}, Cu^{2+}, and Co^{2+} in transition metal ion FRET [52]. Furthermore, these dyes can be linked to ligands (as cAMP, cGMP, or calmodulin) [40] or toxins (charybdotoxin, agitoxin) [57] that bind to the channel and so investigate the accessibility, movement and the structural changes induced by these.

3.6 Noncanonical Amino Acids

In order to minimize the limitations encountered by the use of big-sized dyes (such as Fluorescent Proteins), recently developed alternatives in protein labeling have begun to be used in conjunction with PCF. These involve the incorporation of noncanonical amino acids in the sequence of the protein, especially fluorescent amino acids. This is achieved by the use of an evolved orthogonal pair of aminoacyl-tRNA synthetase and a tRNA from bacterial or yeast [58, 59].

One example of this family of probes is the fluorescent noncanonical amino acid that is a chemically derived from prodan (*N*,*N*-dimethyl-6-propionyl-2-naphthylamine), which receives the name of ANAP (3-(6-acetylnaphtalen-2-ylamino)-2-aminopropanoic acid) [60]. This probe is very sensitive to polar environments, having an emission range around 420–490 nm depending on solvent polarity [61]. To introduce ANAP into the structure of ion channels, a mutation to an amber codon (TAG) is inserted in a chosen site. ANAP has been recently used as a donor in transition metal ion FRET in conjunction with patch-clamp fluorometry experiments [42].

4 Whole-Cell Patch-Clamp Fluorometry

Whole-cell patch clamp recording has been extremely useful in the study of the intrinsic electrophysiology of many cells types, from neurons [62] and myocytes [63] to immune system cells [64], to mention a few examples. Also, this technique is practical in the study of ion channels expressed in heterologous systems using mammalian cell lines [65]. With this configuration it is possible to directly measure the membrane potential of a cell type under current-clamp and evaluate the currents across complete cells under voltage-clamp and even record neurotransmitter secretion [66].

To achieve this modality, we begin with the "cell-attached" configuration, where a high resistance seal is made. Afterward, the researcher must break the membrane inside the pipette tip without disturbing the gigaseal. Once this is achieved, the pipette solution and the cytoplasm make contact [16, 67] and the intracellular solutions is washed out and substituted by the pipette solution, which could represent an advantage (control the components of cytoplasm) or a disadvantage (reduction of the concentration of metabolites and signaling molecules that could be important to the subject of study, either ion channels, neurotransmitter receptors and electrogenic ion carriers) [68].

This important electrophysiological tool has merged with fluorescence techniques as a way to diversify the study of the relations between structure and function of membrane proteins and has been used in the study of voltage-gated potassium ion channels [69], TRPV1 channels [70], voltage-gated Ca^{2+} channels [71], and P_2X_2 receptors [72]. In this section, we discuss the materials and the procedures necessary to achieve the whole-cell patch fluorometry technique in cultured mammalian cells.

4.1 Materials

In whole-cell patch fluorometry experiments, most of the equipment mentioned early for PCF in the inside-out modality is used, so in this section we will focus in the materials and procedures to prepare this kind of experiments. We will focus on HEK cells but these procedures should work equally with other cell lines.

4.1.1 Preparation of HEK 293 Cells

One of the most common biological models used is the cell line derived from human embryonic kidney, HEK 293 [73], these cells are cultured and maintained in an incubator (3110 Water-Jacketed CO_2 Incubator, Forma Scientific) with controlled environment that contains 95% of air, 5% of CO_2 and a constant temperature of 37 °C. The medium normally used for culturing HEK293 cells is DMEM (Sigma-Aldrich). Around 24 or 48 hours before the experimental procedure, a plasmid containing the ion channel of interest is introduced by a method called transfection. There are several protocols to do transfection which use chemical and physical methods to introduce a sequence of nucleic acid inside the cell. In our laboratory, we use the polycationic transfection agent JetPei (Polyplus transfection) that forms complexes with DNA and is endocytosed by the cells. Usually 3–5 μL of transfection agent and 100–600 ng of genetic material are required, both of which are dissolved in 100 μL of a solution of 150 mM of NaCl. Then are softly mixed and incubated for 30 minutes. Later, the mixture is carefully placed in the culture dish that contains the HEK 293 cells with fresh medium. Cells are incubated until the day of the experiment.

HEK 293 cells could be cultured in many types of dishes (*see* **Note 14**). Usually, bottom glassed culture dishes are used (FluoroDish, World Precision Instruments). This kind of dish is useful because it makes it easier to work with high NA objectives (*see* **Note 1**).

Two types of HEK 293 with different growth modes are used in the laboratory: adherent and nonadherent (called "normal" for practical purposes). This feature is important because voltage control is more accurate in rounded cells. The adherent HEK 293 cells tend to be flatter and have more cellular processes and elongations than normal HEK cells. This makes sealing and correct voltage-clamp more difficult. To avoid these two setbacks, a treatment with 300 μL 0.05% trypsin–EDTA solution (Gibco) is applied to transfected HEK 293 cells after removal of DMEM for 4–5 min. Incorporating 300 μL of DMEM interrupts the action of trypsin. After this treatment, cells detach and become rounded and float in the culture dish. Cells are now transferred to another culture dish previously prepared with fresh DMEM. This treatment is applied 2–3 h prior to the experiments (in adherent HEK cells, in the case of normal HEK cells, this span may be 4 or 5 h) to allow HEK 293 cells to partially adhere and permitting recordings in rounded cells.

4.1.2 Whole Cell Recording Solutions

Prior to patching in transfected HEK 293 cells, DMEM is slowly replaced by bath solution with a Pasteur pipette to prevent cells from detaching from the glass. It is beneficial to remember that in this patch-clamp configuration the bath solution is extracellular and the pipette solution is intracellular. Both solutions must have the same osmolarity that the medium solution and if possible, the same temperature to avoid cellular damage. The requirements of other components of solutions can be read in the previous section "Solutions" in this chapter.

4.1.3 Electrophysiology

The electrophysiological recording equipment needed in patch-clamp fluorometry is also used in whole-cell experiments. The most notable difference between the whole-cell experiments and oocyte patch-clamping is the size of pipette tip. For the whole-cell configuration, these pipettes must have a diameter of 2–4 μm. They are made out using capillaries, as mentioned, but programming the pipette puller to obtain pipettes with tips around 5 μm. Then, pipette tips are fire-polished with the microforge to the diameter needed. It is relevant to mention that a bigger tip diameter helps to reduce the pipette access resistance. If the pipette tip is too narrow, the access resistance will slow down voltage-clamping and generate large voltage errors, which will distort the properties of recorded currents. This problem is more serious for inward currents because it results in completely deformed currents. It is worth mentioning that the quantity of ions, especially chloride, in the pipette solution contributes to access resistance. Thus, a larger concentration of ions reduces the access resistance.

4.2 Methods

In this section, we explain the steps to follow in order to obtain whole-cell recordings in HEK 293 cells. As in the "methods" section of inside-out modality of PCF, the steps described apply to channels that where labeled with genetically encoded fluorescent proteins.

4.2.1 Election of Experimental Cell

HEK 293 cells that are useful to perform whole-cell patch-clamp are chosen by a couple of features which facilitates making a seal, voltage-clamping and the fluorescence recording. An ideal cell is one having a diameter approximately of 15–20 μm, is as round as possible and its membrane fluorescence can be distinguished from cytoplasmic fluorescence. This selection is made with the 60x objective to better appreciate the cell conditions. When a suitable cell is located, it is centered in the vision field and the objective is changed to one with less power, usually 40x. Next, the pipette is moved into the bath solution (*see* **Note 15**). Again, before contacting the bath surface, positive pressure is applied to the pipette. Once in the bath, the pipette is located by its shadow and lowered at the same time that focus is adjusted. The pipette is positioned as near the cell as possible; then, the 60x objective is newly elected to do the final approach. Also, the microscope optical path is turned from ocular to camera. If this final change is made after making the seal, it is likely that the seal will be lost because of the vibration caused by the internal mechanism of the microscope during optical path changing.

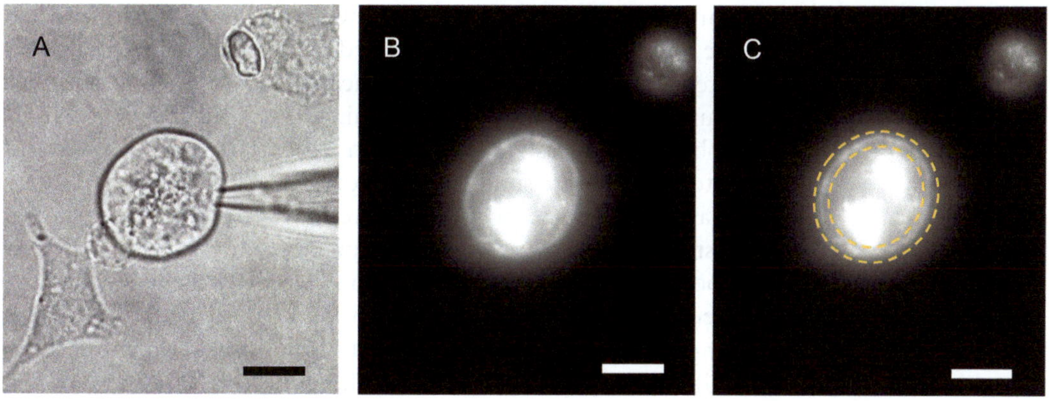

Fig. 5 Whole-cell patch-clamp fluorometry in a HEK 293 cell. Cells were expressing a fusion protein that contains a voltage-gated proton channel (Hv1) and a fluorescent protein (Venus). (**a**) Cell and pipette in light field. (**b**) Fluorescent image of the cell in A excited at 488 nm. The membrane fluorescence can be easily distinguished. (**c**) Region of interest (ROI) designed to quantify only the membrane fluorescence and discard the fluorescence of cell organelles. Scale bars, 10 μm

4.2.2 Making a Seal

Guided by the camera live image in the computer screen, contact of the pipette tip with the cell membrane could be made in two ways. In the first one, more traditional, contact is made from above the cell, placing the blurry silhouette of the pipette tip over the focused image of the cell. Later, the tip is slowly descended until the pipette resistance in the oscilloscope begins to change, indicating cell contact. The movement of the tip is stopped and positive pressure is released while a gently suction is immediately applied. The pipette resistance is monitored until the resistance reaches a value larger a 1 GΩ. It is important to remember that the current of the pipette must be zeroed before cell contact to provide a zero-current level reference.

The other form of contacting the cell membrane with the pipette tip is laterally (Fig. 5a). In this case, the tip is carefully focused in the same plane of the cell but a few microns away from the cell. Then, the tip is moved toward the membrane until the latter begins to deform. At this point, the positive pressure is released and a soft suction is immediately applied. This method is useful when the cells are still spherical or mostly rounded. When the gigaseal is reached, it is advisable to wait for a couple of minutes for it to stabilize, until the resistance value achieves several gigaohms.

4.2.3 Reaching the Whole-Cell Configuration

After seal stabilization, the patch membrane inside the pipette must be broken without losing the seal, to achieve the whole-cell configuration. This can be attained by applying one or several suction pulses with the mouth, strong enough to rupture the membrane. Another way is by using a syringe, applying negative pressure very slowly and waiting to damage the membrane inside the tip. The formation of whole-cell modality will be revealed as a sudden increment of the capacitive current, which is caused by the increase of the membrane area being clamped. This increment depends on the cell size and the access of the pipette to the intracellular medium. Suction should be maintained until the capacitive transient becomes as fast as possible (*see* **Note 16**). If both bath and pipette solutions are similar to physiological extracellular and intracellular solutions, respectively, current recording can be started almost immediate. However, if the pipette solution has a different composition for experimental purposes, it is reasonable to wait between 5–7 min for the pipette solution to dialyze the cell. Also, waiting a couple of minutes before starting the experiment helps to stabilize the seal and whole-cell configuration.

4.2.4 Recording Fluorescence

When the whole-cell configuration is completed, all sources of light in the room are turned off and the cell is excited at the appropriate wavelength (Fig. 5b). The cell's fluorescence is imaged with the camera. This should be done as fast as possible to avoid photobleaching of the fluorophore. In order to record fluorescence and current simultaneously, the same components and protocols for coordinating between electrophysiology software (Patchmaster) and the image software (Micromanager) previously described (*see* Subheading 3.3) are used. These protocols can be changed as required by the experimental procedure.

4.3 Quantification of Cell Membrane Fluorescence Intensity

The images of cell membrane fluorescence obtained with the protocols are initially processed creating a specific ROI that encompasses only the membrane region (Fig. 5c). This "donut-shaped" ROI allows for exclusion of the fluorescence of vesicles and organelles, which do not contribute to membrane fluorescence. This improves the signal to noise ratio of the fluorescence recordings. As was mentioned earlier, the background fluorescence should be subtracted from the brightness value recorded in the ROI to obtain absolute fluorescence, which is the base for more complex analysis.

4.4 An Actual Experiment

As an example of a whole-cell patch clamp fluorometry experiment, we present recordings in HEK 293 cells of the fluorescence and currents of the genetically encoded voltage sensor (GEV) Arclight [74]. HEK 293 cells were transfected with 600 ng of Arclight containing plasmid (obtained from Addgene, www.addgene.org) 2 days before experiments and prepared according to the protocols presented here. After obtaining the whole-cell patch clamp configuration, the cell was held at a holding potential of −70 mV and then stepped to depolarizing voltages lasting for 500 ms. The whole cell currents (Fig. 6d) contain capacitive, leak and the sensing current of the voltage-sensing domain of Arclight. Since no linear component subtraction was applied in these recordings, it is difficult to observe

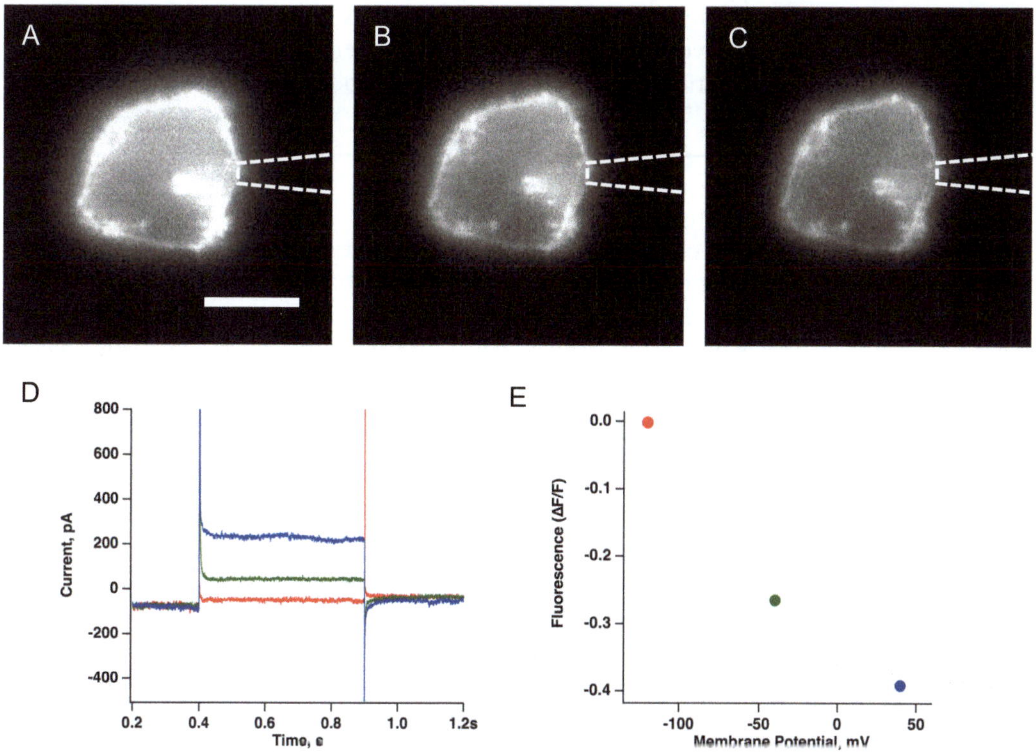

Fig. 6 Whole-cell fluorometry in a HEK 293 cell expressing the fluorescent voltage sensor protein ArcLight. The fluorescence of the same cell clamped to different voltages is imaged: (**a**) −120 mV, (**b**) −40 mV and (**c**) 40 mV. The excitation was achieved with the 488 nm line of an Argon ion laser and excitation was collected with a band pass filter centered at 525 nm. The outline of the patch pipette is shown by the dotted line. (**d**) Unsubtracted whole-cell currents are shown color coded to indicate the clamp voltage as follow. Red, −120 mV, green −40 mV and blue 40 mV. Depolarization pulses cause the voltage-sensing domain of ArcLight to change its conformation and decrease the fluorescence of the pHluorin fluorescent protein attached to it. Membrane fluorescence is quantified in (**e**) and shown as $\Delta F/F = ((F-F_0)/F_0)$ where F is the fluorescence in the test pulse and F_0 is the fluorescence before the test pulse). Scale bar, 10 μm

the sensing currents, but they are discernible as changes in the kinetics of the capacitive transient. Fluorescence was imaged in the same conditions at all voltages, using a 300 ms exposure time. Before each positive voltage a negative pulse of 500 ms to −120 mV was applied and images were acquired during both voltage pulses. A donut-shaped ROI was drawn around the membrane area in the images in A, B and C and the total fluorescence counts were measured in ImageJ. ArcLight's fluorescence is quenched upon depolarization, which is readily visible from the cell fluorescence images. To quantify fluorescence, it is plotted as the difference between the fluorescence at each voltage (F) with respect to the fluorescence at the more negative voltage (F_o), normalized to the value F_o. This form of presenting fluorescence data ($F-F_o/F_o = \Delta F/F$) allows comparisons between different cells or different experimental conditions. In the case of ArcLight, a negative value of ΔF/F indicates quenching of the fluorescence as a function of positive voltages, as expected [75].

5 Concluding Remarks

Patch-clamp fluorometry, in both inside-out and whole-cell configurations, is a powerful technique that allows the experimenter to probe the dynamics of ion channel function. As many techniques, it has benefitted from advances in other areas of research, such as the development and improving of fluorescent proteins, synthetic chemistry, and more recently, the expansion of the genetic code by the use of noncanonical amino acids. PCF is an approach that is seeing more and more use and that in the future will become a standard tool in the neuroscience arsenal.

6 Notes and Advice

1. Numerical aperture (NA). A high NA is preferred since it improves light collection. These objectives are usually 60 or 100× magnification, which also allows for better spatial patch resolution. High NA objectives have reduced working distances of a few micrometers, which means the patch should be close to the cover glass in the bottom of the recording chamber
2. Another alternative for light recording are photomultiplier tubes or photodiodes, instead of CCD cameras. These devices sacrifice imaging capabilities for the ability to record fast signals with good signal to noise ratios. Detectors like these can also be connected to beam splitting units with dichroic mirrors to allow simultaneous detections of at least two wavelengths for ratiometric fluorescence measurements.
3. Most research buildings should have a high quality grounding circuit. If a good ground does not exist, a connection to an independent physical ground or to drainage pipes of the building is suitable. It is important to emphasize the use of very low resistance material, like thick, short, copper wires. Also, it is advisable to provide a single ground point for all equipment and the use of a Faraday cage. These can be big enough to cover the whole table and everything on it or alternatively, only cover the stage of the microscope. This configuration allows easy access to all the optical components which are normally positioned on the vibration isolation table.
4. Pipettes for patch-clamp fluorometry in *Xenopus laevis* oocytes need a smooth shape and a wide aperture. To achieve this

shape, the pipette puller program should loop around eight or ten times to result in pipettes with 40–80 μm diameter, which should be heavily polished in the microforge to achieve a diameter around 10 μm.

5. Removal of the follicular cell layer. A common procedure for enzymatic removal of the follicular cell layer is to use up to 2 mg/mL of collagenase. This high concentration might be detrimental to some batches of oocytes. It is recommended to standardize the concentration beginning with 1.2 mg/mL and incubating during 30 minutes (longer times tend to reduce oocyte viability, regardless of the concentration). If after this time, approximately 50% of ovary segments are not disaggregated, the concentration should be increased. After collagenase treatment, the second wash should mostly remove the remaining follicular cells by mechanical force during agitation.
6. Variant of OR2 solution. OR2 solution has an osmolarity of 172 mOsm and lacks a carbon source. This could be detrimental to some oocytes. Sometimes OR2 can be replaced for an ND96 solution without $CaCl_2$ with an osmolarity of 199 mOsm, reducing the likelihood of oocyte deterioration when they are returned to ND96 after enzyme treatment. This absence of Ca^{2++} is essential to reduce the activation of proteases that can compromise the viability of oocytes.
7. Oocyte maintenance. It is known that during summertime the oocyte quality is reduced. We increase their lifetime a couple of days by adding 5% of fetal bovine serum to the ND 96 solution as well as adding antibiotic solution (Penicillin-Streptomycin) at 1%. This solution should be changed daily.
8. mRNA injection. The mRNA should remain at −20 °C until the very moment of injection. Previously, defolliculated oocytes will be placed inside a 30 mm plastic petri dish with a scraped bottom in order to reduce oocyte mobility during injection. The microinjector with the injection pipette is aligned within the field of vision of the stereomicroscope. The tube of mRNA is then heated by hand and, when liquid, about 1–2 μL is placed on a clean surface, usually a small piece of Parafilm (Sigma). The drop of mRNA is immediately aspirated into the injection pipette and is ready to inject. Gently, the oocyte membrane is perforated with the pipette and 30–50 nL is injected inside the oocyte. After injection, oocytes should be kept in individual wells of a multiwell plate.
9. Removal of the vitelline membrane. The first step before placing the oocyte into the recording chamber is removal of the vitelline membrane. This membrane is transparent in almost all oocytes. Many times it can only be discovered by its absence in

small regions of the oocyte surface where the membrane makes a little protrusion. Using fine tweezers, the edge of the vitelline membrane is taken just within protrusion limits. Sometimes no edge of this membrane can be observed so a fast and delicate movement with the tweezers can be used to push into the oocyte until snagging a piece of the vitelline membrane. If this membrane is adhered to the cell membrane, this is an indicator that the oocyte is nearly dead.

10. Pipette Resistance. The resistance of the pipette in the bath solution is used as a proxy of the tip diameter. However, we must consider that the resistivity of the solution plays a major role in determining pipette resistance. Solutions with overall low electrolyte concentration or with low chloride concentration translate into a larger pipette resistance for a given size opening. Therefore, the size of the pipette tips to be used for giant patch recordings in PCF should be established using a calibrated ocular. The size of the pipettes we use is between 7 and 15 μm.
11. Obtaining a seal. This is the part that needs more practice and patience, especially with oocytes. After removing the vitelline membrane, it is important to try to obtain the patch as fast as possible. After releasing the positive pressure, suction needs to be gently but continuously applied, without losing sight of the resistance of the seal, which must rise continuously and faster as it reaches higher values. For example, at the beginning of seal establishment, resistance will rise to about 0.2–1 MOhm per second, then, after the seal reaches approximately 15 or 20 MOhm, it will increase to 2–5 MOhm per second and when the seal reaches around 50 MOhms, resistance will climb 8–10 MOhm per second, and so on until the gigaseal (larger than 1 GOhm) is obtained. Sometimes the maximum resistance is about 600 to 900 MOhm, and a gigaseal can be encouraged by changing the membrane potential, between −40 and −60 mV. If the resistance stops at lower values, then there is practically no chance of obtaining a gigaseal. The most important thing is to not lose patience and practice, practice, practice.
12. Dealing with intracellular content. It is frequent that cytosolic content near the membrane remains in the patch, even in the *inside-out* configuration. This material may represent an interference factor for fluorescence measurements, because the vesicle membrane is not subject to voltage clamp. It also results in an increase in the background signal. Therefore, the patch must be as clean as possible. When the gigaseal is achieved in the *cell-attached* configuration, the invasion of the patch by

intracellular vesicles can be observed. If this happens, an *inside-out* patch may be achieved by withdrawing the pipette very slowly. In this fashion, at the moment that the membrane is broken around the tip, intracellular content will be pulled out of the patch. This sometimes results in losing the patch because cytoplasmic contents are too attached; nonetheless, it is worth getting a clean patch. An alternative is to very gently blow into the patch trying to expel these vesicles, but the probability of breaking the seal is very high.

13. Photobleaching. Because the membrane patch will be exposed repeatedly to laser or other types of illumination, it is important to evaluate the level of photobleaching produced by it. With that intention, the extent of photobleaching for a given illumination intensity should be determined in some control patches. In addition, photobleaching of the particular fluorophore in use can be estimated by exposing oocyte membrane sheets adhered to coverslips previously treated with poly-D-lysine. If considerable photobleaching is present, a neutral density filter (ND) and low illumination intensity are chosen in order to obtain as much emission light with minimal photo destruction.
14. HEK 293 cells are to be cultured in normal plastic culture dishes while growth and transfection takes place and then sown into glass bottom culture dishes after the trypsin treatment is applied. An alternative to seeding HEK293 cells after trypsin treatment is to use round coverslips which are treated with poly-D-lysine solution to improve attachment of cells to the glass [76]. These coverslips are mounted in a special recording chamber. Round coverslips are more economic and permit higher experimental throughput.
15. In the beginning, centering the pipette tip upon the cell elected to do whole cell is difficult. To solve this, one can turn on the laser beam for a few seconds and move the pipette until it illuminates completely with the laser. When the pipette descends, only a few adjustments will be necessary to center the pipette.
16. Sometimes the membrane can be resealed under the pipette. This increases the access resistance to the cell and is undesirable. Resealing is observed as a slowdown and a reduction of capacitive current amplitude. To solve this problem, it is imperative to apply soft suction again until the capacitive current get faster and larger.

Acknowledgments

Work in the lab of LDI is supported by a grant from CONACYT No.252644; CONACYT-Fronteras de la Ciencia No.77 and DGAPA-PAPIIT IN209515.

References

1. Hille B (2001) Ion channels of excitable membranes, 3rd edn. Sinauer, Sunderland, MA; [Great Britain]
2. Catterall WA, Wisedchaisri G, Zheng N (2017) The chemical basis for electrical signaling. Nat Chem Biol 13(5):455–463. https://doi.org/10.1038/nchembio.2353
3. Rasmussen T (2016) How do mechanosensitive channels sense membrane tension? Biochem Soc Trans 44(4):1019–1025. https://doi.org/10.1042/BST20160018
4. Zheng J, Trudeau MC (2015) Handbook of Ion channels, 1st edn. CRC Press, Boca Raton, FL. https://doi.org/10.1201/b18027
5. Ahern CA, Payandeh J, Bosmans F, Chanda B (2016) The hitchhiker's guide to the voltage-gated sodium channel galaxy. J Gen Physiol 147(1):1–24. https://doi.org/10.1085/jgp.201511492
6. Doyle DA, Morais Cabral J, Pfuetzner RA, Kuo A, Gulbis JM, Cohen SL, Chait BT, MacKinnon R (1998) The structure of the potassium channel: molecular basis of K+ conduction and selectivity. Science 280 (5360):69–77
7. Hodgkin AL, Huxley AF (1952) A quantitative description of membrane current and its application to conduction and excitation in nerve. J Physiol 117(4):500–544
8. Guan B, Chen X, Zhang H (2013) Two-electrode voltage clamp. In: Gamper N (ed) Ion channels: methods and protocols. Humana Press, Totowa, NJ, pp 79–89. https://doi.org/10.1007/978-1-62703-351-0_6
9. Stefani E, Bezanilla F (1998) Cut-open oocyte voltage-clamp technique. Methods Enzymol 293:300–318
10. Rudokas MW, Varga Z, Schubert AR, Asaro AB, Silva JR (2014) The Xenopus oocyte cut-open vaseline gap voltage-clamp technique with fluorometry. J Vis Exp (85). https://doi.org/10.3791/51040
11. Neher E, Sakmann B, Steinbach JH (1978) The extracellular patch clamp: a method for resolving currents through individual open channels in biological membranes. Pflugers Arch 375(2):219–228
12. Sakmann B, Neher E (1984) Patch clamp techniques for studying ionic channels in excitable membranes. Annu Rev Physiol 46:455–472. https://doi.org/10.1146/annurev.ph.46.030184.002323
13. Auerbach A, Sachs F (1984) Patch clamp studies of single ionic channels. Annu Rev Biophys Bioeng 13:269–302. https://doi.org/10.1146/annurev.bb.13.060184.001413
14. Islas L (2015) Patch clamping and single-channel analysis. In: Handbook of ion channels. CRC Press, Boca Raton, FL, pp 71–81. https://doi.org/10.1201/b18027-9
15. Hamill OP, Marty A, Neher E, Sakmann B, Sigworth FJ (1981) Improved patch-clamp techniques for high-resolution current recording from cells and cell-free membrane patches. Pflugers Arch 391(2):85–100
16. Neher E, Sakmann B (1992) The patch clamp technique. Sci Am 266(3):44–51
17. Gorostiza P, Isacoff EY (2008) Nanoengineering ion channels for optical control. Physiology (Bethesda) 23:238–247. https://doi.org/10.1152/physiol.00018.2008
18. Pless SA, Kim RY, Ahern CA, Kurata HT (2015) Atom-by-atom engineering of voltage-gated ion channels: magnified insights into function and pharmacology. J Physiol 593 (12):2627–2634. https://doi.org/10.1113/jphysiol.2014.287714
19. Lueck JD, Mackey AL, Infield DT, Galpin JD, Li J, Roux B, Ahern CA (2016) Atomic mutagenesis in ion channels with engineered stoichiometry. Elife 5. https://doi.org/10.7554/eLife.18976
20. Yuchi Z, Van Petegem F (2016) Ryanodine receptors under the magnifying lens: insights and limitations of cryo-electron microscopy and X-ray crystallography studies. Cell Calcium 59(5):209–227. https://doi.org/10.1016/j.ceca.2016.04.003
21. Fujiyoshi Y (2011) Electron crystallography for structural and functional studies of membrane proteins. J Electron Microsc (Tokyo) 60(Suppl

1):S149–S159. https://doi.org/10.1093/jmicro/dfr033
22. Payandeh J, Scheuer T, Zheng N, Catterall WA (2011) The crystal structure of a voltage-gated sodium channel. Nature 475(7356):353–358. https://doi.org/10.1038/nature10238
23. Kaplan M, Pinto C, Houben K, Baldus M (2016) Nuclear magnetic resonance (NMR) applied to membrane-protein complexes. Q Rev Biophys 49:e15. https://doi.org/10.1017/S003358351600010X
24. Blasic JR, Worcester DL, Gawrisch K, Gurnev P, Mihailescu M (2015) Pore hydration states of KcsA potassium channels in membranes. J Biol Chem 290(44):26765–26775. https://doi.org/10.1074/jbc.M115.661819
25. Kulleperuma K, Smith SM, Morgan D, Musset B, Holyoake J, Chakrabarti N, Cherny VV, DeCoursey TE, Pomes R (2013) Construction and validation of a homology model of the human voltage-gated proton channel hHV1. J Gen Physiol 141(4):445–465. https://doi.org/10.1085/jgp.201210856
26. Randolph AL, Mokrab Y, Bennett AL, Sansom MS, Ramsey IS (2016) Proton currents constrain structural models of voltage sensor activation. Elife 5. https://doi.org/10.7554/eLife.18017
27. Blunck R (2015) Investigation of ion channel structure using fluorescence spectroscopy. In: Handbook of ion channels. CRC Press, Boca Raton, FL, pp 113–133. https://doi.org/10.1201/b18027-12
28. Taraska JW, Zagotta WN (2010) Fluorescence applications in molecular neurobiology. Neuron 66(2):170–189. https://doi.org/10.1016/j.neuron.2010.02.002
29. Mannuzzu LM, Moronne MM, Isacoff EY (1996) Direct physical measure of conformational rearrangement underlying potassium channel gating. Science 271(5246):213–216
30. Gandhi CS, Olcese R (2008) The voltage-clamp fluorometry technique. Methods Mol Biol 491:213–231. https://doi.org/10.1007/978-1-59745-526-8_17
31. Zheng J, Zagotta WN (2003) Patch-clamp fluorometry recording of conformational rearrangements of ion channels. Sci STKE 2003 (176):PL7. https://doi.org/10.1126/stke.2003.176.pl7
32. Zheng J, Zagotta WN (2000) Gating rearrangements in cyclic nucleotide-gated channels revealed by patch-clamp fluorometry. Neuron 28(2):369–374
33. Kusch J, Zifarelli G (2014) Patch-clamp fluorometry: electrophysiology meets fluorescence. Biophys J 106(6):1250–1257. https://doi.org/10.1016/j.bpj.2014.02.006
34. Taraska JW, Zagotta WN (2007) Cyclic nucleotide-regulated ion channels: spotlight on symmetry. Structure 15(9):1023–1024. https://doi.org/10.1016/j.str.2007.08.004
35. Wang S, Lee SJ, Heyman S, Enkvetchakul D, Nichols CG (2012) Structural rearrangements underlying ligand-gating in Kir channels. Nat Commun 3:617. https://doi.org/10.1038/ncomms1625
36. Islas LD, Zagotta WN (2006) Short-range molecular rearrangements in ion channels detected by tryptophan quenching of bimane fluorescence. J Gen Physiol 128(3):337–346. https://doi.org/10.1085/jgp.200609556
37. Liu C, Xie C, Grant K, Su Z, Gao W, Liu Q, Zhou L (2016) Patch-clamp fluorometry-based channel counting to determine HCN channel conductance. J Gen Physiol 148 (1):65–76. https://doi.org/10.1085/jgp.201511559
38. De-la-Rosa V, Suárez-Delgado E, Rangel-Yescas GE, Islas LD (2016) Currents through Hv1 channels deplete protons in their vicinity. J Gen Physiol 147(2):127–136
39. Miranda P, Giraldez T, Holmgren M (2016) Interactions of divalent cations with calcium binding sites of BK channels reveal independent motions within the gating ring. Proc Natl Acad Sci U S A 113(49):14055–14060. https://doi.org/10.1073/pnas.1611415113
40. Trudeau MC, Zagotta WN (2003) Calcium/calmodulin modulation of olfactory and rod cyclic nucleotide-gated ion channels. J Biol Chem 278(21):18705–18708. https://doi.org/10.1074/jbc.R300001200
41. Biskup C, Kusch J, Schulz E, Nache V, Schwede F, Lehmann F, Hagen V, Benndorf K (2007) Relating ligand binding to activation gating in CNGA2 channels. Nature 446 (7134):440–443. https://doi.org/10.1038/nature05596
42. Aman TK, Gordon SE, Zagotta WN (2016) Regulation of CNGA1 channel gating by interactions with the membrane. J Biol Chem 291 (19):9939–9947. https://doi.org/10.1074/jbc.M116.723932
43. Zheng J (2006) Patch fluorometry: shedding new light on ion channels. Physiology (Bethesda) 21:6–12. https://doi.org/10.1152/physiol.00041.2005
44. Geibel S, Kaplan JH, Bamberg E, Friedrich T (2003) Conformational dynamics of the Na+/K+-ATPase probed by voltage clamp fluorometry. Proc Natl Acad Sci U S A 100

(3):964–969. https://doi.org/10.1073/pnas.0337336100
45. Stuurman N, Amdodaj N, Vale R (2007) Micro-manager: open source software for light microscope imaging. Microscopy Today 15(3):42–43
46. Molleman A (2003) Requirements. In: Patch clamping. John Wiley & Sons, Ltd., Chichester, pp 43–93. https://doi.org/10.1002/0470856521.ch3
47. Goldin AL (1992) Maintenance of Xenopus laevis and oocyte injection. Methods Enzymol 207:266–279
48. Soreq H, Seidman S (1992) Xenopus oocyte microinjection: from gene to protein. Methods Enzymol 207:225–265
49. Brown AL, Johnson BE, Goodman MB (2008) Patch clamp recording of ion channels expressed in Xenopus oocytes. J Vis Exp (20):936. https://doi.org/10.3791/936
50. Holmgren M, Liu Y, Xu Y, Yellen G (1996) On the use of thiol-modifying agents to determine channel topology. Neuropharmacology 35 (7):797–804
51. Lundblad RL (2004) Techniques in protein modification, 2nd edn. CRC Press, Boca Raton, FL
52. Taraska JW, Puljung MC, Olivier NB, Flynn GE, Zagotta WN (2009) Mapping the structure and conformational movements of proteins with transition metal ion FRET. Nat Methods 6(7):532–537
53. Taraska JW, Zagotta WN (2007) Structural dynamics in the gating ring of cyclic nucleotide-gated ion channels. Nat Struct Mol Biol 14(9):854–860. https://doi.org/10.1038/nsmb1281
54. Loots E, Isacoff EY (2000) Molecular coupling of S4 to a K(+) channel's slow inactivation gate. J Gen Physiol 116(5):623–636
55. Puljung MC, Zagotta WN (2011) Labeling of specific cysteines in proteins using reversible metal protection. Biophys J 100 (10):2513–2521. https://doi.org/10.1016/j.bpj.2011.03.063
56. Kusch J, Biskup C, Thon S, Schulz E, Nache V, Zimmer T, Schwede F, Benndorf K (2010) Interdependence of receptor activation and ligand binding in HCN2 pacemaker channels. Neuron 67(1):75–85. https://doi.org/10.1016/j.neuron.2010.05.022
57. Posson DJ, Ge P, Miller C, Bezanilla F, Selvin PR (2005) Small vertical movement of a K+ channel voltage sensor measured with luminescence energy transfer. Nature 436 (7052):848–851. https://doi.org/10.1038/nature03819
58. Liu CC, Schultz PG (2010) Adding new chemistries to the genetic code. Annu Rev Biochem 79:413–444. https://doi.org/10.1146/annurev.biochem.052308.105824
59. Serfling R, Coin I (2016) Incorporation of unnatural amino acids into proteins expressed in mammalian cells. Methods Enzymol 580:89–107. https://doi.org/10.1016/bs.mie.2016.05.003
60. Lee HS, Guo J, Lemke EA, Dimla RD, Schultz PG (2009) Genetic incorporation of a small, environmentally sensitive, fluorescent probe into proteins in Saccharomyces cerevisiae. J Am Chem Soc 131(36):12921–12923. https://doi.org/10.1021/ja904896s
61. Chatterjee A, Guo J, Lee HS, Schultz PG (2013) A genetically encoded fluorescent probe in mammalian cells. J Am Chem Soc 135(34):12540–12543. https://doi.org/10.1021/ja4059553
62. Segev A, Garcia-Oscos F, Kourrich S (2016) Whole-cell patch-clamp recordings in brain slices. J Vis Exp (112). https://doi.org/10.3791/54024
63. Li S, Deng Z, Wei L, Liang L, Ai W, Shou X, Chen X (2011) Reduction of large-conductance Ca^2(+) -activated K(+) channel with compensatory increase of nitric oxide in insulin resistant rats. Diabetes Metab Res Rev 27(5):461–469. https://doi.org/10.1002/dmrr.1196
64. Cherny VV, Murphy R, Sokolov V, Levis RA, DeCoursey TE (2003) Properties of single voltage-gated proton channels in human eosinophils estimated by noise analysis and by direct measurement. J Gen Physiol 121 (6):615–628. https://doi.org/10.1085/jgp.200308813
65. Gandini MA, Sandoval A, Felix R (2014) Whole-cell patch-clamp recording of recombinant voltage-sensitive Ca2+ channels heterologously expressed in HEK-293 cells. Cold Spring Harb Protoc 2014(4):396–401. https://doi.org/10.1101/pdb.prot073213
66. Sontheimer H, Olsen ML (2007) Whole-cell patch-clamp recordings. In: Walz W (ed) Patch-clamp analysis: advanced techniques. Humana Press, Totowa, NJ, pp 35–68. https://doi.org/10.1007/978-1-59745-492-6_2
67. Cahalan M, Neher E (1992) Patch clamp techniques: an overview. Methods Enzymol 207:3–14
68. Molleman A (2003) Basic theoretical principles. In: Patch clamping. John Wiley & Sons, Ltd., Chichester, pp 5–42. https://doi.org/10.1002/0470856521.ch2

69. Kobrinsky E, Stevens L, Kazmi Y, Wray D, Soldatov NM (2006) Molecular rearrangements of the Kv2.1 potassium channel termini associated with voltage gating. J Biol Chem 281 (28):19233–19240. https://doi.org/10.1074/jbc.M601231200
70. Yang F, Cui Y, Wang K, Zheng J (2010) Thermosensitive TRP channel pore turret is part of the temperature activation pathway. Proc Natl Acad Sci U S A 107(15):7083–7088. https://doi.org/10.1073/pnas.1000357107
71. Kobrinsky E, Tiwari S, Maltsev VA, Harry JB, Lakatta E, Abernethy DR, Soldatov NM (2005) Differential role of the alpha1C subunit tails in regulation of the Cav1.2 channel by membrane potential, beta subunits, and Ca2+ ions. J Biol Chem 280(13):12474–12485. https://doi.org/10.1074/jbc.M412140200
72. Fisher JA, Girdler G, Khakh BS (2004) Time-resolved measurement of state-specific P2X2 ion channel cytosolic gating motions. J Neurosci 24(46):10475–10487. https://doi.org/10.1523/JNEUROSCI.3250-04.2004
73. Graham FL, Smiley J, Russell WC, Nairn R (1977) Characteristics of a human cell line transformed by DNA from human adenovirus type 5. J Gen Virol 36(1):59–74. https://doi.org/10.1099/0022-1317-36-1-59
74. Jin L, Han Z, Platisa J, Wooltorton JRA, Cohen LB, Pieribone VA (2012) Single action potentials and subthreshold electrical events imaged in neurons with a fluorescent protein voltage probe. Neuron 75(5):779–785
75. Han Z, Jin L, Chen F, Loturco JJ, Cohen LB, Bondar A, Lazar J, Pieribone VA (2014) Mechanistic studies of the genetically encoded fluorescent protein voltage probe ArcLight. PLoS One 9(11):e113873
76. Heuser J (2000) The production of 'cell cortices' for light and electron microscopy. Traffic 1 (7):545–552. https://doi.org/10.1034/j.1600-0854.2000.010704.x

Chapter 8

Modeling Early-Onset Postischemic Seizures in Adult and Aging Mice

Liang Zhang, Chiping Wu, and Justin Wang

Abstract

Stroke is the most common cause of seizures in the adult and elderly population. The development of seizures after stroke is known to be associated with poorer prognoses including higher rates of disability and mortality. Currently, the pathophysiology of how these post-stroke seizures develop and why they occur in certain patients but not in others is not well understood. The use of prophylactic antiepileptics/anticonvulsants after stroke and its unproven role in preventing late-onset poststroke seizures remains a contentious issue. Our recent experiments have focused on examining early-onset postischemic seizures in adult and aging mice using a middle cerebral artery occlusion model and a hypoxia–ischemia model. We focused on electroencephalographic recordings, characterization and detection of convulsive and nonconvulsive seizures, assessment of histological brain injury, and the role of anticonvulsant treatment. Here we discuss these experiments in the hopes that our experience may further future investigations in mouse models of postischemic seizures and other neurological diseases.

Key words Aging, Anticonvulsant, Convulsion, EEG, Epilepsy, Ischemia, Mouse model, Seizure, Stroke

1 Introduction

Stroke (both ischemic and hemorrhagic) is a common cause of seizures in the adult and elderly population, with poststroke seizures reported in 3–10% of all stroke cases [1–9]. Cerebral ischemia is thought to be a major contributor, and global ischemia following cardiac arrest has also been demonstrated to similarly lead to seizures [10–13]. Regardless of the causative event, the development of seizures following stroke or cardiac arrest is known to be associated with a poorer prognosis including higher rates of extended hospitalization, long-term disability, and death [13–18]. Despite this link between seizure development and poorer patient outcomes, specific treatment guidelines for post-stroke seizures have yet to be established and may depend on a better understanding of seizure pathogenesis [6–8, 19, 20]. Where these seizures arise, why

Nicholas J. D. Wright (ed.), *Basic Neurobiology Techniques*, Neuromethods, vol. 152,
https://doi.org/10.1007/978-1-4939-9944-6_8,

they develop in certain patients but not in others and how they contribute to mortality and disability are all questions that remain to be answered.

Seizures that arise within 2 weeks of the initial stroke or cardiac arrest are generally categorized as "early-onset." Early-onset seizures are primarily observed within 24 h of the initial insult and considered a medical emergency as life-threatening status epilepticus may follow [10, 11, 14, 21]. Early-onset seizures can be generalized convulsive seizures (CS) or focal/generalized nonconvulsive seizures (NCS) that may require electroencephalography (EEG) for diagnosis [10, 11, 22–25]. Stroke severity and the degree of cortical involvement have been recognized as risk factors for seizure development following ischemic stroke [1–9], however the underlying pathophysiology remains not well understood.

Animal models have contributed greatly to our understanding of postischemic seizures and our continued exploration of potential treatment strategies that may be eventually translated to humans. Previous studies have modeled early-onset CS and NCS in adult rats following a unilateral middle cerebral artery occlusion (MCAO) [26–31], photothrombotic ischemia [32, 33], and cardiopulmonary bypass resuscitation [34, 35]. Other studies have examined late-onset epileptic seizures in aged rats following a MCAO, and photothrombotic ischemia [36, 37]. Mouse models of stroke have been shown to produce similar infarcts compared to rat models but with the added advantages of being better characterized models with easier animal handling and lower costs [38]. Despite this, studies of postischemic seizures in mice are limited. Our experiments focused on early-onset seizures in both the adult and aging mice using a MCAO model and a hypoxia-ischemia (HI) model [39–41]. Subsequent sections detail our experimental methodology which focuses on EEG recordings, CS and NCS detection, seizure treatment, and classifying the severity of ischemia.

2 Materials

Table 1 (below) lists the main materials and equipment that were used for EEG electrode construction, EEG electrode implantation, EEG recordings, brain ischemia induction, brain histology, and anticonvulsant treatments in our mouse models

3 Methods

3.1 Animals

Inbred C57 black mice (C57BL/6, Charles River Laboratory, Senneville St-Constant, Quebec, Canada), a widely used strain for neurobiology research and transgenic mice models, were utilized

Table 1
Materials and agents used for inducing brain iscehmia and detecting seizures in mouse models

Items	Product number and vendors
EEG electrode construction	
Wires for construction of EEG electrodes	Polyimide-insulated stainless steel wires, outer diameter 0.12 and 0.20 mm; part # 005SW and 008SW, Plastics One, VA, USA
Fine cutters for hard wires	Part # 7248 and # 7250, Tronex Technology Inc., CA, USA; Part #10826-R, Aven Tools, Digikey, Canada
Male pins for implanted headset	Detached from IC sockets, part # ED3014-ND, Digikey, Canada
Multimeter	Tektronix DMM912, OR, USA
Phosphoric acid	Part # p5811, Sigma-Aldrich, Canada
Ethyl alcohol	Part # 64-17-5, Sigma-Aldrich, Canada
Electrode implantation surgery	
Isoflurane	Baxter cooperation, Ontario, Canada
Anesthesia system	Part # 72-6468, Harvard Apparatus, Quebec, Canada
Stereotaxic frame with two micromanipulators	Part # 72-6343, Harvard Apparatus, Quebec, Canada
Plastic weighting boats	Part # 08-732-115, Fisher Scientific, Canada
Glue (Insta-cure+)	Part # BSI-106C, Bob Smith Industries, CA, USA
Motorized drill	Model # FM3545, Foredom Electric, Bethel, CT, USA
Mini drill bit	Part # 115603, Ball Mills Carbide, CircuitMedic, MA, USA
Dental acrylic	Part # 1404, Lang Dental Mfg. Co., Inc., Wheeling, IL, USA
EEG recordings and video monitoring	
Male connection pins	Part # ED90335-ND, Digikey, Canada
Female connection pins	Part # ED1031-ND, Digikey, Canada
Cables (5 conductors)	Part # 78106/SL005-ND, Digikey, Canada
Cables (5 conductors)	Part # Nmuf 5/30-4046SJ, Cooner Wires, CA, USA
Cable (24 conductors)	Part # CWB300, Cooner Wires, CA, USA
Tethered recordings amplifier	Two channel AC microelectrode amplifiers, model 1800 with extended head-stages, A-M systems, WA, USA
Tethered recordings amplifier	One channel AC microelectrode amplifiers, model 3000 with extended head-stages, A-M systems, WA, USA
Digitizer	Digidata 1300 or 1400, Molecular Devices, CA, USA
Electrophysiology software	pCLAMP, version 10/11, Molecular Devices, CA, USA
Telemetric transmitter	Part # ETA-F10, Data Scientific International, MN, USA
Telemetric receiver	Part # RPC-1, Data Scientific International, MN, USA

(continued)

Table 1
(continued)

Items	Product number and vendors
Telemetric data acquisition system	Data exchange matrix 2.0, Data Scientific International, MN, USA
Webcam	Part # C615, Logitech, Canada
Slip ring (6 wires)	Part # 1528-1152-ND, Digikey, Canada
Slip ring (6 wires)	Part # 1528-1157-ND,, Digikey, Canada
Stimulator	Model S88, with isolation units, Grass Medical Instruments, Warwick, RI, USA
Cursor autoclick program	Mini mouse macro program, http://www.turnssoft.com/mini-mouse-macro.html
Induction of ischemia	
Surgical microscope	Zeiss OPMI 6-S with fiber optical source
Silicon-coated suture	part # 6, Doccol Corporation, Redlands, CA, USA
Electrocoagulator	Bovie, FL, USA
Temperature control	DC temperature system, FHC Inc., ME, USA
Hypoxic chamber	Modular incubator chamber 101, Billups-Rothenberg, CA, USA
Thermometer	Part # ZM-86460-03, Digi-Sense, Cole Parmer
Rubber water bottle	MABIS rubber water bottle, Amazon.CA
Compressed air tank	K-size, Praxair, Ontario, Canada
Air tank (8% O_2–92% N_2)	K-size, Praxair, Ontario, Canada
Air tank regulator	2-stage general air regulators, Praxair, Ontario, Canada
Silk suture (USP 5.0)	Braintree Scientific, Inc. Braintree, MA, USA
Laser Doppler system	PF5010, Perimed, New York, USA
Laser Doppler probe	MTB 500, Perimed, New York, USA
Brain histology	
Tissue embedding box	Part # 2219, Fisher Scientific, Canada
Embedding medium	Part # 1310TS; Fisher Scientific, Canada
Cryostat	Model # CM1950, Leica, Canada
Microscope glass slides	Part # 12-550-15, Fisher Scientific, Canada
Micro cover glasses	CA48393-092-1, VWR, Canada
Mounting medium	Part # 8311-4, Richard-Allan Scientific, CA, USA
Cresyl violet	Part # C5042, Sigma-Aldrich, Canada
Fluoro-Jade C	Histo-Chem, Inc., Jefferson, Arkansas, USA
Microscope	DMRN upright microscope, Leica, Canada

(continued)

Table 1 (continued)

Items	Product number and vendors
Image software	Image J, National Institute of Health, USA; https://imagej.nih.gov/ij/download.html
Anticonvulsant drugs	
Diazepam	Clinically available injectable form at 5 mg/2 ml, Sandoz Canada Inc., Quebec, Canada
Lorazepam (Ativan)	Clinically available injectable form at 2 mg/1 ml, Sandoz Canada Inc., Quebec, Canada
Phenytoin	Clinically available injectable form at 50 mg/1 ml, Sandoz Canada Inc., Quebec, Canada
Fosphenytoin (Cerebyx)	Clinically available injectable form at 50 mg/1 ml, ERFA Canada Inc., Quebec, Canada

for all experiments. C57 black mice aged 3–6, 10–14 and 18–24 months may correspond roughly to humans aged 20–30, 38–47 and 56–69 years respectively [42]. However, mice designated as "aged" commonly refer to animals of ≥24 months-old. These "aged" mice are known to frequently encounter many health-related complications including skin lesions, ear/eye infections, and tumors which can confound outcome assessment [42]. Instead, our experiments were conducted in "aging" C57 black mice 16–20 months-old, allowing us to model ischemic stroke in a rodent model corresponding to an older human age while minimizing the confounding health complications common in older mice.

Due to the relative lack of availability of aging mice directly obtainable from commercial sources, we obtained 6–9-month-old C57 black mice (retired breeders) from Charles River Laboratory and housed them in our local animal facility for ≥7 months prior to use in our experiments. This ensues availability of aging mice in a predictable time frame and allows sufficient baseline EEG and behavioral monitoring in aging mice. When animals were caged in group (up to 4 mice per cage), the cost of animal housing may be comparable to that needed to obtain individual aging mice from commercial sources.

3.2 EEG recordings

EEG recordings are essential for assessing the electrographic activity of neuronal populations and for detecting seizure activity in the brain. Invasive EEG recordings in rodent models may be carried out using microelectrodes that are stereotactically implanted in specific brain areas. These implanted electrodes are traditionally cemented onto the skull using multiple anchoring screws and

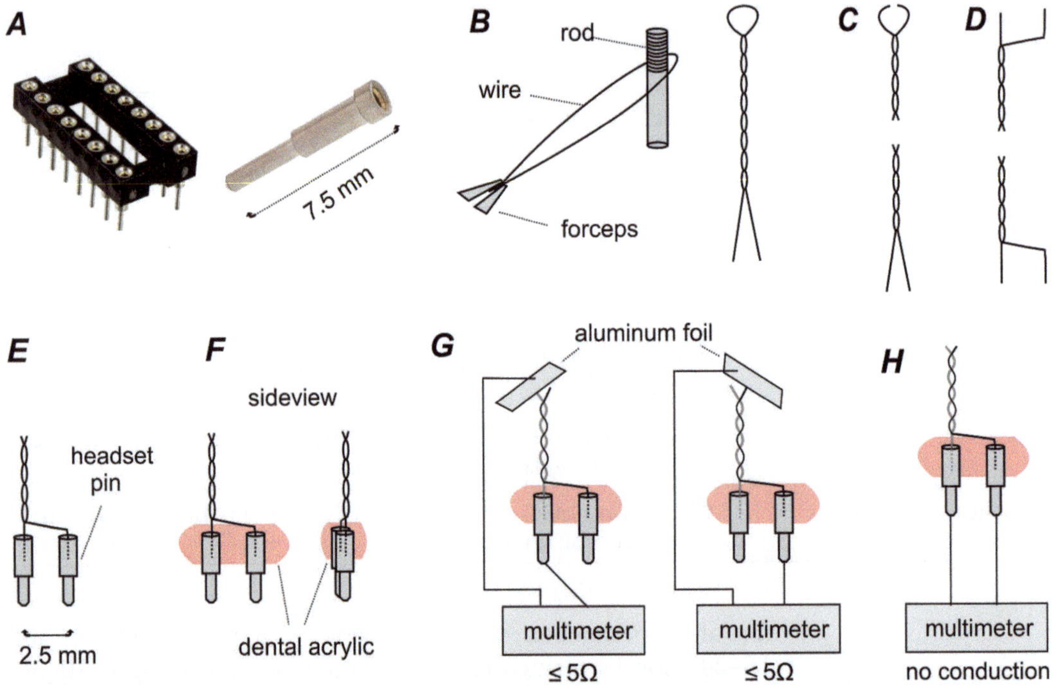

Fig. 1 Procedures for constructing twisted bipolar electrodes. (**a**) Images taken from an IC socket (left) and a "male" pin removed from the socket (right). Similar pins were used to construct EEG electrodes schematically presented in **b–f. (b, c)** A piece (50 mm in length) of polyimide-insulated stainless steel wires were bent and held by forceps along a threaded rod (~7 mm diameter), before being twisted by rotating the rod 8–10 complete turns. (**d**) Twisted wires were cut at middle and at the looped end, producing two shorter twisted wires. (**e**) The shorter wires were bent to form an "h" shape at one end using a pair of pliers. (**f**) The polyimide insulation layer was stripped from the "h" end at the tip and then soldered to the headset pins. (**g**) The base of the headset pins were embedded with dental acrylic. After the dental acrylic hardened, the twisted wires were further cut to a desired length. (**h**) A multimeter and a sheet of aluminum foil (20 mm × 20 mm) were used to test the conductivity of a twisted bipolar electrode. When one tip of the electrode contacted the aluminum foil sheet, the resistance between the aluminum foil sheet and a headset pin was expected to be $\leq 5\ \Omega$. When the electrode was tested alone, no conduction was expected between the two headset pins

dental acrylic. However, the thinness of the mouse skull is suboptimal for the placement of anchoring screws to be able to create a secure attachment. Our lab therefore chose to utilize a novel screw-free, glue-based method of securing EEG electrodes onto free moving mice [43–45].

3.2.1 EEG Electrode Construction

The procedure for constructing twisted bipolar electrodes is outlined schematically in Fig. 1. Monopolar electrodes were similarly constructed except without wire twisting. "Male" pins detached from standard IC sockets (Table 1; Fig. 1a) were used in our implanted headsets. We chose these pins for their ideal length and durability in addition to the wide commercial availability of IC sockets. We found that these pins maintained their integrity for several months despite the wear and tear of frequent usage and impacts against the cage elements. We used polyimide-insulated stainless steel wires (outer diameter 0.12 or 0.20 mm; Table 1) for our EEG electrodes as the polyimide insulation layer is relatively thin but protects well against mechanical and chemical damage.

For construction of twisted bipolar electrodes, a piece of wire (~50 mm in length) was bent, held with forceps around a threaded metal rod (~7 mm diameter) and turned 8–10 complete times (Fig. 1b). The twisted wires were cut in the middle and at the looped end, to produce two shorter twisted wires (Fig. 1c). The open ends of each shorter wire were then shaped into an "h" by a pair of pliers (Fig. 1d). The insulation layer was then stripped at tips of the h's, and the stripped wire tips were soldered to the headset pins described above (Fig. 1e). A solution containing 1:1 phosphoric acid and alcohol (Table 1) was used as a soldering flux. Fine tweezers were used to drop a small amount of this solution onto the base of the headset pin before soldering. After soldering, the base of the pins was embedded with dental acrylic (Table 1) to strengthen the electrode construct and facilitate implantation (Fig. 1f). Once the dental acrylic completely hardened, the twisted wires were further cut to their desired length (e.g., ~3 mm for targeting the middle hippocampal CA3 area in adult mice) using a fine hard-wire cutter (Table 1). Following the final cut, the wire tips were usually separated ~100 μm apart.

A multimeter (Table 1) and a piece of aluminum foil (20 mm × 20 mm) were used to test the conductivity of the twisted bipolar electrodes. When an electrode tip achieved good contact with the aluminum foil, a low resistance connection (≤5 Ω) between the aluminum foil and corresponding headset pin would be expected if the construction was successful (Fig. 1g). If a twisted bipolar electrode was tested alone, there should be no conduction between the two headset pins if the insulation layer remained intact and the exposed electrode tips did not contact each other (Fig. 1h). The electrodes that passed the multimeter test were cleaned with 75% alcohol and stored in a sterilized glass vial until usage.

3.2.2 EEG Electrode Implantation

The procedure for implanting intracranial EEG electrodes in mice was adopted from our previous studies [43–45] and outlined schematically in Fig. 2. Briefly, the animal was anaesthetized with isoflurane (initially at 5% and then maintained at 2%) and placed on a stereotaxic frame equipped with two micromanipulators (Table 1). After skin incision and exposure of the skull, the tip of a mini drill bit (Table 1) was aimed at bregma using the micromanipulator. After determining the position of bregma, the drill bit was moved up but its X–Y position corresponding to the bregma remained unchanged. A plastic base matching the size of the skull exposure was cut from the curved portion of a polystyrene weighing dish

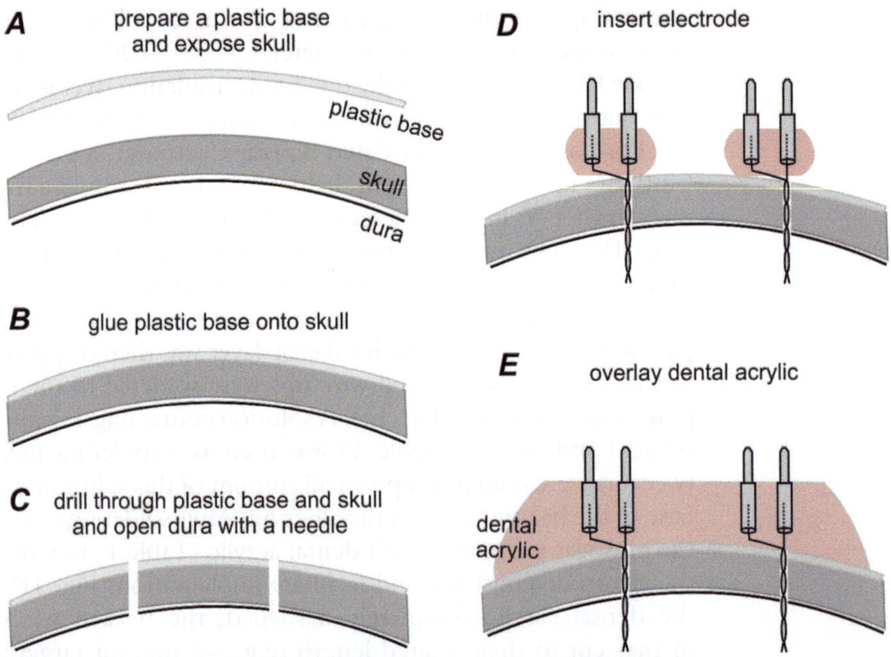

Fig. 2 Schematic illustration of electrode implantation procedures. (**a**) A plastic base cut from a polystyrene weighing dish was size matched to the exposed skull surface. (**b**) The plastic base was glued onto the skull surface. (**c**) Small holes were drilled through the plastic base and skull, and the dura underneath was opened with a fine needle. (**d**) Preassembled bipolar electrodes were inserted into the brain using micromanipulators. (**e**) Dental acrylic overlaid onto the plastic base covered the inserted electrodes

(0.2 mm thickness; Table 1; Fig. 2a) and glued onto the skull surface using a cyanoacrylate-based glue (Table 1; Fig. 2b).

After glue drying, small holes (≤0.5 mm in diameter) were drilled through the plastic base and the skull using a motorized drill (Table 1) and in accordance to the stereotaxic coordinates of targeted brain areas. The stereotaxic coordinates were bregma −2.5 mm, lateral 3.0 mm, and depth of 2.5 mm for hippocampal CA3; bregma −2.3 mm, lateral 2.0 mm, and depth 2.0 mm for hippocampal CA1; bregma −0.8 mm, lateral 1.8 mm, and depth 1.5 mm for parietal cortex [46]. A reference electrode was positioned at bregma +1.0 mm, lateral 1.0 mm, and depth of 0.5 mm.

After the dura was opened using a fine needle (Fig. 2c), preassembled bipolar electrodes were individually inserted into the brain using micromanipulators (Fig. 2d). Dental acrylic was overlaid onto the plastic base covering the inserted electrodes (Fig. 2e). Once the dental acrylic hardened, the electrodes were released from the micromanipulators but remained secured onto the skull by the dental cement. The incised skin was glued to the dental cement to prevent potential infection.

The glue-based method described above is successful because the dental acrylic denatures the plastic base and therefore upon hardening, bonds the base to the underlying skull to secure the implanted electrodes in place. Although the hardening time of the dental acrylic we used is 6–9 min, practice may be needed to overlay the acrylic onto the plastic base and cover the implanted electrodes promptly. In our studies, implanted mice rarely encountered local infections or mechanical failure of the electrodes and underwent repeated EEG recordings for months with the same construct [47–55].

3.3 Tethered Recordings in Free Moving Mice

Microelectrode amplifiers with extended head-stages (model 3000 or 1800, A-M systems, Table 1) were used for EEG recording in free-moving mice. We used these amplifiers as they accommodate low- and high-impedance electrodes and can operate in differential or monopolar recording mode. The head-stages with flexible cables can be conveniently placed near the animal to minimize cable/movement-related artifacts. The model 3000 amplifier also allows TTL-gated switches between recording and stimulation mode, therefore enabling the delivery electrical stimuli and the recording of subsequent discharges using a single bipolar electrode. To collect EEG signals in a wide frequency range and high amplitude resolution, the input frequency band of these amplifiers was set in a range of 0.1–1000 Hz and amplification gain at 1000×. Amplifier's output signals were digitized at 5000 Hz. A digitizer (Digidata 1300 or 1400, Molecular Devices; Table 1) and electrophysiology software (pCLAMP, version 10, Molecular Devices; Table 1) were used for EEG data acquisition, storage and analysis. We used this system as it is versatile and compatible with our in vitro recording setups.

The amplifier, digitizer, a desktop computer, and a platform for animal recording were all mounted on a mobile equipment rack (Fig. 3a) as a compact, movable recording unit. For EEG recordings concurrent with webcam monitoring, the animal was placed in a large glass beaker or bowl-shaped container and allowed to freely access hydrogel and food pellets provided on floor.

"Female" connection pins (part # ED1031-ND, Digikey; Table 1) were used to connect to the implanted headset pins. These pins could be used repeatedly without loosening of their connection. Tweezers with threaded tips were used to connect and disconnect the female pins individually. For animals implanted with two bipolar electrodes and a single reference electrode, the female pins were soldered to a soft 5-conductor cable (Table 1) or to 5 individual cables detached from a 24-conductor cable (Table 1). For the electrode connection or disconnection, we suggest hold the animal in air by gripping its neck skin with thumb and forefinger and securing its tail via little finger and lower thumb but avoid press the animal against the hard surface of a table or bench as the latter may cause death due to respiratory suppression.

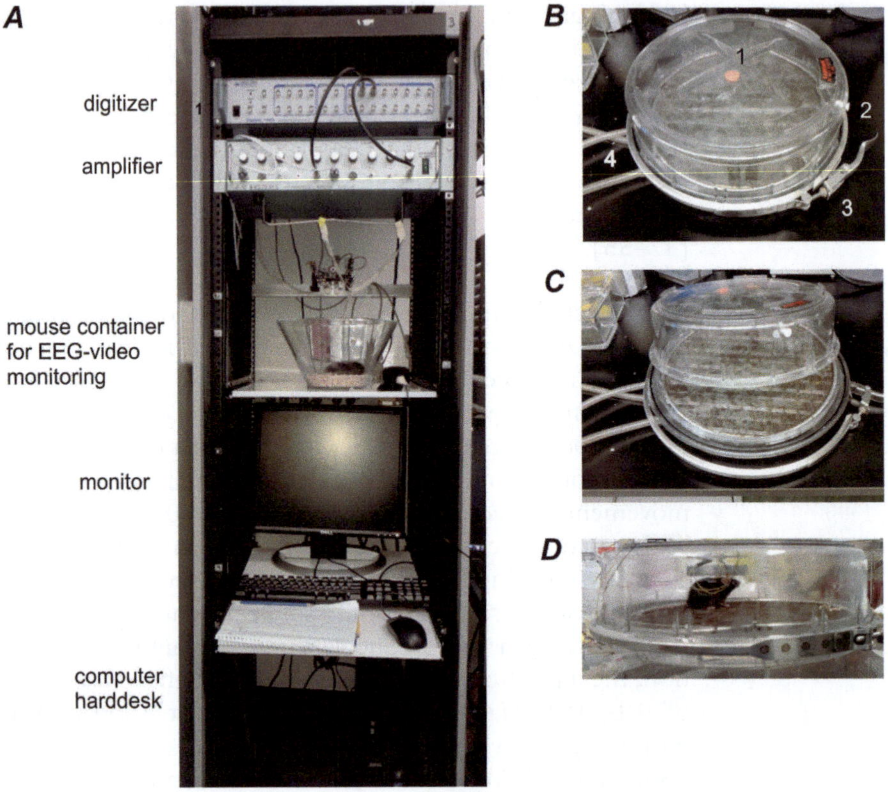

Fig. 3 EEG recording setup and hypoxia chamber used for mouse models. (**a**) An image taken from the tethered EEG recording unit. This unit consisted of (from top to bottom) a digitizer, an amplifier, amplifier-associated head-stages, a glass container with a mouse to be recorded, a webcam, and essential components of a desktop computer. (**b**, **c**) Bird's eye view of the airtight hypoxia chamber we used. The chamber has a top cover and a base, and the two parts are held tightly together by a metal clamp and handle [3]. Input–output air passages [4] are at the base. Two holes were drilled in the top cover of the chamber for passing the EEG wires [1] and thermometer probe [2]. (**d**) A lateral view of the closed chamber with a mouse placed inside

Our experience showed that local differential recordings using twisted bipolar microelectrodes were preferable to monopolar recordings in mouse models. As local differential recordings detect signal differences between adjacent microelectrode tips, they are more effective in sampling local circuitry activity and less susceptible to movement and signal artifacts relative to monopolar recordings that detect signal difference between recording and reference electrodes. However, if electrode contamination precludes the possibility of the local differential recordings, monopolar recordings can be performed using the same amplifiers (A-M systems; Table 1). For monopolar recordings, one input of the head-stage was connected to a pin of a bipolar electrode, and another input was grounded as per the manual of A-M amplifiers.

4 Brain Ischemia

4.1 Middle Cerebral Artery Occlusion

Middle cerebral artery occlusion (MCAO) using either intraluminal suture insertion or electrocoagulation has been widely used in rodent models for the induction of focal brain ischemia [38, 56–60]. We utilized both methods while varying the occlusion time to study the effect of ischemia duration and subsequent reperfusion on seizure occurrence and outcome [40]. Animals were operated on under isoflurane anesthesia, with their rectal temperatures maintained between 35.5 and 36.5 °C using an automatic heating device (Table 1). A silicon-coated fine suture (#6, Doccol Corporation; Table 1) was used to conduct the intraluminal suture insertion. A clinical-grade cautery (Bovie; Table 1) was used to conduct permanent MCAO via electrocoagulation. The procedure of the intraluminal suture insertion or electrocoagulation was guided under a Zeiss surgical microscope (Table 1).

We adopted the Koizumi's method [56] to induce a permanent MCAO. The suture was inserted into the common carotid artery and advanced through the internal carotid artery with its tip 7–8 mm distal to the carotid bifurcation. The common carotid artery and the inserted suture were then permanently ligated using a fine silk suture (Table 1). We adopted the Longa's method [56] to induce a reversible MCAO. The silicon-coated suture was inserted through the external carotid artery and advanced through the internal carotid artery as described above. After suture placement, the external carotid artery was temporarily ligated with one knot and the skin wound was loosely sutured. Animals recovered from anesthesia with the suture in for either 15 or 90 min, and were then reanesthetized to remove the inserted suture. To induce a permanent MCAO via electrocoagulation (MCAO-e), a skin incision was made to expose the skull between the right ear and eye. A small hole was drilled through the skull, and the main ascending branch of the MCA was cauterized. Animals that underwent surgical exposure of the common carotid artery without ligation or underwent skin incision without MCA electrocoagulation were used as sham controls.

4.2 Hypoxia-Ischemia

Both the permanent and reversible MCAO using the intraluminal suture insertion method were more effective than the electrocoagulation method for inducing early-onset seizures. However, a potential confounder of the intraluminal suture insertion method was subarachnoid hemorrhage due to vessel perforation [61], which may cause early-onset seizures independently. Therefore, we also utilized the alternative hypoxia-ischemia (HI) model in a separate experimental series [39, 41].

The HI model requires a permanent ligation of the common carotid artery followed by systemic hypoxia for a prespecified amount of time [62]. Advantages to this model over the MCAO model include the absence of subarachnoid hemorrhage as a possible confounder, and a shorter, less complex surgical procedure that minimizes confounding anesthetic effects, particularly on more susceptible older animals [60, 63]. However, limitations of this model include the possibility of altering cardiorespiratory activity due to the systemic hypoxic episode [57]. We addressed this by telemetrically monitoring the animals' core body temperature and electrocardiogram, in addition to bilateral EEG, and regional cerebral blood flow.

For the HI model, the animal was anesthetized with isoflurane and the right common carotid artery was isolated and ligated with a fine silk suture (Table 1). Care was taken to avoid damage to the vagus nerve and other surrounding vessels. Systemic hypoxia was conducted 1.5 h after the right common carotid artery ligation, at which time point the animal had recovered from anesthesia and exhibited "normal" behaviors such as exploring and grooming. The animal was placed in an airtight plastic chamber (Billups-Rothenberg; Table 1; Fig. 3b) and exposed to a hypoxic episode for 30 min, while its EEG signals and behavior were continuously monitored [39, 41, 47]. Sham control animals had surgical exposure of their common carotid artery without ligation and were exposed to compressed air for 30 min.

The hypoxic chamber we used had two openings for air passage (input and output) and a total volume of 5.89 cm^3. The chamber consisted of a base and a top cover. The two parts were sealed using with a releasable metal clamp (Fig. 3b–d). The chamber was filled with compressed air for baseline monitoring and then with 8% O_2–92% N_2 for the hypoxic challenge. Air tanks (K size, medical grade) with two stage regulators (Table 1) were connected to the input passage of the chamber to allow airflow at low pressure ($\leq$2 psi). The output passage of the chamber was controlled by an adjustable clamp to regulate airflow rate. A 7-mm hole was drilled into the top cover of the chamber to allowing EEG wires (or a Laser Doppler probe, *see* below) to pass through and connect to the implanted animal. A rubber cork surrounding the EEG wires was used to seal the hole and prevent air leakage. Another 2-mm hole was drilled on the side of the top cover to allow a fine thermometer probe (Table 1) pass through and record the inner chamber temperature. This hole and inserted probe were similarly sealed to prevent air leaks. A rubber bottle filled with hot water was placed underneath the chamber to warm it. During the hypoxic episode, the air temperature of the inner chamber was continuously monitored and maintained between 35–35.5 °C to prevent hypoxia-induced hypothermia [47, 62]. The goal was a 30-min hypoxic episode for each animal. However, hypoxia was terminated immediately if animals exhibited observable respiratory distress or vigorous convulsions, in keeping with our animal care guidelines.

5 Assessments of Ischemic Brain Injury

5.1 Ischemic EEG Suppression

EEG signal suppression was used as a sensitive measure of brain ischemia [22, 23]. In both the MCAO and HI models, bilateral hippocampal-cortical EEG signals were recorded before ischemia induction as a baseline, and then postischemia at different time points (hours to weeks). To quantify EEG changes over time, we used the root mean square (RMS) of the EEG signals as this has been shown to be a sensitive measure of ischemic EEG suppression in adult and aging mice [39–41]. RMS calculations were made from 30-s EEG segments collected while animals were immobile as these data segments were minimally contaminated by movement-related artifacts. Power spectra were then generated with 50% window overlap at a spectral resolution of 0.3 Hz. The RMS of the EEG power spectrum was automatically calculated in pCLAMP and normalized to corresponding baseline values for individual animals. In our experiments, a decrease in the RMS of the ipsilateral hippocampal EEG signal to $\leq$50% of the baseline at 1-h postischemia strongly correlated with the development of early-onset CS and histological brain injury [39–41].

An example of ischemic EEG suppression is shown in Fig. 4, in which ipsilateral and contralateral (in reference to the unilateral common carotid artery ligation) EEG signals collected from an aging mouse before and following the HI episode. Ipsilateral hippocampal EEG signals, but contralateral cortical EEG signals, were greatly decreased following hypoxia, suggesting that the HI model induces brain ischemia primarily in the ipsilateral hemisphere.

5.2 Monitoring Regional Cerebral Blood Flow

We used a Laser Doppler system (PF5010, Perimed; Table 1) to monitor regional cerebral blood flow (rCBF) in mouse models [39–41]. This system consisted of a fiber glass probe (MTB 500, 0.5 mm in diameter and 150 mm in length), a sensor head with extended cables, and a software-controlled central unit. rCBF signals could be collected and displayed continuously, averaged at desired time intervals, and then stored for further analysis. In addition, analog rCBF signals could be exported to a digitizer with a $\geq\pm5$ V input range, and stored and analyzed independently. In some pilot experiments, we simultaneously collected rCBF and local cortical EEG signals using the digitizer and pCLAMP system described above.

To measure CBF, the animal was anesthetized with isoflurane and its skull was surgically exposed. One end of the fiber glass probe was glued onto the skull surface using a vendor-provided kit. The probe was positioned overlying an area corresponding to the parietal cortex (bregma −2.2 to −2.8 mm and lateral 1.5–2.2 mm) ipsilateral to the MCAO or carotid occlusion (HI model). The other end of the fiber glass probe was then inserted into the sensor

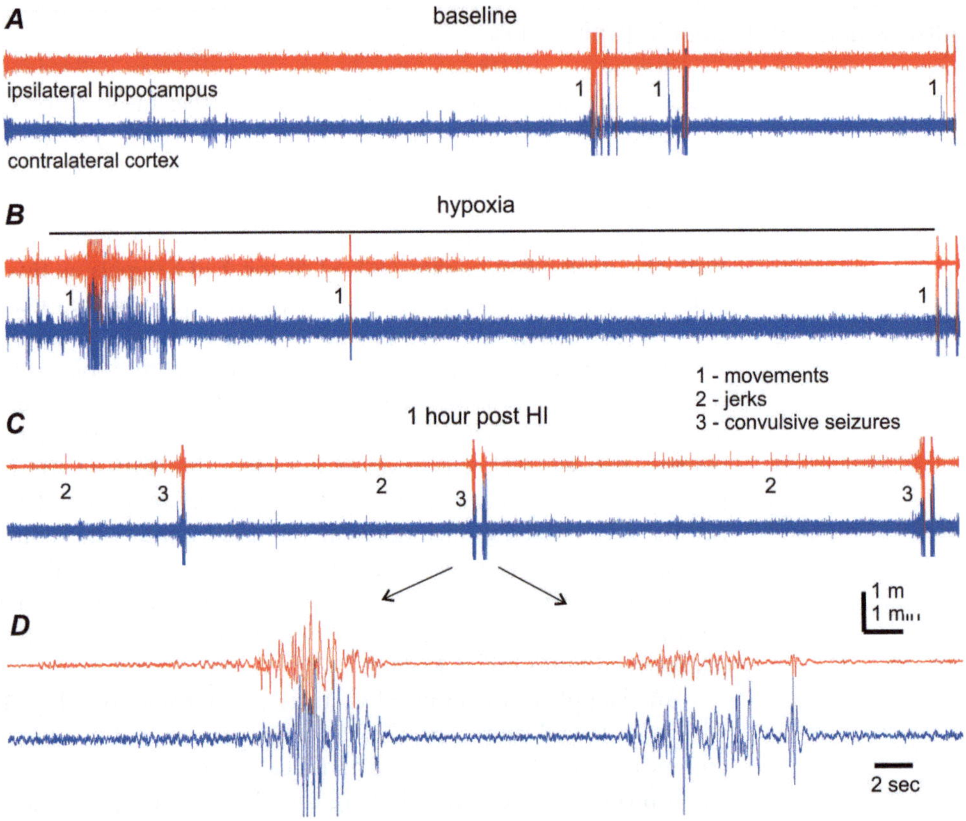

Fig. 4 Post-HI EEG suppression. EEG recordings were made from ipsilateral hippocampal (red) and contralateral cortical (blue) areas in an aging mouse. Ipsilateral and contralateral were in reference to the common carotid artery occlusion. (**a**) Representative EEG traces collected during baseline monitoring. (**b**) Traces collected 1.5 h after the common carotid occlusion and during the period of hypoxic episode (8% O_2 for 30 min). (**c**) Traces collected at 1 h post HI. (**d**) Expanded traces corresponding to indicated convulsive seizures. Note in B and C that ipsilateral EEG signals were greatly decreased following the HI episode. Note in D large convulsion-related artifacts but no evident discharge in hippocampal and cortical recordings

head. rCBF signals were averaged every 10 s to remove cardiorespiratory fluctuations. Baseline monitoring was conducted ≥1 h after probe placement when the animal had fully recovered from anesthesia. Signals collected over a 5–10-min period of behavioral immobility were used as the baseline rCBF.

To examine rCBF changes following a permanent MCAO, CBF signals were collected for 2 h following termination of the intraluminal suture insertion and data were normalized to the baseline rCBF. To examine rCBF signal changes following HI, CBF signals were collected before, during, and following hypoxia. These data were normalized to prehypoxic signals to minimize the confounding effect of compensatory changes in rCBF following the unilateral common carotid artery occlusion alone. Our data suggest a decrease in ipsilateral rCBF to ≤50% of baseline 1-h following ischemia correlates strongly with ischemic brain injury and development of postischemic seizures in both models.

5.3 Brain Histology

Brain histological assessments were conducted as previously described [39–41, 47]. The animal was anesthetized with an intraperitoneal injection of sodium pentobarbital (70 mg/kg) and transcardially perfused with 10% neutral buffered formalin solution. The brain was removed and further fixed in 10% formalin with 20% sucrose. Frozen coronal sections (30 μm thick) were obtained throughout the brain using a cryostat microtome (Table 1), and brain sections were air-dried for at least 1 week before cresyl-violet staining (Table 1). We used cresyl-violet staining in most animals to examine gross brain injury at different postischemic time points [60, 64]. Brain sections were treated sequentially with anhydrous chloroform, 100%, 95%, 90% and 75% alcohol for 3 min, and then exposed to 0.1% cresyl-violet for 2–3 min. Stained sections were briefly treated with 75% alcohol to reduce background staining and then covered by cover glasses (Table 1). Images were obtained using a Leica (DMRN) upright microscope and analyzed using the Image J software (National Institute of Health, USA). Quantitative analysis were made from brain sections at 8 different coronal levels (bregma 1.9, 1.2, 0.5, −0.2, 1.1, 1.5, 2.4 and −3.2 mm respectively) [39–41].

At 24–48 h postischemia, gross ipsilateral brain injury was denoted by regions of weak or hypochromic cresyl-violet staining (a surrogate measure of cell injury), and by an enlarged ipsilateral to contralateral hemispheric area (indicative of ipsilateral edema). At 4–5 weeks postischemia, gross brain injury was recognized as structural deformities, cavities, atrophy, and dark areas of scar tissue in the ipsilateral hemisphere, which we collectively referred to as cystic infarctions. These infarctions were associated with a decreased ipsilateral to contralateral hemispheric due to loss of brain tissue.

To quantify regions with hypochromic staining or cystic infarctions, the brightness, contrast, and color balance of the Leica microscope captured images were adjusted to clearly demarcate the boundaries of these regions. The injured ipsilateral brain areas were normalized as a percentage of the total ipsilateral hemispheric area at eight coronal levels, and the means from multiple coronal levels were presented for each animal. To differentially quantify both ipsilateral edema and tissue loss, the area of the total ipsilateral hemisphere was normalized as a percentage of the corresponding contralateral hemisphere at the same coronal level. All analyses at multiple coronal levels were averaged for each animal in order to adequately demonstrate the extent and variability of the brain injury observed.

Fluoro-Jade positive signals, a marker of degenerating cells [65], were examined 24-h postischemia [41]. Fluoro-Jade C

(Histo-Chem, Inc.; Table 1) staining was carried out in accordance with the manufacturer's instructions and protocols established in our previous work [39, 47]. Fluoro-Jade positive cells were counted in the striatal, dorsal hippocampal, and caudolateral cortex at three coronal levels (bregma 0.5 mm, −1.5 mm and −3.2 mm) and cell counts from each level were averaged for each animal. These areas were selected based on the greater degree of brain injury consistently observed on cresyl-violet staining. Images were acquired under 20× magnification, resulting in individual imaging fields of 0.13 mm^2 each.

The brain histological assessments in cresyl-violet staining sections also served to confirm the tracks of implanted EEG electrodes in their correct neuroanatomical locations (Fig. 5c, d). One of the limitations of our histology protocol however was the extensive time and labor required to perform frozen sectioning, staining, image acquisition, and analysis.

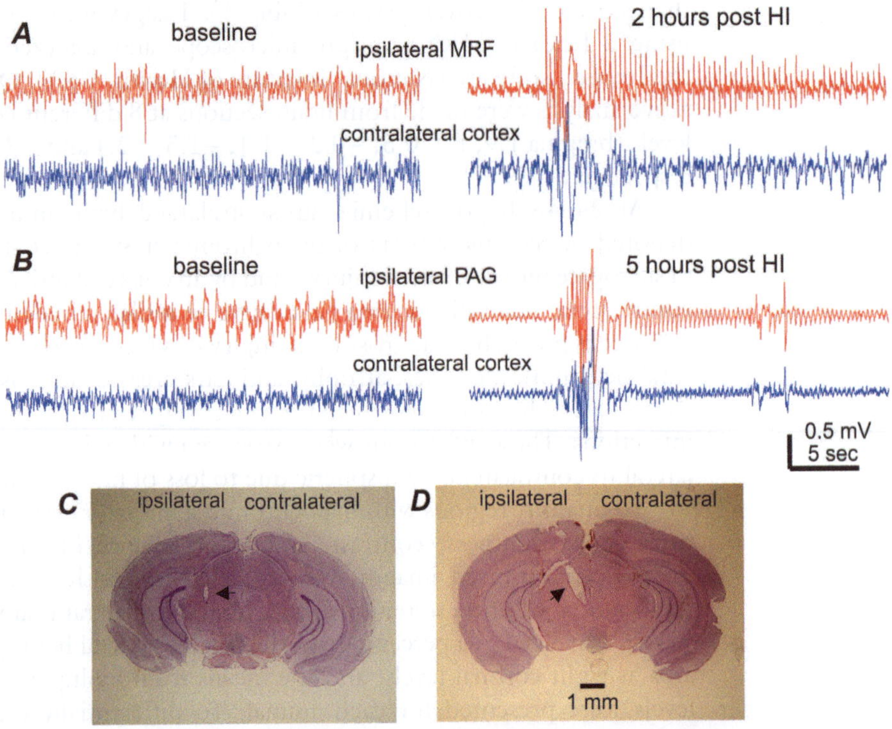

Fig. 5 CS-concurrent brainstem EEG discharges. EEG traces were collected from two adult mice during baseline monitoring and 2–5 h following HI episodes. (**a**) EEG recordings were made from ipsilateral midbrain reticular formation (MRF, bregma −3.52 mm, lateral 1.5 mm, and depth 3.0 mm; red) and contralateral cortex (blue). (**b**) recordings made from ipsilateral periaqueductal gray matter (PAG; bregma −3.5 mm, lateral 0.6 mm, and depth 2.6 mm) and contralateral cortex. Note that post-HI discharges were evident in the MRF and PAG areas but not in corresponding cortical recordings. (**c**, **d**) Images of cresyl violet-stained sections were taken from other 2 adult mice. The tracks of implanted MRF and PAG electrodes were indicated by arrows

6 Detection of Early-Onset Seizures

6.1 Seizure Monitoring Protocols

We utilized both visual behavioral state monitoring and EEG recordings to detect early-onset convulsive seizures (CS) and non-convulsive seizures (NCS) following ischemia [39–41]. Animals were initially placed under continuous visual surveillance by experimenters with concurrent EEG recordings for the first 4–6 h immediately post-MCAO or post-HI, the time window during which they were most likely to develop seizures. This initial period was also optimal for anticonvulsant treatment under specific experimental protocols (*see* below).

After this initial period of close monitoring, animals were monitored overnight using high-resolution video surveillance for an additional 10–14 h. We utilized a Logitech C615 webcam with autofocus, adjustable screen resolution up to 1280 × 720 pixels, built-in microphone, and a tripod base for these recordings (Table 1). The overnight video monitoring was sensitive for detecting delayed onset CS or recurrent CS following a seizure-free period after anticonvulsant treatment, but with a drawback of lacking concurrent EEG monitoring.

Following this period of overnight video monitoring, live monitoring was resumed the following day and then again at predetermined serial time points later when animal survival permitted. Although these subsequent monitoring periods were essential for assessing the animals' overall activity level and state of health, they were insufficient to reliably detect late-onset seizures, which was one of limitations of our studies. Future experiments that conduct continuous EEG and video monitoring from postischemic animals may be of great help to address this issue (*see* below).

6.2 Features of Postischemic CS

The CS observed were virtually identical in morphology between adult and aging mice and in the MCAO model versus the HI model. CS behavior included repeated jumping, fast running, barrel-rolling ($\geq$360° rotations along the trunk axis), and/or falling (loss of the righting reflex) with limb spasms. Most animals developed their first 2–3 CS within 3-h of ischemia termination, and CS tended to recur without anticonvulsant treatment. Interestingly, in several pilot experiments, CS were not accompanied by any evident EEG discharges from the hippocampal or cortical areas (Fig. 4b–d) but rather with discharges from specific brainstem areas (unpublished data; Fig. 5).

Animals with CS had a significantly higher rate of acute mortality, defined as spontaneous death or mandatory euthanasia within 48-h of ischemia termination, compared to those that remained seizure free. On later histological analysis, all animals noted to have had CS also demonstrated severe brain injury in the ipsilateral hemisphere as described above. Based on these

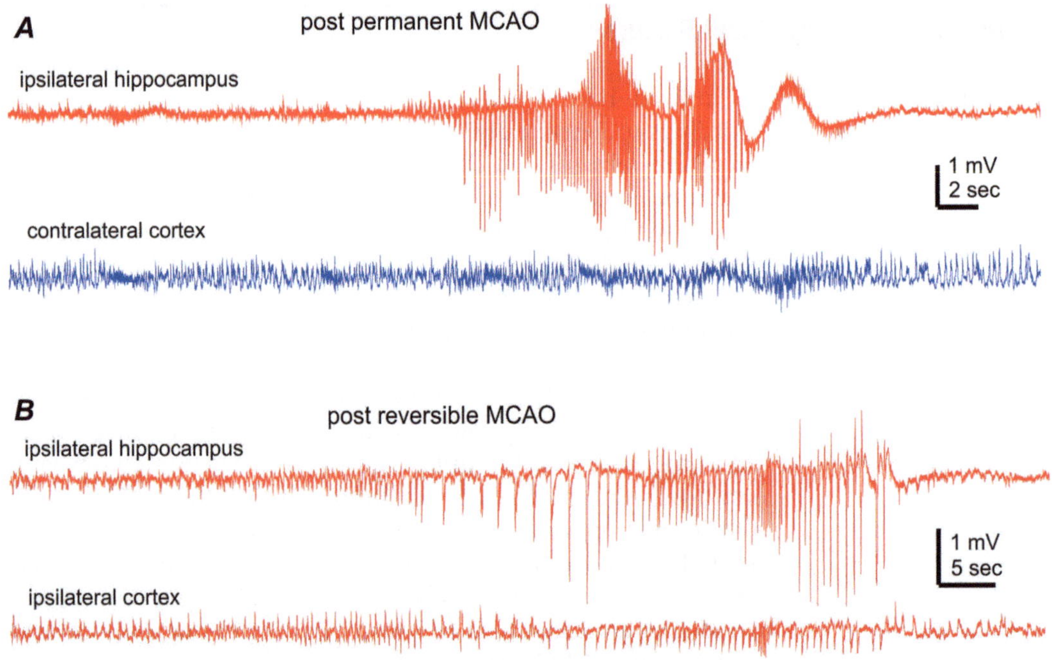

Fig. 6 MCAO-induced nonconvulsive EEG discharges. EEG traces were collected from two aging mice following permanent (**a**) and reversible (**b**) MCAO episodes via intraluminal suture insertion. (**a**) Recordings were made from ipsilateral hippocampal and contralateral cortical areas. Ipsilateral and contralateral were in reference to the intraluminal suture insertion. Note that robust ipsilateral discharge was not concurrent with contralateral cortical discharge. (**b**) Recordings made from ipsilateral hippocampal and cortical areas. Note that discharges were concurrent in ipsilateral hippocampal and cortical recordings. In both cases, these discharges were recorded while the animals were immobile or in a "freezing" behavior

observations, it logically follows that CS may be an early consequence of severe brain ischemia and a poor prognostic factor. Furthermore, seizure genesis may arise initially from subcortical structures with secondary spread. A potential site of seizure genesis within the brainstem may contribute to the disruption of cardiorespiratory centers and lead to some of the sudden deaths observed shortly after CS.

6.3 Features of Postischemic NCS

NCS were also observed in both adult and aging mice following MCAO [40] (Fig. 6). These occurred during behavioral immobility and presented with ictal-like EEG discharges with repetitive single and poly-spike waveforms lasting ≥ 5 s at an amplitude ≥ 2 times that of the baseline signal. These EEG discharges were detectable in all microelectrode leads including the bilateral hippocampus and cortex. However, ipsilateral hippocampal discharges were sometimes observed in the absence of any cortical discharges (unpublished data; Fig. 6a). This is altogether not unsurprising since the hippocampus is known to be a seizure prone structure and its blood supply from the posterior cerebral artery has been shown to be affected in mouse models of MCAO [66].

Interestingly, all animals who developed NCS also developed subsequent CS, although not all animals with CS had preceding NCS. The animals that eventually developed CS had worse outcomes than those with NCS alone. These findings suggest that the hippocampal circuitry may play an important role in the development of NCS, which in turn may be predictive of eventual CS development. Antiepileptic treatment following detection of NCS prior to any CS development may prevent subsequent CS and provide some measure of neuroprotection in adult and aging mice post-MCAO (*see* below). However, this type of treatment regimen needs to be further evaluated for feasibility and reproducibility.

Interestingly, early-onset NCS were not observed in the HI model [39, 41]. The greater degree of EEG suppression in the HI model immediately following systemic hypoxia with 8% O_2 may have suppressed any subsequent ictal discharges [39, 41]. Future studies should investigate whether NCS appear if a more moderate degree of hypoxia (12–15% O_2) was utilized instead.

7 Anticonvulsant Treatments

7.1 Treatments After CS

In our studies [39–41], animals with CS following either HI or MCAO were treated with clinically approved anticonvulsant medications as per our animal care ethics guidelines. Adult mice were treated with diazepam and phenytoin while aging mice were treated with lorazepam and fosphenytoin as per clinical recommendations for the treatment of status epilepticus [6, 67, 68]. Both diazepam and lorazepam are allosteric potentiators of the GABAa receptor. Diazepam acts more rapidly but lorazepam has a longer duration of effect. Phenytoin and fosphenytoin on the other hand both act to block the voltage gated Na^+ channels, but the latter has fewer side effects clinically. These drugs were chosen to be used in combination for their synergistic effect to simultaneously enhance GABAergic inhibition while reducing Na^+ channel mediated hyperexcitability to achieve seizure control. Diazepam, lorazepam, phenytoin, and fosphenytoin were obtained in clinically available injectable forms (Sandoz Canada Inc., and Erfa Canada Inc., Quebec, Canada; Table 1). These medications were diluted in saline and administered via intraperitoneal injections at dosages of 1.5 mg/kg for diazepam and lorazepam, 20 mg/kg for phenytoin and 30 mg/kg for fosphenytoin.

In both our MCAO and HI experiments, a combination of either diazepam and phenytoin or lorazepam and fosphenytoin was administered immediately following the first two observed CS. CS-free animals remained untreated. The purpose of using the combination of anticonvulsants instead of a single agent was to ensure complete seizure suppression to better assess the effect of seizure suppression on animal outcome, including mortality. However, while treatment given after CS provided temporary seizure control for approximately 4–6 h, CS invariably recurred, usually during overnight video monitoring. The acute mortality rate of these treated mice, both adult and aging, was 63–90%, the same as their untreated counterparts.

7.2 Prophylactic Treatments

In a separate cohort of adult and aging mice, we utilized prophylactic anticonvulsant treatment to determine if there would be any difference in outcome [39, 41]. In these experiments, the same combination of either diazepam and phenytoin or lorazepam and fosphenytoin at the same doses as above were administered within 5-min of hypoxia termination, prior to any CS occurring. It was found that this type of prophylactic treatment effectively prevented seizure development and reduced acute mortality. CS were rarely observed in the first few hours postischemia and acute mortality for this cohort of animals was in the range of 13–58%, significantly lower than for the animals treated only after having developed CS. These observations are in keeping with previous studies in adult rats [27, 28] and may suggest that inhibition of postischemic seizures prophylactically may have a neuroprotective effect. However, due to the well-known side effects of anticonvulsants/antiepileptics and the lack of evidence for the prevention of late-onset post-stroke seizures, prophylactic treatment is not currently recommended clinically [6–8, 19, 20].

8 Notes

1. Limitations of modeling postischemic seizures in C57BL/6 mice

   The posterior communicating arteries are often underdeveloped or hypoplastic in C57BL/6 mice. This inherited cerebrovascular defect may contribute to brain injury of greater severity in these mice following periods of global or focal ischemia due to decreased collateral flow from the posterior circulation in the face of a compromised anterior circulation [39–41, 69–74]. Since the risk of developing postischemic seizures increases with stroke severity, this specific cerebrovascular defect may contribute to a higher incidence of severe brain injury and postischemic seizures in our experiments [39–41] than what has been observed clinically [1–9].

   Our animal models were focused on evaluating early-onset postischemic seizures, but may be inadequate or inappropriate for assessing late-onset seizures. From our experience, animals that survived several weeks after MCAO or HI and remained in good physical condition rarely exhibited any spontaneous EEG discharges or late-onset CS during intermittent monitoring [39–41]. In a study evaluating late-onset seizures following neonatal HI, a separate cohort of C57BL/6 mice underwent HI on postnatal day 7 and was monitored visually and with EEG for seizure development intermittently until 10–11 months of age [54]. Only 10% of all animals had observable CS under continuous video monitoring (24 h daily for 10 consecutive days at a time). However, the intermittent EEG recording and video monitoring utilized in our experiments may miss seizure events during periods where animals are not monitored and therefore is methodologically inadequate to reliably detect late-onset, postischemic seizures. Previous studies have also suggested that C57BL/6 mice may be less vulnerable than other strains to kainite-induced hippocampal injury and subsequently less susceptible to the development of late-onset seizures and epilepsy [75–77]. Future experiments using uninterrupted EEG and video monitoring over an extended period (78, 79; *see* also below) in addition to experimentation with different mouse strains are needed to further investigate postischemic, late-onset epileptic seizures.

2. Issues about EEG monitoring of seizure activity in mouse models

   Telemetric EEG recordings

   A major disadvantage of tethered EEG recordings in free moving animals is movement-related artifacts. Interpretation of EEG signals during CS events was sometimes difficult due to contamination by these artifacts. The use of a physical restrainer for mice reduced movement artifacts related to running and jumping but did not eliminate them altogether and did not affect artifacts caused by barrel rolling [39–41]. To address these concerns, we supplemented tethered recordings with telemetric EEG recording in adult and aging mice using a mouse-specific transmitter (ETA-F10) and data acquisition system from Data Scientific International (Table 1) [39, 41, 49]. This transmitter had a built-in temperature sensor, a motion sensor and a biopotential channel, and permitted continuous data collection for ~60 days. The transmitter was implanted in the animal's intraperitoneal cavity and wires leading to cortical microelectrodes were tunneled to the head subcutaneously from the site of transmitter implantation. The telemetric EEG signals were minimally affected by movement artifacts, even during CS [41, 48]. These observations are in keeping with previous studies in rats [78, 79] that recommend telemetric EEG recording and concurrent video monitoring as the ideal combined modality for seizure detection. However, the transmitter we used was relatively large, weighed 2 grams, and produced a low signal transmission rate (200 Hz). These factors may restrict its ideal use to capturing preferentially higher frequency EEG signals in mice with a high body weight. Future technological advances that will miniaturize the implantable transmitter and increase both the biopotential channels available and the signal transmission rate will be helpful.

3. Continuous tethered EEG recordings

   Another complication of tethered EEG recordings in free moving animals is that wires become tangled due to animal movement. Wire tangling can disrupt EEG recordings and/or restrict an animal's normal behavior. Commutators or swivels that act as rotating electrical connectors are commonly used to overcome or minimize wire tangling. EEG commutators for rodent models are commercially available but they are manufactured to match specific recording apparatus at relatively high costs. Therefore, we used a small 6-wire slip ring (part # 1528–1152-ND, Digikey; Table 1; Fig. 7a) as an EEG commutator for our recording system.

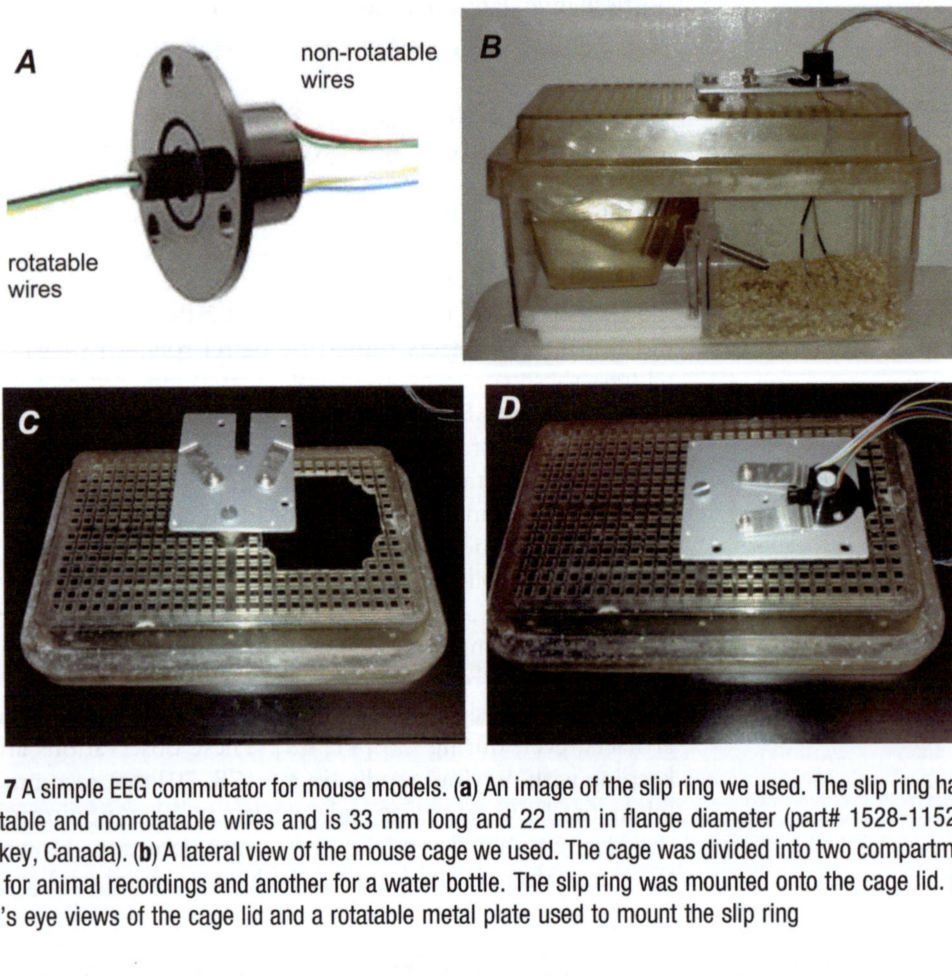

Fig. 7 A simple EEG commutator for mouse models. (**a**) An image of the slip ring we used. The slip ring has six rotatable and nonrotatable wires and is 33 mm long and 22 mm in flange diameter (part# 1528-1152-ND, Digikey, Canada). (**b**) A lateral view of the mouse cage we used. The cage was divided into two compartments, one for animal recordings and another for a water bottle. The slip ring was mounted onto the cage lid. (**c**, **d**) Bird's eye views of the cage lid and a rotatable metal plate used to mount the slip ring

Rotatable wires of the slip ring were soldered to female connecting pins (Table 1) that connected to animal's implanted headset and corresponding nonrotatable wires were connected to the EEG amplifier head-stage. Five rotatable and nonrotatable wires were used for animals that were implanted with two twisted bipolar electrodes and one reference electrode. Unused wires were cut to reduce noise. Since the original wires of the slip ring were relatively rigid, only one original rotatable wire was used and the others were replaced with soft wires to avoid restriction on cage activities of the animals. The soft wires were detached from a 24-conductor cable (Table 1).

A standard mouse cage was modified for continuous EEG recordings. The cage was divided by a plastic glass into two compartments, one for animal housing/recording and another for placing a water bottle (Fig. 7b). A narrow cut was made in the plastic glass allowing the water bottle outlet to pass through. Food pellets were provided on floor of the recording compartment. An opening (about 60 mm × 50 mm) was made in a cage lid to allow placement of an animal. A plastic glass plate, screwed on the cage lid but movable horizontally, was used to cover the lid opening and to accommodate the slip ring (Fig. 7c, d). The plastic glass plate had a rectangular cut (10 mm × 40 mm) which supported the flange of the slip ring and allowed rotatable wires of the slip ring to pass through. Two small metal bars, screwed onto the plastic glass plate but movable horizontally, were used to secure the slip ring by pressing onto its flange.

To set up continuous EEG monitoring, an implanted animal was connected to rotatable wires of the slip ring and placed into the recording compartment through the cage lid opening. The plastic glass plate screwed onto the lid was moved back to cover the lid opening and the slip ring was then secured onto the plastic glass plate by the two small metal bars. Nonrotatable wires of the slip ring were connected to amplifier's head-stages, which were placed on top of the cage lid. The webcam described above (Table 1) was placed near the recording compartment of the cage for monitoring of animal motor behaviors. Webcam monitoring in the light-off period was done under a dim light condition.

EEG recordings were conducted similarly as described above. To avoid complications of saving and analyzing large data files, EEG and video data were saved every 2 h using the cursor autoclick function of Mini Mouse Macro program (Table 1). A 10-s interval between recording sessions was intentionally given to ensure proper data storage. Data collection was stopped approximately 30 min daily for the purpose of animal care.

We tested the above apparatus in four C57 black mice (8–9 months of ages and 25–34 g in body weight). An extended kindling protocol [80] was adopted to induce spontaneous recurrent seizures in these mice. Unilateral hippocampal stimulation was conducted using a Grass stimulator through an isolation unit (Table 1). Daily electrical stimulation consisted of repetitive square current pulses at 60 Hz for 2 s, with pulse duration of 0.5 ms and intensity at 125% of initial after-discharge threshold as determined using an ascending protocol [52]. The four animals experienced 95–110 daily stimulation prior to continuous EEG-webcam monitoring.

Spontaneous recurrent seizures were observed from all 4 extendedly kindled mice. When monitored 24 h daily for 5–9 consecutive days, these animals exhibited 3–8 spontaneous seizures per day but no apparent restriction on their baseline cage activity. These seizures featured hippocampal discharges lasting 30–80 s (Fig. 8) and concurrent convulsive behaviors included rearing, forelimb clonus, and/or falling. EEG signals

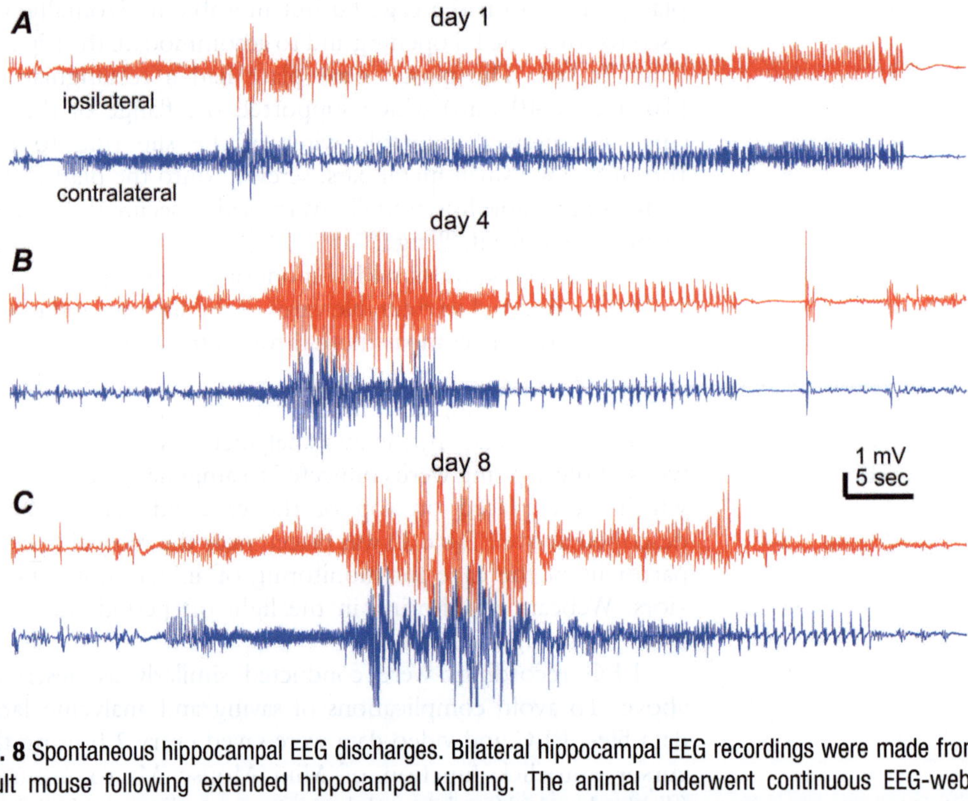

Fig. 8 Spontaneous hippocampal EEG discharges. Bilateral hippocampal EEG recordings were made from an adult mouse following extended hippocampal kindling. The animal underwent continuous EEG-webcam monitoring 24 h daily for 9 consecutive days. (**a**–**c**) One episode of bilateral discharges collected in the 1st, fourth and eighth day respectively. “Ipsilateral” (red) and “contralateral” (blue) were in reference to the site of hippocampal kindling stimulation. These hippocampal discharges were concurrent with convulsive behaviors including head nodding, forelimb clonus, tail erection and/or fall with forelimb clonus and hind-limb extension. Note large convulsion-related large artifacts in the middle of discharges in **b**, **c**

collected through the slip ring did not appear to be associated with high electronic noise but movement-related artifacts were evident in these EEG signals. Moreover, 1–2 of the rotatable wires became disconnected from the implanted electrodes in three animals after 5–9 days of recordings. Further improvements of the above described apparatus are needed to minimize movement-related artifacts and achieve more stable connections between slip ring inputs and implanted electrodes. Additionally, a smaller slip ring (part# 1528–1157-ND, Digikey; Table 1) remains to be tested for continuous EEG recordings in mice with lower body weights.

4. Optimization of ischemic episodes and anticonvulsant treatment for aging mice

   Our previous experiments [40, 41] demonstrated that aging mice with early-onset CS following either MCAO or HI episodes had a high rate of acute mortality despite anticonvulsant treatment after observation of seizures. Considering the increased vulnerability of aging animals to ischemic brain injury, other health related complications that may increase their susceptibility to brain ischemia [60, 63], and the observed beneficial effect of prophylactic anticonvulsant medication [27, 28, 39, 41], it may have been possible that the severity of ischemia produced disproportionately worse outcomes. Therefore, ischemia of lesser severity or a shorter period of ischemia may have been sufficient to reproduce early-onset seizures, without the higher attendant risk of mortality and poor outcomes. For example, a reversible MCAO (via intra luminal suture insertion) for $\leq$30 min or an HI episode with 12–15% O_2 may have sufficed. Furthermore, considering that NCS, whenever they were present, were a predictor of CS development and preceded CS by 5–30 min [40], it may be prudent to establish a future protocol with anticonvulsants/antiepileptics that are given immediately after the first 1–2 NCS events. Successfully suppressing the development of CS may help to decrease mortality in this group of animals while maintaining the ability to examine NCS.

Acknowledgments

This research was supported by the Canadian Institute of Health Research, the Natural Sciences and Engineering Research Council of Canada and the Epilepsy Research Program of the Ontario Brain Institute. The authors thank Ms. Nila Sivanenthiran for editing the manuscript.

References

1. Bladin CF, Bornstein N (2009) Post-stroke seizures. In: Fisher M (ed) Stroke Part II: Clinical manifestations and pathogenesis. Handbook of clinical neurology, vol 93. Elsevier B. V., pp 613–621
2. Brodie MJ, Elder AT, Kwan P (2009) Epilepsy in later life. Lancet Neurol 8:1019–1030
3. Menon B, Shorvon SD (2009) Ischaemic stroke in adults and epilepsy. Epilepsy Res 87:1–11
4. Chen RL, Balami JS, Esiri MM et al (2010) Ischemic stroke in the elderly: an overview of evidence. Nat Rev Neurol 6:256–265
5. Balami JS, Chen RL, Grunwald IQ et al (2011) Neurological complications of acute ischaemic stroke. Lancet Neurol 10:357–371
6. Gilad R (2012) Management of seizures following a stroke. What are the options? Drugs Aging 29:533–538
7. Guekht A, Bornstein NM (2012) Seizures after stroke. In: Stefan H, Theodore WH (eds) Epilepsy. Handbook of clinical neurology, vol 108. Elsevier (3rd series), pp 570–583
8. Procaccianti G, Zaniboni A, Rondelli F et al (2012) Seizures in acute stroke: incidence, risk factors and prognosis. Neuroepidemiology 39:45–50
9. Guth JC, Gerard EE, Nemeth AJ (2014) Subarachnoid extension of hemorrhage is associated with early seizures in primary intracerebral hemorrhage. J Stroke Cerebrovasc Dis 23:2809–2813
10. Mani R, Schmitt SE, Mazer M et al (2012) The frequency and timing of epileptiform activity on continuous electroencephalogram in comatose post-cardiac arrest syndrome patients treated with therapeutic hypothermia. Resuscitation 83:840–847
11. Rittenberger JC, Popescu A, Brenner RP et al (2012) Frequency and timing of nonconvulsive status epilepticus in comatose post-cardiac arrest subjects treated with hypothermia. Neurocrit Care 16:114–122
12. Knight WA, Hart KW, Adeoye OM et al (2013) The incidence of seizures in patients undergoing therapeutic hypothermia after resuscitation from cardiac arrest. Epilepsy Res 106:396–402
13. Hofmeijer J, Tjepkema-Cloostermans MC, Blans MJ, Beishuizen A et al (2014) Unstandardized treatment of electroencephalographic status epilepticus does not improve outcome of comatose patients after cardiac arrest. Front Neurol 5:39
14. Waterhouse EJ, Vaughan JK, Barnes TY et al (1998) Synergistic effect of status epilepticus and ischemic brain injury on mortality. Epilepsy Res 29:175–183
15. Vespa PM, O'Phelan K, Shah M et al (2003) Acute seizures after intracerebral hemorrhage: a factor in progressive midline shift and outcome. Neurology 60:1441–1446
16. Szaflarski JP, Rackley AY, Kleindorfer DO et al (2008) Incidence of seizures in the acute phase of stroke: a population-based study. Epilepsia 49:974–981
17. Burneo JG, Fang J, Saposnik G et al (2010) Impact of seizures on morbidity and mortality after stroke: a Canadian multi-centre cohort study. Eur J Neurol 17:52–58
18. Huang CW, Saposnik G, Fang J et al (2014) Influence of seizures on stroke outcomes: a large multicenter study. Neurology 82:768–776
19. Kulhari A, Strbian D, Sundararajan S (2014) Early onset seizures in stroke. Stroke 45: e249–e251
20. Sykes L, Wood E, Kwan J (2014) Antiepileptic drugs for the primary and secondary prevention of seizures after stroke. Cochrane Database Syst Rev (1):CD005398
21. Mecarelli O, Pro S, Randi F et al (2011) EEG patterns and epileptic seizures in acute phase stroke. Cerebrovasc Dis 31:191–198
22. Jordan KG (2004) Emergency EEG and continuous EEG monitoring in acute ischemic stroke. J Clin Neurophysiol 21:341–352
23. Claassen J, Jetté N, Chum F et al (2007) Electrographic seizures and periodic discharges after intracerebral hemorrhage. Neurology 69:1356–1365
24. Silverman IE, Restrepo L, Mathews GC (2014) Post stroke seizures. Arch Neurol 59:195–201
25. Chung JM (2014) Seizures in the acute stroke setting. Neurol Res 36:403–406
26. Hartings JA, Williams AJ, Tortella FC (2003) Occurrence of nonconvulsive seizures, periodic epileptiform discharges, and intermittent rhythmic delta activity in rat focal ischemia. Exp Neurol 179:139–149
27. Williams AJ, Tortella FC, Lu XM et al (2014) Antiepileptic drug treatment of nonconvulsive seizures induced by experimental focal brain ischemia. J Pharmacol Exp Ther 311:220–227
28. Williams AJ, Bautista CC, Chen RW et al (2006) Evaluation of gabapentin and ethosuximide for treatment of acute nonconvulsive seizures following ischemic brain injury in rats. J Pharmacol Exp Ther 318:947–955
29. Karhunen H, Nissinen J, Sivenius J et al (2006) A long-term video-EEG and behavioral follow-

up after endothelin-1 induced middle cerebral artery occlusion in rats. Epilepsy Res 72:25–38
30. Lu XC, Si Y, Williams AJ et al (2009) NNZ-2566, a glypromate analog, attenuates brain ischemia-induced non-convulsive seizures in rats. J Cereb Blood Flow Metab 29:1924–1932
31. Cuomo O, Rispoli V, Leo A et al (2013) The antiepileptic drug levetiracetam suppresses non-convulsive seizure activity and reduces ischemic brain damage in rats subjected to permanent middle cerebral artery occlusion. PLoS One 8:e80852
32. Karhunen H, Bezvenyuk Z, Nissinen J et al (2007) Epileptogenesis after cortical photothrombotic brain lesion in rats. Neuroscience 148:314–324
33. Paz JT, Davidson TJ, Frechette ES et al (2013) Closed-loop optogenetic control of thalamus as a tool for interrupting seizures after cortical injury. Nat Neurosci 16:64–70
34. Han F, Boller M, Guo W et al (2010) A rodent model of emergency cardiopulmonary bypass resuscitation with different temperatures after asphyxial cardiac arrest. Resuscitation 81:93–99
35. Boller M, Jung SK, Odegaard S et al (2011) A combination of metabolic strategies plus cardiopulmonary bypass improves short-term resuscitation from prolonged lethal cardiac arrest. Resuscitation 82(Suppl 2):S27–S34
36. Kelly KM, Kharlamov A, Hentosz TM et al (2001) Photothrombotic brain infarction results in seizure activity in aging Fischer 344 and Sprague Dawley rats. Epilepsy Res 47:189–203
37. Kelly KM, Shiau DS, Jukkola PI et al (2011) Effects of age and cortical infarction on EEG dynamic changes associated with spike wave discharges in F344 rats. Exp Neurol 232:15–21
38. Fluri F, Schuhmann MK, Kleinschnitz C (2015) Animal models of ischemic stroke and their application in clinical research. Drug Des Devel Ther 9:3445–3454
39. El-Hayek YH, Wu CP, Chen R et al (2011) Acute post-ischemic seizures are associated with increased mortality and brain damage in adult Mice. Cereb Cortex 21:2863–2875
40. Wu CP, Wang J, Peng J et al (2015) Modeling early-onset post-ischemic seizures in aging mice. Exp Neurol 271:1–12
41. Wang J, Wu CP, Peng J et al (2015) Early-onset convulsive seizures induced by brain hypoxia-ischemia in aging mice: effects of anticonvulsive treatments. PLoS One 10(12):e0144113
42. Flurkey K, Currer JM, Harrison DE (2007) Mouse models in aging research. In: Fox JG, Davisson MT, Quimby FW et al (eds) The mouse in biomedical research. Academic Press, New York, pp 637–672
43. Wu CP, Wais M, Sheppy E et al (2008) A glue-based, screw-free method for implantation of intra-cranial electrodes in young mice. J Neurosci Meth 171:126–131
44. Wu CP, Wais M, Zahid T et al (2009) A reliable method for intracranial electrode implantation and chronic electrical stimulation in the mouse brain. Behav Res Meth 41:736–741
45. Jeffrey M, Lang M, Gane J et al (2013) An improved screw-free method for electrode implantation and intracranial electroencephalographic recordings in mice. BMC Neurosci 14:82
46. Franklin KBJ, Paxinos G (1997) The mouse brain in stereotaxic coordinates. Academic Press, San Diego, CA
47. Wais M, Wu CP, Zahid T et al (2009) Repeated hypoxic episodes induce seizures and alter hippocampal network activities in mice. Neuroscience 161:599–613
48. D'Cruz JA, Wu CP, Zahid T et al (2010) Alterations of cortical and hippocampal EEG activity in MeCP2-deficient mice. Neurobiol Dis 38:8–16
49. Wither RG, Colic S, Wu CP et al (2012) Daily rhythmic behavior and thermoregulatory patterns are disrupted in female MeCP2-deficient mice. PLoS One 7:e35396
50. El-Hayek YH, Wu CP, Zhang L (2011) Early suppression of intracranial EEG signals predicts ischemic outcome in adult mice following hypoxia-ischemia. Exp Neurol 231:295–303
51. El-Hayek YH, Wu CP, Ye H et al (2013) Hippocampal excitability is increased in aged mice. Exp Neurol 247:710–719
52. Jeffrey M, Lang M, Gane J et al (2014) Novel anticonvulsive effects of progesterone in a mouse model of hippocampal electrical kindling. Neuroscience 257:65–75
53. Lang M, Wither RG, Colic S et al (2014) Rescue of behavioral and EEG deficits in male and female Mecp2-deficient mice by delayed Mecp2 gene reactivation. Hum Mol Genet 23:303–318
54. Peng J, Li R, Arora N et al (2015) Effects of neonatal hypoxic-ischemic episodes on late seizure outcomes in C57 black mice. Epilepsy Res 111:142–149
55. Zhang L, Wither RG, Lang M et al (2016) A role for diminished GABA transporter activity in the cortical discharge phenotype of MeCP2-

deficient mice. Neuropsychopharmacology 41:1467–1476
56. Durukan A, Tatlisumak T (2007) Acute ischemic stroke: overview of major experimental rodent models, pathophysiology, and therapy of focal cerebral ischemia. Pharmacol Biochem Behav 87:179–197
57. Hossmann KA (2008) Cerebral ischemia: models, methods and outcomes. Neuropharmacology 55:257–270
58. Sicard KM, Fisher M (2009) Animal models of focal brain ischemia. Exp Transl Stroke Med 1:7
59. Howells DW, Porritt MJ, Rewell SS et al (2010) Different strokes for different folks: the rich diversity of animal models of focal cerebral ischemia. J Cereb Blood Flow Metab 30:1412–1431
60. Liu FD, McCullough LD (2011) Middle cerebral artery occlusion model in rodents: methods and potential pitfalls. J Biomed Biotechnol. https://doi.org/10.1155/2011/464701
61. Schmid-Elsaesser R, Zausinger S, Hungerhuber E et al (1998) A critical reevaluation of the intraluminal thread model of focal cerebral ischemia: evidence of inadvertent premature reperfusion and subarachnoid hemorrhage in rats by laser-Doppler flowmetry. Stroke 29:2162–2170
62. Vannucci SJ, Willing LB, Goto S et al (2001) Experimental stroke in the female diabetic, db/db, mouse. J Cereb Blood Flow Metab 21:52–60
63. Popa-Wagner A, Buga AM, Kokaia Z (2011) Perturbed cellular response to brain injury during aging. Ageing Res Rev 10:71–79
64. Zille M, Farr TD, Przesdzing I et al (2011) Visualizing cell death in experimental focal cerebral ischemia: promises, problems, and perspectives. J Cereb Blood Flow Metab 32:213–231
65. Schmued LC, Stowers CC, Scallet AC et al (2005) Fluoro-Jade C results in ultra high resolution and contrast labeling of degenerating neurons. Brain Res 1035:24–31
66. El-Amki M, Clavier T, Perzo N et al (2015) Hypothalamic, thalamic and hippocampal lesions in the mouse MCAO model: potential involvement of deep cerebral arteries? J Neurosci Meth 254:80–85
67. Meierkord H, Boon P, Engelsen B et al (2010) EFNS guideline on the management of status epilepticus in adults. Eur J Neurol 17:348–355
68. Shorvon S (2011) The treatment of status epilepticus. Curr Opin Neurol 24:165–170
69. Fujii M, Hara H, Meng W et al (1997) Strain-related differences in susceptibility to transient forebrain ischemia in SV-129 and C57black/6 mice. Stroke 28:1805–1811
70. Kitagawa K, Matsumoto M, Yang G et al (1998) Cerebral ischemia after bilateral carotid artery occlusion and intraluminal suture occlusion in mice: evaluation of the patency of the posterior communicating artery. J Cereb Blood Flow Metab 18:570–579
71. Ozdemir YG, Bolay H, Erdem E et al (1999) Occlusion of the MCA by an intraluminal filament may cause disturbances in the hippocampal blood flow due to anomalies of circle of Willis and filament thickness. Brain Res 822:260–264
72. Majid A, He YY, Gidday JM et al (2003) Differences in vulnerability to permanent focal cerebral ischemia among 3 common mouse strains. Stroke 31:2707–2714
73. McColl BW, Carswell HV, McCulloch J et al (2004) Extension of cerebral hypoperfusion and ischaemic pathology beyond MCA territory after intraluminal filament occlusion in C57Bl/6J mice. Brain Res 997:15–23
74. Adhami F, Liao GH, Morozov YM et al (2006) Cerebral ischemia-hypoxia induces intravascular coagulation and autophagy. Am J Pathol 169:566–583
75. Schauwecker PE, Steward O (1997) Genetic determinants of susceptibility to excitotoxic cell death: implications for gene targeting approaches. Proc Natl Acad Sci U S A 94:4103–4108
76. McKhann GM, Wenzel HJ, Robbins CA et al (2003) Mouse strain differences in kainic acid sensitivity, seizure behavior, mortality, and hippocampal pathology. Neuroscience 122:551–561
77. McLin JP, Steward O (2006) Comparison of seizure phenotype and neurodegeneration induced by systemic kainic acid in inbred, outbred, and hybrid mouse strains. Eur J Neurosci 24:2191–2202
78. Kadam SD, White AM, Staley KJ et al (2010) Continuous electroencephalographic monitoring with radio-telemetry in a rat model of perinatal hypoxia-ischemia reveals progressive post-stroke epilepsy. J Neurosci 30:404–415
79. Rakhade SN, Klein PM, Huynh T et al (2011) Development of later life spontaneous seizures in a rodent model of hypoxia-induced neonatal seizures. Epilepsia 52:753–765
80. Brandt C, Ebert U, Löscher W (2004) Epilepsy induced by extended amygdala-kindling in rats: lack of clear association between development of spontaneous seizures and neuronal damage. Epilepsy Res 62(2–3):135–156

Chapter 9

Modeling Psychological Trauma in Rats

Samina Salim

Abstract

Stress is an unavoidable part of our lives. While some types of stress are easily managed by most, extremely stressful situations such as those resulting from experiencing or witnessing of traumatic events can have a long lasting negative impact on our health, often causing serious psychiatric illnesses such as post-traumatic stress disorder (PTSD). Therapeutic interventions to treat such complex ailments are rather limited and are associated with serious side effects. Better interventions are clearly needed. Understanding the biological impact of stressful life events is central to this need. Animal models offer useful insights. Here we describe a method, which simulates direct as well as vicariously acquired form of psychological stress in male Sprague-Dawley rats. The role of social support considered as a critical component of stress-coping mechanisms were also modeled. Extensive details are provided to promote reproducibility.

Key words PTSD, Stress, Trauma, Animal behavior, Anxiety, Depression

1 Introduction

Psychological stress is known to play a central role in the pathogenesis of several neuropsychiatric disorders including post-traumatic stress disorder (PTSD) [1]. The National Institute of Mental Health reports that the lifetime prevalence of PTSD in adults from the United States (U.S.) is 6.8%. The individuals that have acquired PTSD can exhibit symptoms of anxiety and/or depression. PTSD is not limited to adults that have experienced a traumatic event. According to the Centers for Disease Control and Prevention, approximately 14% of children in the USA between the ages of 2 and 8 years old are diagnosed with mental, behavioral, or developmental disorder. The source of psychological stress varies from being social in nature [2], to being extremely traumatic, as observed in the cases of victims of violent crimes, accidents, or natural disasters [3–5]. Interestingly, not just experiencing but witnessing of traumatic events can also be very damaging. This is supported by reports suggesting that in North America approximately 30% of individuals who witnessed a traumatic event

Nicholas J. D. Wright (ed.), *Basic Neurobiology Techniques*, Neuromethods, vol. 152,
https://doi.org/10.1007/978-1-4939-9944-6_9,

developed PTSD [6]. Thus, PTSD can be acquired not only from physically experiencing traumatic events but also from vicariously witnessing of traumatic events [6]. This is especially relevant considering early life trauma exposures. It is well known that adverse experiences during early life can contribute to development of psychiatric conditions later in life. In fact, adults with a history of traumatic experiences of either childhood abuse or other traumatic events are considered to be at increased risk of developing depression, cognitive deficits and even post-traumatic stress disorder (PTSD), in their later life. Relevant to this, in the United States alone, 1 in 58 children experience maltreatment (Fourth National Incidence Study of Child Abuse and Neglect), highlighting increased vulnerability of children to exposure of early life stress. While examining the link between early life stress and later life behavioral and cognitive well-being is important, conducting such studies in children are difficult to carry out. Therefore, animal models are valuable in studying the behavioral consequences of early life stress across the developmental course. Furthermore, currently, the drugs of choice for treatment of anxiety disorders, depression and PTSD in children and adults alike, are traditional-antidepressants and benzodiazepines, with modest efficacy and major side effects. Therefore, alternative therapies with higher efficacy and lesser side effects are needed, which can be delivered only when mechanistic underpinnings are clear. Animal models are valuable in this regard, as they provide us with the convenience of modeling direct as well as indirect traumatic stress that resembles traumatic occurrences and also the ability to study the neurobiological basis of trauma. Furthermore, examining the effect of early life stress on later life behaviors has direct translational relevance considering the high human vulnerability to early life stress exposures. The concept of intervening with key biological processes before the full and chronic expression of an illness has many precedents in non-CNS disorders, including those in cardiovascular disease and diabetes.

We have developed an experimental method, which mimics trauma by simulating direct as well as indirect stress [7, 8]. Social defeat is a resident–intruder paradigm, which was originally developed by Miczek [9]. This model employs social conflict between two rodents, and believed to induce psychological stress in animals without causing physical harm or habituation [10, 11]. Thus it is considered as an ethologically valid and robust model of stress-induced neuropsychopathologies including PTSD, depression and anxiety [12, 13]. In social defeat model, either a male rodent (intruder) is placed into the home cage of the aggressive dominant male (resident), or a cohabitating female is placed as the intruder in the resident cage. Agonistic encounters occur between the resident and the intruder. Following the exposure to aggressive encounters by the resident, the intruder displays repeated signs of distress and exhibits freezing behavior along with assuming submissive supine position [12, 13]. The resident and the intruder is separated by a transparent perforated plexiglass partition following the defeat exposure to prevent physical contact but enable auditory, olfactory and visual interactions between the resident and the intruder for the remainder of the session [12, 13]. Socially defeated animals also are known to exhibit increased heart rate and blood pressure, and elevated corticosterone levels, indicating heightened levels of stress [12, 13]. Furthermore, social defeat exposure also causes behavioral changes, and reported to cause decreased locomotor and exploratory activity [13–15], increased anxiety-like behavior [13, 16, 17], reduced aggression, low mating behavior [15], social avoidance, and depression-like behavior [13, 18–20]. These behaviors are long lasting, persisting at least four weeks following social defeat stress [19, 21], and reported to be reversible with chronic antidepressant administration [19, 21]. In our method, we used the modified version of the resident–intruder paradigm originally developed by Miczeck et al. [9]. Briefly, two Sprague-Dawley rats were housed together during acclimatization period. Later, one Sprague-Dawley rat considered as an intruder, and designated as socially defeated (SD), rat was introduced into the cage of the resident Long-Evans (LE) rat, and the cage mate of the SD rat was placed in the Plexiglas enclosure surrounding the cage where both LE and SD rats were present (Fig. 1). Putting an intruder rat into the home cage of a resident rat results in intruder subordination, or social defeat. The cage mate of intruder is positioned to witness the event and is designated as, the trauma witnessing (TW) rat. Three exposures of social defeat per day, lasting 7 consecutive days, were witnessed by the TW rat. After daily exposures, the TW rat and their corresponding cage mate SD rat were housed together. At the end of the social defeat witnessing/control exposures protocol, anxiety- and depression-like behavior as well as memory tests were conducted. We observed that experiencing as well as witnessing social defeat, both caused behavioral (anxiety-like and depression-like behaviors) and cognitive impairments (long-term memory deficits) in rats [7]. The behavioral phenotype lasted until 1 month after the last social defeat session [8].

The inverse relationship between PTSD and social support is one of the most consistent associations observed in stress/trauma research [22, 23]. It is well known that social support in humans is an important component of PTSD coping strategy [24, 25]. Our method examines not only the effects of experiencing or witnessing traumatic events but also examines the issue of social support and PTSD in this model.

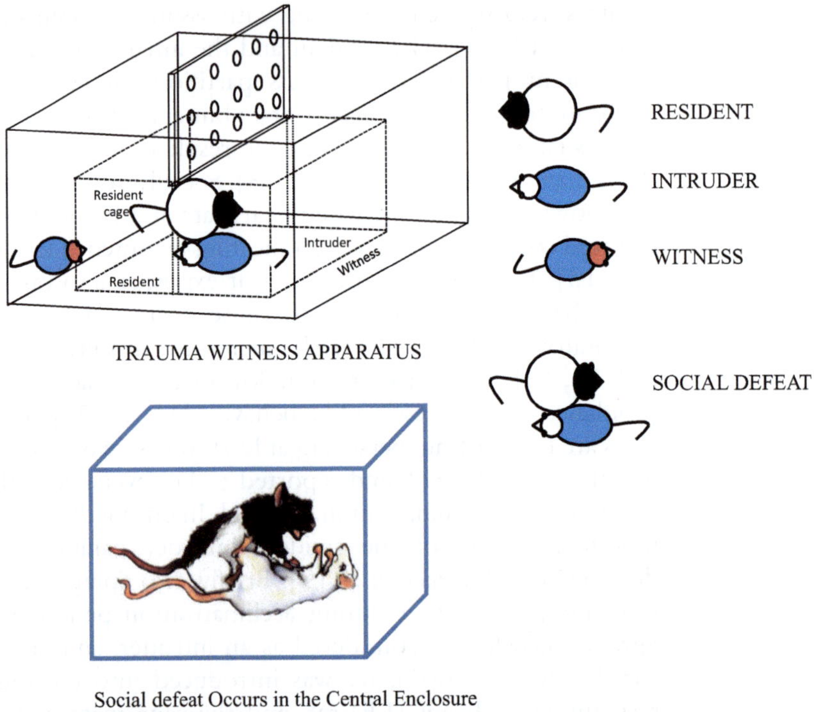

Fig. 1 Depiction of the trauma witness apparatus. Two Sprague-Dawley rats were housed together to allow bonding (social support) during acclimatization period. Later, one Sprague-Dawley rat considered as an *intruder* was introduced into the cage of the *resident* LE rat. This resulted in a typical social defeat behavior indicated by the intruder surrendering, when attacked by the resident LE rat. After defeat, a perforated Plexiglas partition was placed for 10 min, between the resident and the intruder to avoid injury to the intruder. The partition allowed visual, auditory, and olfactory interactions. The cage mate of the intruder designated as the trauma witnessing (TW) rat was put in an enclosure surrounding the cage and witnessed social defeat of its cage mate. This initiated a freezing response in the TW rat. Two more bouts of social defeat were performed with 5-min separation, in order to reinforce the visual stress in the TW rat

2 Materials

Trauma witness: Male Sprague-Dawley rats (225–250 g) were used as controls or intruders, and male Long-Evans (LE) retired breeders (400–500 g) served as resident aggressors (Charles River, Wilmington, MA).

Maternal trauma witness: Sprague-Dawley pups arrived on post-natal day (PND) 14 along with their dams. They were placed with the dams for 4 days for acclimatization. Before PND 18, pups and dams were housed together to allow bonding. Then age-matched pups of the same gender were reared in littermate groups. Female Sprague-Dawley rats (225–250 g) were used as controls or intruders, and male Long-Evans (LE) retired breeders (400–500 g) served as resident aggressors (Charles River, Wilmington, MA).

Female social defeat: Female Sprague-Dawley rats (225–250 g) were used as controls or witnesses, and male Long-Evans (LE) retired breeders (400–500 g) served as resident aggressors, while male Sprague-Dawley rats (225–250 g) served as intruders (Charles River, Wilmington, MA).

Upon arrival at the animal facility, rats were housed on a 12-h light/dark cycle in a climate-controlled room with food and water provided ad libitum. Experiments with rats were conducted in accordance with the NIH guidelines using protocols approved from the University of Houston Animal Care and Use Committee.

3 Methods

3.1 Trauma Witness Model

The social defeat model originally developed by Miczek [9], which involves aggressive encounters between a large Long-Evans (LE) male rat (resident) and a smaller Sprague-Dawley male rat (intruder) is an accepted rodent model of stress and is used as a model of PTSD [12, 13, 26–28]. We have modified the resident–intruder model to create a trauma witness model (TWM). The detrimental effects due to loss of social support and the benefits of continued social interaction, which are core components of PTSD coping strategy [24, 25], were also modeled.

Two Sprague-Dawley were housed together to allow bonding (social support) during acclimatization period. Later, one Sprague-Dawley rat considered as an intruder and designated as socially defeated (SD) rat was introduced into the cage of the resident LE rat, and the cage mate of the SD rat was placed in the Plexiglas enclosure surrounding the cage where both LE and SD rats were present. The cage mate is therefore the trauma-witnessing (TW) rat. Introduction of the SD intruder rat into the cage of the resident LE rat results in a typical social defeat behavior of the intruder SD rat, indicated by the intruder surrendering, when attacked by the resident LE rat. After defeat, a perforated Plexiglas partition was placed in between the LE and SD rats to avoid injury to the intruder SD rat. This partition allowed visual, auditory, and olfactory interactions for the remainder of the 30-min session (Fig. 1). The TW rat present outside the cage in the enclosure witnesses the social defeat and subordination of the SD rat, initiating a freezing response in the TW rat. Two more bouts of social defeat were performed with 5-min separation, in order to reinforce the visual stress in the TW rat.

3.2 Screening of aggressor Long-Evans (LE) rats

The 3-day screening process for selection of appropriate LE rats was dependent on the LE rats that exhibited consistent levels of aggression. Such steady dominant behaviors of LE rats are vital for inducing chronic social defeat stress in rats. The aggressive behavior among all male retired breeder LE rats differed in the intensity, extent and quality. Only those aggressors were included in the social defeat paradigm that met the following criteria: aggressive resident performing a defeat as demonstrated by the intruder attaining supine position for approximately 3 s. If the residents were to be used for multiple defeats, all aggressors were subjected to single screening session before the initiation of consecutive social defeat exposures [29].

3.3 Trauma Witness Apparatus

The apparatus comprises of a central cage which houses LE and SD rats during social defeat sessions. The social defeat cage measures 24 × 41 × 50 cm in size and the external enclosure measures 24 × 41 × 50 cm in size. The social defeat cage was designed to be placed in the center of the apparatus, with the enclosure surrounding it. A perforated plexiglass comprises of holes that is inserted to stop social defeat. During the social defeat sessions, the entire apparatus is covered with a perforated plexiglass cover. An experimenter was seated nearby hidden from the animals to avoid distraction. The experimenter was positioned to be able to view social defeat sessions clearly with easy reach to the apparatus to allow physical interruption of social defeat by putting the perforated separator in between the SD and LE rats to avoid any physical injury or fighting. Another experimenter records movement and freezer behavior patterns of the TW rat.

3.4 Trauma Witness Model Modifications

An important and interesting feature of our trauma witness model is that it allows for researchers to make numerous modifications to ask interesting questions. For example, we simulated maternal trauma witness stress using this model with the purpose of creating a rat model in which we are able to study the effect of maternal trauma witness on pups. By exposing pups to witnessing social defeat of their mothers, one could use this model to study later life behaviors in rats. Description of this model is provided below.

3.5 Maternal trauma witness stress

One female Sprague-Dawley rat (intruder) was introduced into the cage of an aggressive Long–Evans male rat (resident). During the encounter, the male Long-Evans rat attacked the female Sprague-Dawley rat. Three exposures of social defeat (attacks) were given for 7 consecutive days. Typically, in social defeat paradigm the resident dominates the intruder and fully defeats it with the intruder assuming a supine position. Here, the resident attacks were met with more frequent defensive postures by the intruder, than typically observed in the social defeat paradigm. This can be attributed as protective attempts by the dam to shield the pups. The social defeat traumatic events were witnessed by 6 pups (offspring of the intruder female rat, PND21–27), which were placed in six separate enclosures surrounding the cage. As indicated in Fig. 2, the apparatus was designed to include a central enclosure of social defeat, with six separate small witness enclosures. Pups were natural litters

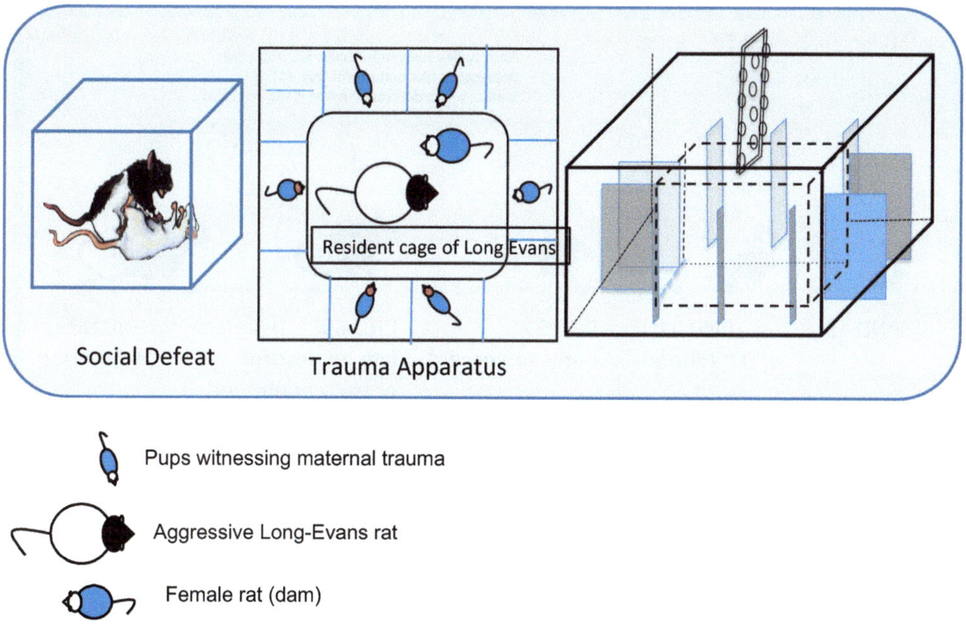

Fig. 2 Illustration of the maternal trauma witness apparatus. The maternal trauma witness apparatus comprising of a central enclosure where social defeat events between the dam and the aggressive LE rat occur. Also exhibited are surrounding enclosures from where the social defeat events are witnessed by the pups

of the dam and were exposed to daily witnessing of repeated social defeat (attacks) of their mother by an unfamiliar aggressive LE male rat for 45 min (3 intervals, 10 min of social defeat followed by 5 min of rest), for 7 consecutive days. Male and female pups were separately placed and their behavioral parameters recorded separately. As indicated in the experimental design (Fig. 3), after conclusion of social defeat witness, pups were left undisturbed for a month in their home cages and their behavior assessment was conducted at PND60 (considered adult). As another segment of this study, 24 h after conclusion of the social defeat procedures, dams were examined for behavioral and cognitive functions and results compared with another set of female rats not subjected to social defeat but exposed to control exposures as previously published [4].

Group designations and controls: (1) pups at PND21 witnessed social defeat of dams (M-TW: maternal trauma witness), (2) pups at PND21 witnessed only LE (LE-W: witness Long Evans rat only), (3) pups at PND21 witnessed dams alone (W-D), (4) pups at PND21 witnessed social defeat of a new female rat which is not a dam (F-TW: female trauma witness), (5) pups at PND21 witnessed female rat only without social defeat (F-W). Thus, several controls were included in the study. We used age-matched pups that did not witness maternal social defeat. Pups were placed into the chambers

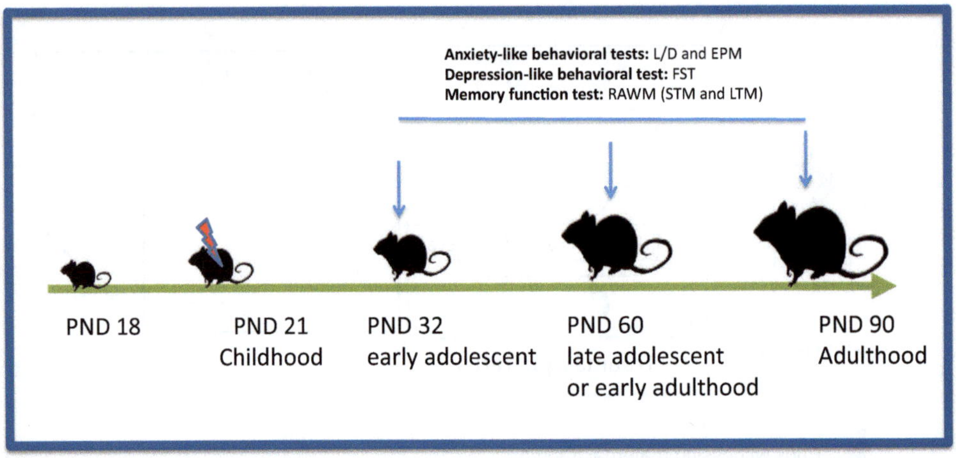

Fig. 3 Schematic representation of the experimental protocol used for examination of early life maternal trauma witness in rats. Pups at PND21 witnessed social defeat (3 exposures per day) of dams for 7 consecutive days. They were then returned to their cages. Behavior assays were conducted when the pups were 32 days old, 60 days and 90 days old

surrounding the resident cage of their mother in the absence of the aggressive male LE rat. This control group involved the natural litters of the dam witnessing their mother in her resident cage for 45 min, for 7 consecutive days (W-D). An additional control group was included to confirm that the changes observed in adult rats were a result of witnessing maternal social defeat, and not a consequence of fear from the sight of the aggressive LE rat. Pups were placed into the chamber surrounding the resident cage of LE rat. This time the dam was not present throughout the procedure (7 days) (LE-W). Dams that underwent social defeat are referred as SD-D (social defeat-dams) while control females undergoing control exposures: (1) Ctrl Dam: the dams that did not undergo social defeat (attacks), (2) Ctrl Female: the non-lactating female rats that did not undergo social defeat (attacks), (3) SD-Female: the non-lactating female rats that underwent social defeat (attacks).

3.6 Female Social Defeat

Re-purposing our trauma witness model to study female social defeat behavior is another highly significant feature of this model. Conducting social defeat in females is a challenging task and considered highly impractical, as social defeat never happens between a male resident LE and a female intruder rat. The male LE rat does not exhibit aggressive behavior towards the female but instead exhibits mating cues and playful behavior. Due to the difficulty associated with conducting social defeat in females, several groups reported utilizing highly territorial Syrian hamsters, California mice, or a lactating female as the aggressive resident [30–33]. Our lab has published a modification to the resident–intruder

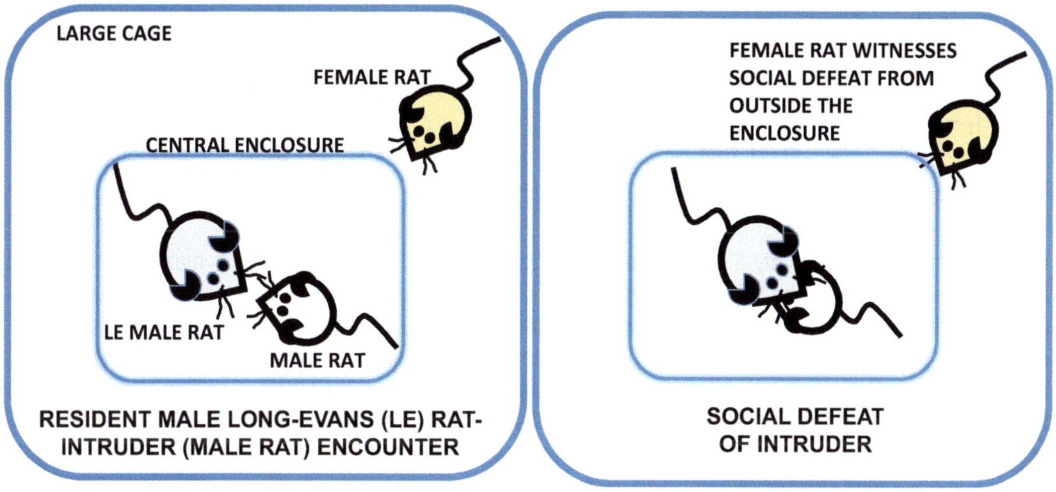

Fig. 4 Illustration of the social defeat paradigm in female rats. Female rats witnessed daily social defeat exposures between a Sprague-Dawley male rat and a male Long-Evans rat for 1 week. Following social defeat witness the female rats were returned to their cages

paradigm, which consists of a rodent bearing witness to a social defeat episode between a resident and an intruder [7, 8]. This model effectively combines social isolation and visual, olfactory, and auditory exposure to social defeat. Using this model, we identified that both physical (intruder) and psychological (witness) social stress results in the emergence of robust anxiety- and depressive-like behaviors [7, 8]. Thus, directly experiencing social defeat or witnessing social defeat both, leads to an identical phenotype (anxio-depressed). These findings were later independently confirmed by other groups [34–37]. More importantly, the effectiveness of witnessing social defeat in inducing depressive- and anxiety-like behaviors was reported in female mice [37]. Therefore, it is reasonable to use the social defeat witness paradigm to investigate stress-induced impairments in female rats. As indicated in Fig. 4, the female rats were allowed to witness 3 daily social defeat events between a male LE and a male SD rat for 7 consecutive days. Using this social defeat paradigm, we observed that social defeat witnessing female rats (SDW) developed anxiety- (Fig. 5a, b) and depression-like behavior (Fig. 5c) as well as exhibited short- and long-term memory deficits (Fig. 5d, e). SDW rats spent significantly greater time in the dark box of the LD test than control rats. Similarly, SDW rats spent significantly less time in the open arms of the EPM test. In the FST test, SDW rats spent significantly greater time being immobile indicating higher depression-like behavior than control female rats and exhibited more errors in the STM and LTM tests than control female rats. Anxiety-like behavior was assessed using light–dark and open-field test while depression-like behavior was evaluated using forced swim test. Learning and

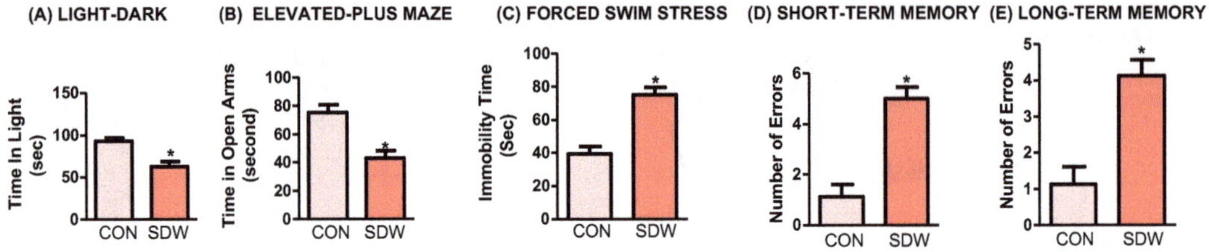

Fig. 5 Examination of anxiety-like behavior using light–dark (**a**), and elevated-plus maze (**b**) tests, and evaluation of depression-like behavior using forced swim test (**c**) and learning-memory function using radial arm water maze test (**d**, **e**). Female rats subjected to social defeat witness spent significantly less time in the open arm of the EPM and lit area of the LD box. Female rats subjected to social defeat also spent significantly increased amount of time immobile and exhibited impaired short and long-term memory as compared to control female rats ∗significantly different from control, $p < 0.05$. Values are mean ± SEM, $n = 10–12$ rats/group. Groups: CON: female rats witnessing only LE rats, SDW: female rats witnessing social defeat between male LE and Sprague-Dawley rats

memory function was assessed using radial-arm-water-maze (RAWM) test to examine short-and long-term learning and memory deficits in rats.

4 Notes

1. It is important that animals are handled by the same personnel as those conducting social defeat sessions, video recordings, and behavior assays. This is important as any novel scent or different handling maneuvers might generate stressful stimuli in rats, which can easily confound results and decrease reproducibility.
2. Experimenter should not wear any strong cologne or perfume and should be very quiet and gentle with movements.
3. The experimenter should use the same color of lab wear (white or grey coats or aprons) throughout the experiment to prevent rats from getting unwanted color cues.
4. Social defeat sessions should be conducted at the same time of the day. We have noted some differences in aggression of LE rats as well as in behavior of SD rats in winter versus summer times. Therefore, this consideration can be taken into account when planning experiments.
5. Social defeat apparatus should be wiped with 70% ethanol and left to air dry at least for 10 min to remove all residual scent/odor of previously tested animals.
6. The social defeat session room should be a quiet area preferably a sound proof room (no sounds of doors closing, no elevator sounds, no hallway noise or no loud talking).

7. The behavior room vents should not be noisy. Air speed can be modified to keep the vent sound to the minimum. Regulated air often causes a sound when the air cycle switches on or when it shuts down. This noise tends to startle the animals especially pups.
8. Cage bedding is important. Cage bedding vendors should not be changed during the experiment.
9. Rodent diet should be kept constant. Do not change vendors.
10. Cage water bottles should always be the same type and preferably the same bottles per cage. Do not use any scented dish detergents for washing these bottles.

4.1 Important Controls

Several controls can be included. In our study, one set of controls included was when one Sprague-Dawley rat was placed in a novel cage and another Sprague-Dawley rat, a cage mate, was put outside the cage, in the enclosure for 30 min daily for 7 days. The resident LE rat was not present. After each 30 min control session/exposure, they were pair-housed. In another set similar controls were included, except that rats were single housed after each 30 min control exposure. An additional control included placing a Sprague-Dawley rat in the enclosure outside the cage of the resident LE rat, but no social defeat session was conducted. This was done to examine whether the changes in behavior are specific to the social defeat process or due to the mere sighting of an aggressive rat. In order to confirm that behavioral effects are primarily through intrinsic emotion-evoked neurocircuitry and that witnessing of traumatic events is required to cause PTSD-like behaviors in rats, role of olfactory and auditory cues was also examined. Basically, an intruder rat was put into the home cage of a resident rat resulting in social defeat of the intruder. During this process occasional vocalization sounds were made by the intruder. Brief sounds were also made when the intruder hit the walls of the central enclosure in an effort to escape attacks and avoid defeat. These sounds were audible to the cage mate (another small male Sprague-Dawley rat) of the intruder rat, but the defeat process was not seen as the central enclosure was covered with opaque black paper. Auditory stress arising from hearing sounds of the social defeat process did not lead to PTSD-like behaviors in rats, suggesting that witnessing traumatic events is essential for development of PTSD-like behaviors in rats [8]. In a similar setup, a small Sprague-Dawley rat was allowed to smell but not see the LE (resident) rat. This was done either by placing the LE rat in a blacked-out central enclosure, or by putting the fur and urine of a LE rat inside the enclosure. No social defeat was carried out. Our data suggest that the olfactory stress of smelling resident odor does not lead to PTSD-like behaviors in rats, suggesting that witnessing traumatic events is required for PTSD-like behaviors. Finally, our

TWM model meets the standard criteria generally accepted for animal models of PTSD, as PTSD-like behaviors observed in TW rats persisted until after a month of witnessing traumatic events [7, 8].

4.2 Behavior Assessments

4.2.1 Qualitative Assessments

(a) *Freezing behavior*: Behavior of the rats after termination of the actual social defeat while still in the separated enclosure was monitored. Motionless posture for more than 1 min was interpreted as freezing (fearful or shock) behavior.

(b) *Cage-behavior*: The behavior of the rats upon return to their home cages with pair housing or in isolation was also carefully noted at least for 1 h. SD rats after undergoing defeat sessions were returned to the home cage and pair housed with the cage mate TW rat (witness of the defeat sessions). Initially, the TW rat remains aloof and restless as evident from its repetitive movements in the cage, but then huddles with the SD rat and spent time licking and/or sniffing. These qualitative assessments are representative of comforting and supporting behavior. SD and TW rats when housed separately following social defeat or witnessing social defeat, led to a freezing behavior by SD and TW rats in their home cages. In addition, neither rat made any effort to drink or eat for a long time.

4.2.2 Quantitative Assessments

Anxiety-like Behavior: We used three tests to quantitatively examine anxiety-like behavior. Anxiety test were conducted back-to-back after a rest period of 24 h of the last day of social defeat witness.

Light–dark (LD) test: The LD box consisted of two compartments: a light compartment (27 × 27 × 27 cm) and a blackened wall dark compartment with dark floor (27 × 18 × 27 cm), separated by a single partition with an opening (7 × 7 cm) to facilitate the movement of rat between the two compartments [38] (Fig. 6). Each rat was given 5 min to explore both compartments and the total time spent in the lit area was recorded manually as previously published [38–40]. This apparatus was situated within a screen enclosed area of behavior core facility room with only one experimenter/observer present in the room at the time of experiment under standard lighting conditions of approximately 700 lx. The observer scored the data using a Microsoft Excel software program that we designed ourselves that enabled the observer to record time by manually scoring the data by pressing computer keys "L" for light and "D" for dark on the keyboard. The experiment began by putting the rat gently at the center of the light compartment facing the opening to the dark chamber. "L" was pressed immediately on the computer keyboard placed next to the apparatus but hidden from the rat by a screen. Pressing the "L" key automatically started the computer clock. Movement

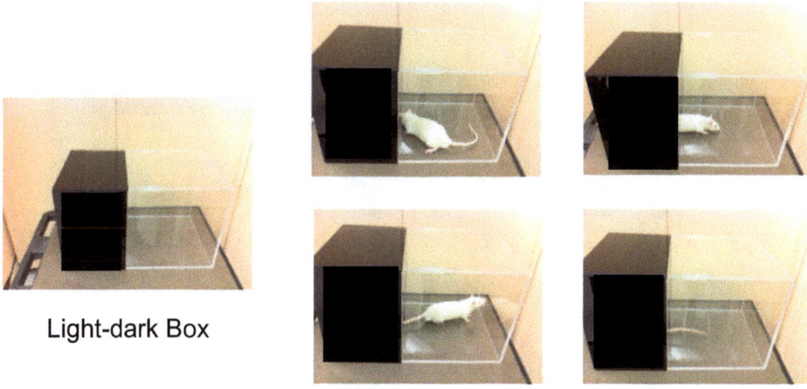

Fig. 6 Illustration of the light–dark apparatus. The light and a dark compartment is separated by a partition with an opening for passage from one compartment to the other. Time spent in the light and dark part is recorded for a 5 min period. Reduced time spent in the light area is indicative of increased anxiety

of the rat to the dark compartment was recorded by pressing "D" on the keyboard. This way, total time spent in the illuminated part was recorded by an observer blinded to treatment for 5 min. A rat was defined to have entered the lit or dark box when both front paws and shoulders were inside the respective compartment. The less time a rat spent in the light compartment versus the dark compartment is an indicator of anxiety-like behavior.

Open Field (OF) test: The OF test is widely used to study anxiety level, exploratory as well as locomotor activity in rodents [13, 38–40]. The OF apparatus consisted of open arena (60 × 40 cm) surrounded by transparent plexiglass walls (Fig. 7). Rats were placed in the center of the arena and allowed to move freely in the arena for 15 min. The infrared light sensors detected the movement, which was quantified using Opto-Varimex Micro Activity Meter v2.00 software (Optomax, Columbus Instruments; OH). The light intensity was kept at 300 lux. The time spent in the center of the arena was analyzed and reported as percentage. The less time a rat spent in the center of the arena versus periphery of the arena is an indicator of anxiety-like behavior. The apparatus was wiped down with alcohol in between each test animal. Activity was quantitated using a computer-operated Opto-Varimex Micro Activity Meter v2.00 system (Optomax, Columbus Instruments; OH) that utilizes sensors containing eight infrared light emitting diodes and eight phototransistors that emit and detect modulated infrared light beams. Sensors were positioned to form two-dimensional cages each with rearing monitoring. Movement was detected by beam breaks and data from three test chambers was recorded simultaneously, one rat per chamber, collected in 3 min intervals over a 15-min test session. The

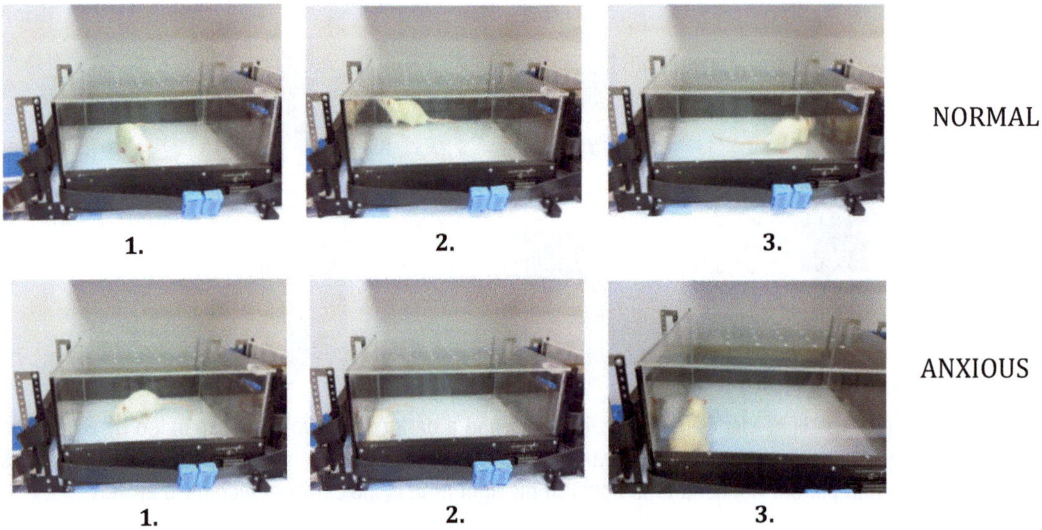

Fig. 7 Illustration of the open-field apparatus. The open-field apparatus is an open arena attached to a software-based analysis to detect movement of the animals. The more time the rat spends in the periphery and the less in the center, the more anxious it is considered to be

program tabulated activity counts, zone entries, zone times, center time and the periphery time, distance travelled and rearing for every cage in the system. For center time analysis, an approximately 25 cm × 25 cm square in the center of the open-field arena was defined as the center zone for data analysis. The total time spent in the center of the arena and rearings were calculated for each group.

Elevated plus maze (EPM) test: The standard rat EPM apparatus consists of four arms (10 cm × 50 cm length) (two open and two closed) intersecting to create a plus shaped path, was obtained from Med Associates Inc., (St. Albans, VT) (Fig. 8). The arms of the EPM were elevated about 60 cm off the ground. The EPM procedure was performed as described [13, 38–40]. Briefly, the rat was placed in the central area facing the open arms of the maze and allowed to explore the maze for 5 min. The movement of the rat was tracked visually to measure the amount of time the rat spent in the open arm. Reduced time spent by a rat in the open arm is an indication of anxiety-like behavior.

Depression-like Behavior—Depression-like behavior was assessed using forced swim stress test (FST). A rat was put in water (25 °C) filled tank (24 cm in diameter and 30 cm high) for 5 min (Fig. 9). Soon after being placed in the water tank, rats exhibit immobile posture and assume motionless floating. The total time spent being immobile was recorded as previously

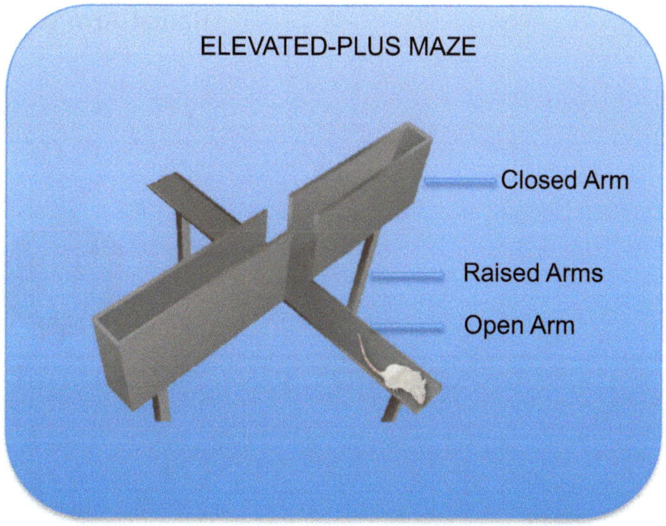

Fig. 8 Illustration of elevated plus maze apparatus (EPM). The EPM is an elevated apparatus comprising of two open and two closed arms. The more time the rat spends in the closed arms and less in the center or the open arms-the more anxious it is considered to be

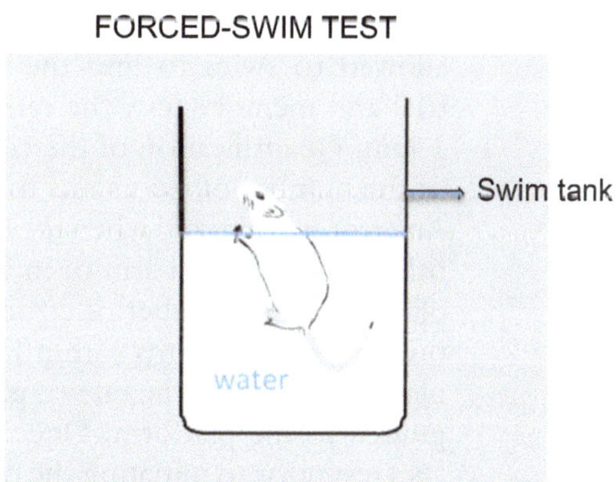

Fig. 9 Illustration of forced swim test apparatus (FST). The FST comprises of a cylinder filled with water. The greater the reluctance of the rat to swim, the higher the depression-like behavior

published by us [40]. The more time a rat spent being immobile in the FST, the higher the depression-like behavior.

Learning-Memory Function—Radial Arm Water Maze (RAWM) test was performed as previously published by us [13, 38–40]. In this test, a circular black water pool with six swimming arms and an open central area was used (Fig. 10). The test was performed in a dimly lit room with visual cues on the surrounding walls. Each rat was randomly assigned a goal arm with a hidden platform

Fig. 10 Illustration of radial arm water maze test apparatus (RAWM). Short and long-term memory is tested in the RAWM test. The more errors made in finding the hidden platform, the greater is the learning-memory deficit

submerged 1 cm under water at the end of the arm. Each rat was released from one of the arms different from the goal arm and allowed to swim to find the hidden platform. In each learning trial and memory test, the rats were allowed a maximum time of 1 min. Quantification of the cognitive function was done in terms of the number of errors rats made in finding the hidden platform. An error was marked when the rat entered the halfway in to any arm other than the goal arm or in the goal arm but failed to reach the platform. The number of errors ranges from 1 to 7, as the rat can only swim into 7 arms within 1 min. When the rat failed to find the platform in 1 min, the rat was given a score of 7 errors and manually guided to the platform. Once on the platform, the rat was given 15 s rest prior to initiating the next learning trial. Each rat was given 12 learning trials with a 5 min rest period between two sets (trials # 1–6) and (trials # 7–12). 30 min after the 12th learning trial the short-term memory was performed. 24 h after the 12th learning trial each rat was subjected to long-term memory test. Each rat was returned to its home cage until ready for long-term memory test.

5 Notes

For obtaining reliable behavior data it is important to be mindful of the following details, whether one is using automated setups or conducting manual recordings.

1. The sequence of conducting behavior analysis on the same group of rats is important, in order to make sure that habituation does not occur and test-induced stress does not happen. Apparatus cleaning in-between tests with alcohol or bleach should not be done. This should be done only once all tests have been completed.
2. First, anxiety-like behavior tests should be conducted followed by forced swim test and behavior and radial arm water maze test.
3. Amongst anxiety-like behavior tests, first, light–dark test should be done, followed by open-field and elevated-plus maze tests. The rationale for choosing this order is that least stressful tests should be conducted first followed by the most stressful test. Light–dark test is considered to be a mild test as it involves only one aspect of anxiogenic stimuli, that is, light. Open-field test involved two anxiogenic stimuli, open areas and lighted areas while elevated-plus maze test involved open, elevated and lit areas, thus involving three stress components simultaneously adding fearful cues. Anxiety-like behavior tests should be followed by tests involving water. Water is a stressful stimulus for rodents and adding stressful stimuli adds stress to the test itself. Therefore, it would become difficult to evaluate if the behaviors exhibited are a result of a pharmacological intervention or psychological manipulation used or test-related stress. Therefore, if RAWM and FST have to be done on the same animals, these test should be done last.
4. The test timing is also important. In our experience, we have obtained reliable results when anxiety tests were conducted in the first half of the day. This might be due to the absence of a variety of stimuli during the early hours of the day, which increases as the day becomes busier with doors closing and opening, more experimenters in adjacent rooms, the conversations in the hallways, elevator noise etc.

5.1 Data Analysis

Data are expressed as mean ± SEM. Significance was determined by unpaired Student's t-test (GraphPad Software, Inc., San Diego, CA). A value of $p < 0.05$ was considered significant. At least 10–12 rats will be needed per group to achieve statistical significance.

6 Conclusion

By developing a modified version of the rodent resident–intruder paradigm, we have provided neurobehavioral researchers a novel tool to study the impact of both direct and indirect trauma in animals. This rodent model offers extensive advantages towards promoting better understanding of the underlying biology of trauma and stress-related impact on the brain. Our animal model is both robust and valid. This model is suitable for conducting studies in both mice and rats.

Using this model, we have established that rats witnessing social defeat of cage-mates for 7 consecutive days developed PTSD-like behavioral (anxiety and depression-like) and cognitive phenotype [7, 8]. The anxious and the cognitive phenotype lasted until 6 weeks of last traumatic event exposure, but these impairments were reversible after 8 weeks of traumatic exposure for both direct as well as indirect exposures [8]. Interestingly, depressive phenotype did not recover in this time frame, further adding an interesting feature to our model. Considering the involvement of a variety of neural circuits and mechanisms in regulation of depression [32], perhaps, depressive recovery is a more time taking process. Further investigation into these observations might particularly be helpful in drug discovery and in identifying novel molecular targets critical for drug discovery and drug design.

Furthermore, inclusion of a variety of control experiments further solidifies the experimental utility and relevance of our animal model. Our observations in control experiments suggest that witnessing is critical for PTSD-like phenotype and cannot be acquired solely via application of auditory or olfactory stresses. Thus, visually witnessing the traumatic event(s) is an important aversive stimulus for the development of PTSD-like symptoms in rats. Our results underline the importance of stress sensitization and associative learning as explained by the dual-branch hypothesis. Finally, we suggest that TWM is an excellent model for studying vicarious trauma. This model can be utilized for examining variety of neurobehavioral effects. This model is also highly suitable for studying social defeat in female rats. Female rats witnessing social defeat of other rats exhibited PTSD-like behaviors. Questions such as the neurobehavioral impact of one versus multiple traumatic event exposures and the effect of social housing versus isolation housing on reversing behavioral and cognitive phenotype remain to be determined. PTSD-like behavior reversal versus rescue effects of a variety of drugs and interventions can be tested using this model.

Translational relevance: The greatest utility of these rodent models is their translational relevance. For example, negative consequences of early life stress, either from experiencing a traumatic event directly or witnessing a traumatic event, are often expressed much later in life as anxiety disorders, depression and in worse cases as post-traumatic stress disorder. In fact, young children who experience stressful traumatic event(s) during early life, a sensitive developmental period, are considered highly vulnerable to psychiatric disorders in adult life. Interestingly, not all children who experience stressful events are equally at risk of developing later life psychiatric disorders. Some are resilient in spite of being exposed to the same risk factors, while others are susceptible. Role of gender in stress-susceptibility is also unclear. Our current understanding of stress susceptibility in females is limited, although overall women are considered to have enhanced susceptibility to psychological consequences of stress. Thus, the relationship between early life stress/trauma exposure and development of later life psychiatric symptoms in males versus females is not fully understood. This knowledge is critical for early intervention. Conducting these studies in children and/or adults is challenging; therefore, animal models are useful tools in addressing some of these questions, which are critical from the standpoint of therapeutic as well as preventive measures.

Acknowledgments

This work was supported by National Institutes of Health grant (2R15MH093918-02) and University of Houston start-up funds awarded to Samina Salim.

References

1. Yehuda R, Resnick H, Kahana B, Giller EL (1993) Long lasting hormonal alterations to extreme stress in humans: normative or maladaptive. Psychosom Med 55:287–297
2. de Ridder D. 2000. Social status and stress. In G. Fink, (ed.) Encyclopedia of stress. Academic Press, San Diego, CA. 3:468-473
3. Kessler RC (2000) Posttraumatic stress disorder: the burden to the individual and to society. J Clin Psychiatry 61(Suppl. 5):4–12. discussion 13–14
4. Kessler RC, Sonnega A, Bromet E, Hughes M, Nelson CB (1995) Posttraumatic stress disorder in the national comorbidity survey. Arch Gen Psychiatry 52:1048–1060
5. Stein MB, Walker JR, Forde DR (2000) Gender differences in susceptibility to posttraumatic stress disorder. Behav Res Ther 38:619–628
6. American Psychiatric Association (ed) (2013) Diagnostic and statistical manual of mental disorders, 5th edn. American Psychiatric Association, Washington, DC
7. Patki G, Solanki N, Salim S (2014) Witnessing traumatic events causes severe behavioral impairments in rats. Int J Neuropsychopharmacol 17(12):2017–2029. https://doi.org/10.1017/S1461145714000923
8. Patki G, Salvi A, Liu H, Salim S (2015) Witnessing traumatic events and post-traumatic stress disorder: Insights from an animal model. Neurosci Lett 600:28–32. https://doi.org/10.1016/j.neulet.2015.05.060
9. Miczek KA (1979) A new test for aggression in rats without aversive stimulation: differential effects of d-amphetamine and cocaine. Psychopharmacology 60:253–259
10. Bjorkqvist K (2001) Social defeat as a stressor in humans. Physiol Behav 73:435–442
11. Rohde P (2001) The relevance of hierarchies, territories, defeat for depression in humans: hypotheses and clinical predictions. J Affect Disord 65:221–230
12. Wood SK, Walker HE, Valentino RJ, Bhatnagar S (2010) Individual differences in reactivity to social stress predict susceptibility and resilience to a depressive phenotype: role of corticotropin-releasing factor. Endocrinology 151:1795–1805
13. Patki G, Solanki N, Atrooz F, Allam F, Salim S (2013) Depression, anxiety-like behavior and memory impairment are associated with increased oxidative stress and inflammation in a rat model of social stress. Brain Res 1539:73–86. https://doi.org/10.1016/j.brainres.2013.09.033. Epub 2013 Oct 3
14. Koolhaas JM, De Boer SF, De Rutter AJ, Meerlo P, Sgoifo A. Social stress in rats and mice. Acta Physiol Scand Suppl 1997;640:69-72. Review
15. Meerlo P, Overkamp GJ, Daan S, Van Den Hoofdakker RH, Koolhaas JM (1996 Jul) Changes in behaviour and body weight

following a single or double social defeat in rats. Stress 1(1):21–32

16. Crawford LK, Rahman SF, Beck SG (2013) Social stress alters inhibitory synaptic input to distinct subpopulations of raphe serotonin neurons. ACS Chem Neurosci 4(1):200–209. https://doi.org/10.1021/cn300238j. Epub 2013 Jan 8
17. Ruis MA, te Brake JH, Buwalda B, De Boer SF, Meerlo P, Korte SM, Blokhuis HJ, Koolhaas JM (1999) Housing familiar male wildtype rats together reduces the long-term adverse behavioural and physiological effects of social defeat. Psychoneuroendocrinology 24 (3):285–300
18. Rygula R, Abumaria N, Flügge G, Fuchs E, Rüther E, Havemann-Reinecke U (2005) Anhedonia and motivational deficits in rats: impact of chronic social stress. Behav Brain Res 162(1):127–134. Epub 2005 Apr 14
19. Berton O, McClung CA, Dileone RJ, Krishnan V, Renthal W, Russo SJ, Graham D, Tsankova NM, Bolanos CA, Rios M, Monteggia LM, Self DW, Nestler EJ (2006) Essential role of BDNF in the mesolimbic dopamine pathway in social defeat stress. Science 311 (5762):864–868
20. Hollis F, Wang H, Dietz D, Gunjan A, Kabbaj M (2010) The effects of repeated social defeat on long-term depressive-like behavior and short-term histone modifications in the hippocampus in male Sprague-Dawley rats. Psychopharmacology 211(1):69–77. https://doi.org/10.1007/s00213-010-1869-9. Epub 2010 May 8
21. Tsankova NM, Berton O, Renthal W, Kumar A, Neve RL, Nestler EJ (2006) Sustained hippocampal chromatin regulation in a mouse model of depression and antidepressant action. Nat Neurosci 9(4):519–525. Epub 2006 Feb 26
22. Brewin CR, Andrews B, Valentine JD (2000) Meta-analysis of risk factors for posttraumatic stress disorder in trauma-exposed adults. J Consult Clin Psychol 68:748–766
23. Ozer EJ, Best SR, Lipsey TL, Weiss DS (2003) Predictors of posttraumatic stress disorder and symptoms in adults: a meta-analysis. Psychol Bull 129:52–73
24. Andrews B, Brewin CR, Rose S (2003) Gender, social support, and PTSD in victims of violent crime. J Trauma Stress 16:421–427
25. Clapp JD, Gayle BJ (2009) Understanding the relationship between PTSD and social support: the role of negative network orientation. Behav Res Ther 47:237–244
26. Bhatnagar S, Vining C (2003) Facilitation of hypothalamic-pituitary-adrenal responses to novel stress following repeated social stress using the resident/intruder paradigm. Horm Behav 43:158–165
27. Bhatnagar S, Vining C, Iyer V, Kinni V (2006) Changes in hypothalamic-pituitary-adrenal function, body temperature, body weight and food intake with repeated social stress exposure in rats. J Neuroendocrinol 18:13–24
28. Wood SK, McFadden KV, Grigoriadis D, Bhatnagar S, Valentino RJ (2012) Depressive and cardiovascular disease comorbidity in a rat model of social stress: a putative role for corticotropin-releasing factor. Psychopharmacology 222:325–336
29. Golden SA, Covington HE III, Berton O, Russo SJ (2011) A standardized protocol for repeated social defeat stress in mice. Nat Protoc 6:1183–1191
30. Huhman KL et al (2003) Conditioned defeat in male and female Syrian hamsters. Horm Behav 44(3):293–299
31. Shimamoto A et al (2015) Individual differences in anhedonic and accumbal dopamine responses to chronic social stress and their link to cocaine self-administration in female rats. Psychopharmacology 232(4):825–834
32. Jacobson-Pick S et al (2013) Social agonistic distress in male and female mice: changes of behavior and brain monoamine functioning in relation to acute and chronic challenges. PLoS One 8(4):e60133
33. Trainor BC et al (2011) Sex differences in social interaction behavior following social defeat stress in the monogamous California mouse (Peromyscus californicus). PLoS One 6 (2):e17405
34. Finnell JE, Lombard CM, Padi AR, Moffitt CM, Wilson LB, Wood CS, Wood SK (2017) Physical versus psychological social stress in male rats reveals distinct cardiovascular, inflammatory and behavioral consequences. PLoS One 12(2):e0172868
35. Iñiguez SD, Riggs LM, Nieto SJ, Dayrit G, Zamora NN, Shawhan KL, Cruz B, Warren BL (2014) Social defeat stress induces a depression-like phenotype in adolescent male c57BL/6 mice. Stress 17(3):247–255
36. Warren BL, Vialou VF, Iñiguez SD, Alcantara LF, Wright KN, Feng J, Kennedy PJ, Laplant Q, Shen L, Nestler EJ, Bolaños-Guzmán CA (2013) Neurobiological sequelae of witnessing stressful events in adult mice. Biol Psychiatry 73(1):7–14
37. Sial OK, Warren BL, Alcantara LF, Parise EM, Bolaños-Guzmán CA (2016) Vicarious social

defeat stress: bridging the gap between physical and emotional stress. J Neurosci Methods 258:94–103

38. Salim S, Sarraj N, Taneja M, Saha K, Tejada-Simon MV, Chugh G (2010) Moderate treadmill exercise prevents oxidative stress-induced anxiety-like behavior in rats. Behav Brain Res 208(2):545–552. https://doi.org/10.1016/j.bbr.2009.12.039. Epub 2010 Jan 12
39. Vollert C, Zagaar M, Hovatta I, Taneja M, Vu A, Dao A, Levine A, Alkadhi K, Salim S (2011) Exercise prevents sleep deprivation-associated anxiety-like behavior in rats: potential role of oxidative stress mechanisms. Behav Brain Res 224(2):233–240. https://doi.org/10.1016/j.bbr.2011.05.010. Epub 2011 May 19
40. Solanki N, Salvi A, Patki G, Salim S (2017) Modulating oxidative stress relieves stress-induced behavioral and cognitive impairments in rats. Int J Neuropsychopharmacol. https://doi.org/10.1093/ijnp/pyx017. (Epub ahead of print)

Chapter 10

Extending the Life Span of Acute Neuronal Tissue for Imaging and Electrophysiological Studies

Yossi Buskila, Alba Bellot-Saez, Orsolya Kékesi, Morven Cameron, and John Morley

Abstract

Slice preparations of neuronal tissue are among the most commonly used experimental approaches in the field of neuroscience. They are employed for a variety of techniques addressing questions across the entire neuroscience spectrum, including immunohistochemical, anatomical, and electrophysiological methods to study the properties of individual, and networks of neurons. In the past decades, slice preparations have provided information that has allowed us to develop our understanding of the central nervous system. Unlike cultures, slice preparations leave the topography of neurons and glia intact and therefore retain a considerable degree of functionality that allows molecular, cellular, and network investigations. However, a major limitation of using acute brain slices is their life span which is limited to 6–8 h due to intrinsic and extrinsic factors. Recently, new technological and methodological modifications have proved efficient in extending the life span of acute neuronal tissue. In this chapter, we will review the mechanisms leading to tissue deterioration and describe in detail the steps required to achieve a significant enhancement in neuronal viability and longevity.

Key words Brain slices, Bacteria, Ischemia, Braincubator, ATP, Electrophysiology, Incubation

1 Introduction

1.1 History of Brain Slices

The goal of the slice preparation is to obtain a thin piece of neuronal tissue containing the cells and circuits of interest, and to maintain the slice in a viable (although artificial) condition that is as close as possible to the in vivo environment. The first studies using tissue slices were published in the early twentieth century by the Nobel laureate Otto Warburg, who studied the metabolism of tumor cells [1]. To avoid anaerobiosis in the centre of the tissue, Warburg calculated that the tissue sample should be no thicker than 200–400 μm. Following Warburg's success, other neurochemists started using brain slices, mainly to study neuronal metabolism during aerobic and anaerobic glycolysis [2, 3]. Slices of neuronal tissue became a popular tool to study cellular neurophysiology two

Nicholas J. D. Wright (ed.), *Basic Neurobiology Techniques*, Neuromethods, vol. 152,
https://doi.org/10.1007/978-1-4939-9944-6_10,

decades later, after the pioneering work of Henry McIlwain's group who showed that brain slices are not only capable of resynthesizing creatine phosphates hours after slicing [4], but that cortical neurons maintained their membrane potential and spiking activity similar to in vivo preparations [5]. McIlwain's group further showed that neurophysiological properties of both individual neurons [6] and neuronal circuits [7] could be measured hours after slicing, and manipulated by alterations of the ionic environment. This seminal work identified brain slices as an ideal tool for metabolic and neurophysiological experiments, especially studies that require complete isolation of the tissue from the body. Moreover, this work was also generalized to other tissues, such as spinal cord and retina.

The use of spinal cord slices to study sensory and motor pathways was initially limited to early postnatal period [8], as recordings from adult lower motorneurons (LMN) in vitro remained a challenge [9]. While LMNs from young animals are fairly resilient to hypoxic conditions, due to low levels of myelination that permits oxygen permeability [10–12], mature LMNs are large cells with a high metabolic rate that makes them highly vulnerable to ischemia and mechanical insults associated with the slicing procedure [9]. Although in vitro preparations of immature mammalian spinal cord can provide valuable information, the developmental status of the spinal circuitry is quite different from that in the adult [13]. To limit reductions in viability of adult spinal cord slices, neuroscientists developed methods of rapid cooling to mitigate the ischemic damage [14, 15], as well as adding glutamate antagonists to reduce excitotoxicity [16]. To reduce the extensive damage caused by the mechanical insult, Carp and colleagues added polyethylene glycol (PEG), which induces rapid refusion of neural processes [14]. Others include in the artificial cerebrospinal fluid (aCSF) ethyl pyruvate, which provides antioxidant protection and a metabolic boost that has been shown to have neuroprotective effects in the postischemic tissue [9, 17].

In vitro recordings from retinal tissue have also added significantly to the study of neuronal circuits. The pioneering work by Svaetchin [18] and Tomita [19], who recorded from single cells in isolated vertebrae retina, revolutionized the study of retinal physiology. In contrast to brain slices, the retina can be excised intact, with only the severance of the optic nerve. This allows the function of a complete neuronal circuit to be analyzed. However, the greatest advantage of the retinal preparation is the ability to input tightly defined patterns of light stimuli and record the output of the circuit. This means the neuronal processing completed by this tissue from input, to action potential output of retinal ganglion cells, can be precisely evaluated.

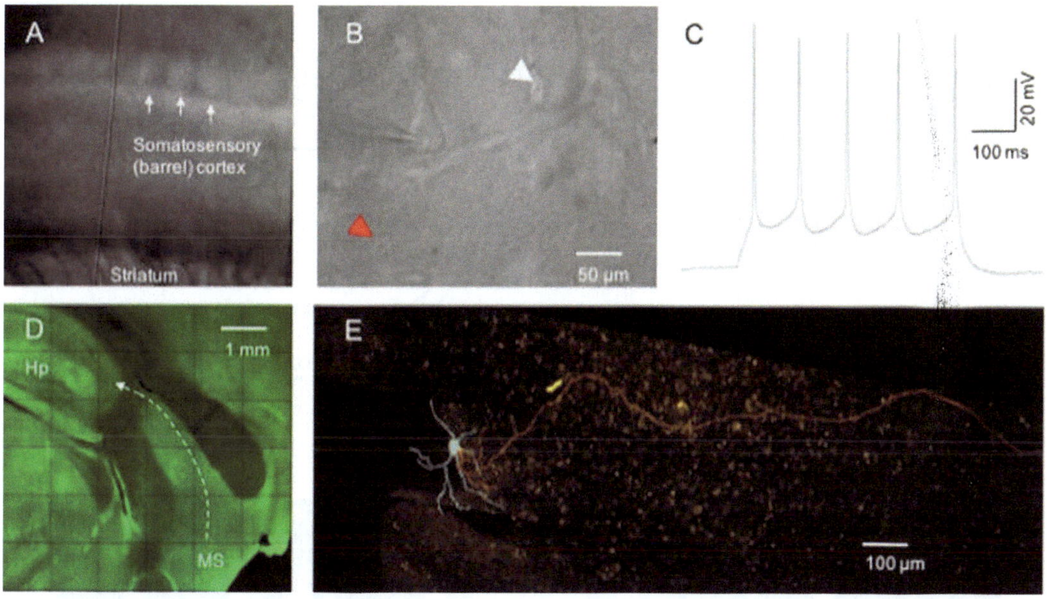

Fig. 1 Acute neuronal preparations allow the study of specific circuits of isolated networks. (**a**) Image of mouse brain slice containing the primary somatosensory (barrel) cortex (arrows) and the striatum, 24 h after slicing and recovery in the Braincubator. (**b**) High magnification DIC image of layer 5 pyramidal neuron, while it is being recorded. The membrane morphology can be used to assess cellular integrity; note the shrunken morphology of a neighboring cell (white arrow) and vacuole of what used to be a cell (red arrow). (**c**) Action potentials recorded from the cell in B, illustrating normal biophysical properties. (**d**) acute slice containing the septohippocampal pathway in $ChAT^+$ mouse, allowing investigation of cholinergic cells in basal forebrain. Note the fluorescent cells originating in the medial septum (MS) and sending their axons to the hippocampus (Hp), as indicated by the arrow. (**c**) High magnification image depicting a $ChAT^+$ neuron in the medial septum filled with biocytin and reconstructed by Neurolucida 360

1.2 Advantages of Acute Neuronal Slices

Slices of neuronal tissue offer several advantages over in vivo and cultured preparations, which have been emphasized over the years through a plethora of manuscripts [9, 20–23]. The fact that they allow stable measurements of metabolic and physiological properties from individual cells (Figs. 1 and 2), without contamination from anesthetics, gives them a significant advantage over in vivo preparations, setting them as the gold-standard model system for cellular and molecular experiments. Moreover, they permit direct access to the neurons (no need to pass the BBB), which allows perfusion of drugs and fast regulation of the external environment. In contrast to cultures or homogenates, acute slices maintain their structural integrity, and therefore allow the study of specific circuits of isolated networks [24–26], such as in the somatosensory or cholinergic system illustrated in Fig. 1. However, it should always be kept in mind that the slice is surrounded by an artificial environment. Many substances that are present in vivo and important in regulating neuronal function, such as trophic factors and amino acids, are not included in the aCSF. Furthermore, although local

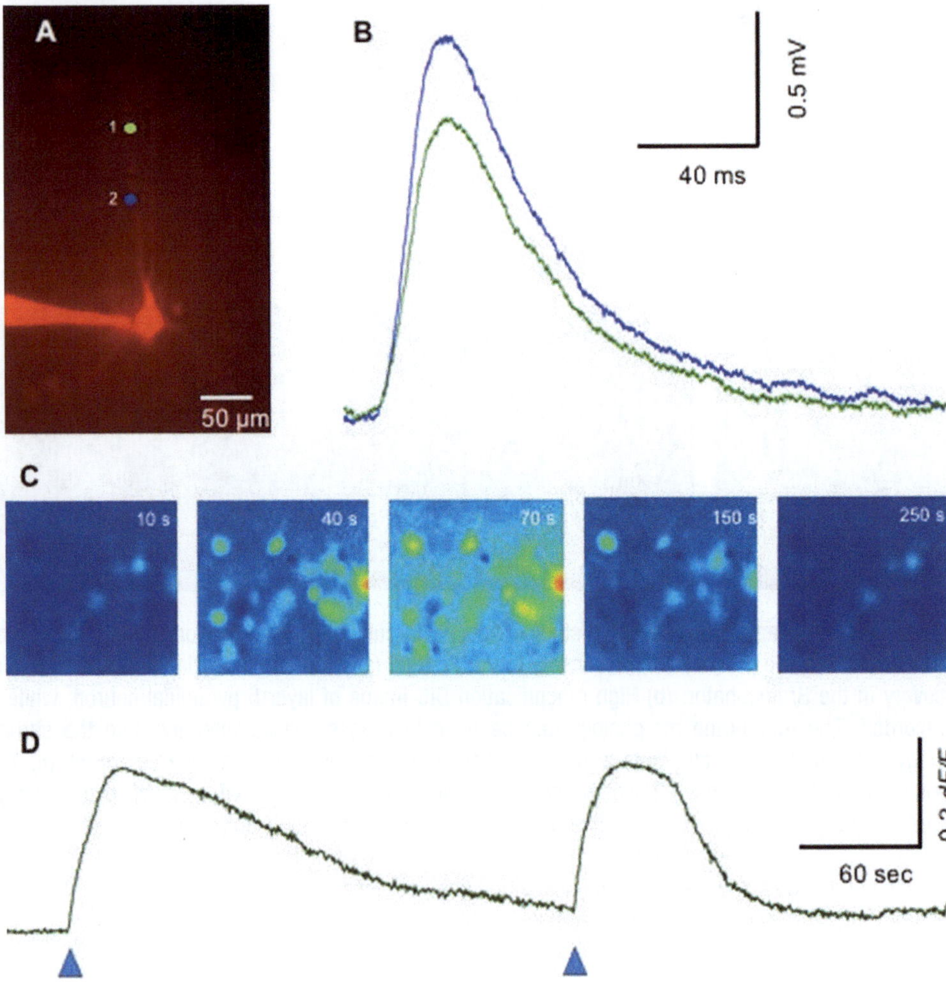

Fig. 2 Slice viability after long term incubation. (**a**) Image of L3 pyramidal cell (filled with Alexa-594 to allow visualization of neuronal compartments) from a YFP-channelrhodopsin-2 expressing mouse after 24 h of incubation in the Braincubator. The blue and green dots indicate regions of photostimulation. (**b**) electrophysiological recordings of the light stimulated synaptic activity in the cell from A. (**c**) Time lapse images of slices loaded with Fluo-4 AM for >24 h, depicting an increase in intracellular calcium levels following brief local application of 30 mM KCl (1 s). (**d**) Average trace of intracellular calcium transients in C following the application of KCl (blue triangles)

neuronal circuits may be preserved, the synaptic connections of the slice to and from other brain regions are interrupted.

1.3 Previous Efforts to Extend Slice Viability

As the slice preparation is often used to study circuits and the impact of pharmacological compounds on networks of neurons and glia, the viability of the slice is of great importance. In past years, scientists worldwide have tried to improve the viability of neuronal slices through various approaches (summarized in Table 1), including pretreating the animal with ketamine to reduce excitotoxicity [36], inducing hypothermia prior to decapitation

Table 1
Methodological approaches to extend slices viability

Methodology	What have they tried to do?	Effects	Ways of measurement	References
Replacement of NaCl with either: 1. High sucrose	Reduce cell swelling and excitotoxicity in slices from adult mice	Improved motoneuron viability	Electrophysiology (intracellular recordings)	[27]
2. NMDG	Increase cortical interneuron viability in slices from adult mice	Enhanced GABAergic interneuron viability	Density of GFP^+ neurons/mm^2 in GAD67-GFP knock-in mice	[28]
3. Glycerol	Reduce cell swelling and lysis to increase slice viability	Reduced cell swelling; increased cell viability in different brain areas and spinal cord	Live/dead ratio (morphology); electrophysiology (intracellular recordings)	[29]
Addition of agents: 1. HEPES (pH buffer)	Reduce cell swelling and excitotoxicity during slicing; increases carbon dioxide transport	Reduced adult slice edema; slowed slice deterioration	IR-DIC optics (morphology); Ca^{2+} imaging; electrophysiology (intracellular recordings)	[30]
2. Antioxidants: (ascorbate; pyruvate; thiourea)	Reduce cell swelling and lysis	Prevented cell swelling and brain edema; improved slice viability from adult animals	Water content; Ca^{2+} imaging (GCAMP3); electrophysiology (intracellular recordings)	[30, 31]
3. Antibiotics:	Increase long-term survival of adult slices	Extended slice life span for 15 hours	Electrophysiology (extracellular and intracellular recordings)	[32, 33]
4. Polyethylene-glycol (PEG)	Improve lower motoneuron's viability in adult rats	Maximizes tissue oxygenation; reseals damaged dendritic processes; enhances the number of viable motorneurons	Electrophysiology (intracellular recordings)	[14]
Technological methods: 1. Vibrotome	Adjustment of vibration amplitude with the vibroprobe	Reduced cellular damage on the slice surface	IR-DIC images; electron microscopy; electrophysiological recordings	[34]
2. Braincubator	Reduce bacteria levels; tight control of aCSF pH and temperature	Extend slices viability >24 h	DAPI/PI method; electrophysiology	[35]

[36], changing the aCSF composition to alleviate the direct damage of the slicing procedure [30], and adding protective antioxidants to reduce cellular deterioration [9]. Recently our group showed that a combination of hypothermia with tight control of the pH, temperature, and bacteria levels of the aCSF in the recovery chamber significantly extends slice viability over a period of more than 24 h [35, 37, 38]. In this chapter, we focus on methodologies developed in recent years that aim to extend and improve slice viability made from both young and old animals. Given the breadth of its usage, improving slice viability and extending life span can considerably reduce the cost, time, and number of animals used.

1.4 Reasons for Deterioration of Slice Viability

Preparation of acute neural tissue leads to considerable cellular damage. Immediately following slicing, neurons die by necrosis due to membrane disruption, excitotoxicity, and calcium influx [39], while a subsequent wave of neuronal death occurs via an apoptotic pathway mediated by mitochondrial dysfunction, ATP depletion, and the release of regulatory proteins [39, 40]. Whittingham and colleagues have shown that after slicing, ATP levels in brain slices drop significantly, probably due to ischemia [41]. Feig and colleagues showed that after 45 min incubation at ~37 °C, ATP levels return to normal [42]. This is likely the main reason for a minimum incubation period before measuring any physiological activity from these preparations.

Recently, we have used a cell death/live assay to measure slice viability over time [35, 38]. Our results indicated that approximately 30% of all cells in the slice die within 1 h after slicing. The reasons for cellular deterioration can be divided to several chronological subcategories: (1) preslicing, (2) slicing, and (3) incubation (short-term and long-term post slicing), as indicated in Table 2. Although some of the cellular decay is inevitable due to the ischemia that occurs prior to slicing and the traumatic stress during slicing, alterations of the experimental procedures can mitigate the cellular deterioration over time.

1.4.1 Preslicing

Neuronal deterioration starts immediately after decapitation, as disruption of the blood supply to the neural tissue leads to ischemia followed by hypoxic processes. The direct impact of the ischemic episode, which starts within seconds after the termination of blood supply, is depletion of adenosine triphosphate (ATP), followed by dysfunction of ATP-dependent processes, mitochondrial dysfunction, and excitotoxicity through massive release of glutamate [45]. The excitotoxicity is mediated by a calcium influx through NMDA receptors, which accumulates in the cytosol to toxic levels. Additionally, depolarization of the cell leads to reversal of the activation of the Na^{+}-Ca^{2+} exchanger, causing more calcium accumulation [46]. Moreover, Ca^{2+} is driven into the mitochondria, leading to inhibition of oxidative phosphorylation, enhanced

Table 2
Improving slice viability at different phases

Phase	The main cause for slice deterioration	How to improve slice viability?	References
Pre-slicing	Ischemia; hypoxia	Perfusion of cold aCSF to induce hypothermia	[43]
Slicing	Ischemia; trauma; dendriotomy; excitotoxicity; swelling	1. Neuroprotective cutting method (high sucrose, NMDG, glycerol)	[27, 29, 30]
		2. Vibroslicer with vibroprobe	[34]
Recovery—short term	Cytotoxicity; cellular deterioration; depletion of glutathione and decrease of ATP	1. Addition of antioxidants, e.g., pyruvate and ascorbate	[44]
		2. 10–12 min incubation at NMDG aCSF (32–34 °C)	[30]
Recovery—long term	Apoptosis; hyperexcitability; lack of normal circuitry and neuromodulation; high bacteria levels	Hypothermia (15–16 °C) and UVC filtration to reduce bacteria growth and improve life span for >24 h (Braincubator™)	[35]

generation of reactive oxygen species (ROS), oxidative stress and lipid peroxidation which results in membrane failure, inflammation, and apoptosis [9, 45, 47].

Neurons are especially vulnerable to the ischemic process, due to their high rate of metabolism and dependence on glucose as an energy source [10, 48]. Although in vivo glycolysis provides only 5% of the energy needs of the neuron, during hypoxia, it is the only source of energy [49]. For that reason glucose levels in aCSF are normally much higher than in the CSF (2–7 Mm), ranging from 10–26 mM [50, 51]. To reduce neural loss, many physiologists perfuse cold aCSF to induce deep hypothermia, which reduces the metabolic rate and thus limits ischemic damage [52].

1.4.2 Slicing

The slicing procedure is damaging by nature, as it ruptures the membrane (cell body and/or processes) of most of the cells within the tissue. Essentially, the viable cells are confined by two layers of significantly damaged tissue. To rectify the cellular damage occurring during slicing, the cellular membrane of individual cells seals itself, to avoid further leakage of cytosolic contents. The damage during the slicing procedure also results from excitotoxicity caused by the spillover of glutamate from ruptured cells, which leads to further calcium cytotoxicity and the detrimental cascade detailed above.

To reduce the mechanical damage elicited by slicing, Jonas's group developed a Vibroslicer with minimal vertical blade vibrations (>1 μM), which reduces the depth of the damaged area on the slice surface [34]. Based on their work, several companies (e.g., Leica, Camden instruments) have produced vibrotomes with a Vibrochek, a device that aligns and reduces the blade deflection in the Z-axis. This reduces the number of cells that are irreparably damaged, and consequently, the resulting depolarization and excitotoxicity (Table 1). Further reduction in cellular damage during the slicing procedure can be achieved by using the "protective cutting method," as described in the Methods section.

1.4.3 Recovery—Incubation Period

The damage during the recovery period can be divided into short and long (>3 h) term. The short-term damage is a direct result of the slicing procedure. K^+ spillover from the leaky neurons and glia results in transient depolarization. Furthermore, cellular deterioration is mediated through secondary processes of the ischemic period (postischemic deterioration), when intracellular signaling such as activation of PKC and apoptotic pathways takes place in order to mitigate the overall damage to the tissue [53]. During this period, the cellular morphology of apoptotic cells will look a bit shrunken (Fig. 1b) and can serve as a way to assess the overall slice viability as detailed in Table 3 [42].

The largest consumer of ATP in the brain is the electrogenic Na/K pump, which maintains the intracellular Na^+ and K^+ levels against their concentration gradients [61]. Therefore, the reduction in ATP levels during the slicing procedure leads to dysfunction of the Na/K ATPase, and accumulation of intracellular Na^+. High levels of intracellular Na^+ promotes passive leak of Cl^- and attraction of water molecules into the cell to reduce osmolarity, resulting in cell swelling and death.

Cellular deterioration during the incubation period also occurs due to lack of synaptic inputs, reorganization of the neuronal network, and lack of efficient waste removal system, causing further toxicity. Moreover, cell death can be triggered by the depletion of cytoplasmic glutathione resulting from the loss of intracellular glutamate due to oxidative stress [62].

Further to the reasons mentioned above, deterioration of slice viability during long-term incubation (>3 h) is also affected by delayed and prolonged processes including pH fluctuations that occur following increased metabolism, and decreased penetration of dissolved oxygen and glucose into deep layers of tissue due to cell swelling on the slice surface [63]. Moreover, as the aCSF provides the ideal conditions for bacterial growth, this leads to increased bacteria levels that release endotoxins, causing neuroinflammation and cell death [38].

Table 3
Methods to assess slice viability

Method	Viability measurement	Advantages	Disadvantages	References
Metabolic activity	Levels of phosphocreatine and inorganic phosphates	The level of metabolites gives information about the quality of the slice	Need homogenates, cannot look into individual cells	[4, 54]
	Mitochondrial function by TTC (2,3,5-triphenyltetrazolium chloride)	The quality of the slice can be monitored during the measurement	The process of staining depends on multiple factors, the assessment is subjective	[55]
	Protein synthesis by protein-incorporated [^{3}H]leucine	Changes in protein synthesis can indicate decreases in cell viability	Can only be assessed post experiment	[56]
Morphological integrity: 1. IR-DIC imaging	Appearance of live cells	Allow visualization of individual cells	Subjective; based on the observer's judgement No information about functional integrity	[42]
2. Fluorescent labeling of living/dead cells: propidium iodide (PI); Hoechst dye; DAPI; GFP-*knock-in* detection)	PI—Selectively stain nuclei of dead cells	Accurate measurement over depth of tissue	No information about functional integrity	[57]
	Hoechst dye fluorescent labeling of living cells		Confocal microscope is required for dual imaging	[58]
	DAPI—Fluorescent staining of all nucleus			[59]
	Cell-counting of GFP-positive neurons (considered to be live cells)			[28]

(continued)

Table 3
(continued)

Method	Viability measurement	Advantages	Disadvantages	References
Functional integrity: 1. Electrophysiology	Intra- and extracellular recordings of membrane potential and spiking activity Light evoked responses that require functional circuitry—lateral inhibition (retina)	In situ measurement of functionality	Difficulties with recordings from deep layers of the tissue	[36, 52]
2. Calcium imaging	Calcium-AM dyes (i.e., Calcein-AM/Fura-2 AM) stain live cells	Function of both individual cells and a large area of the neuronal network can be assessed	Penetration of dyes to deep layers is time consuming. Requires sophisticated light stimulating equipment and expertise	[56, 60]

2 Materials

2.1 Artificial Cerebrospinal Fluid

Artificial cerebrospinal fluid (aCSF) is a solution used to mimic the brain interstitial fluid and provides neurons the necessary ingredients that allow them to survive in vitro for several hours. Although the ideal composition of aCSF mirrors that of the cerebrospinal fluid (CSF) in vivo, investigators often use an aCSF that differs from CSF, to allow better viability of the tissue. Normally, the composition of aCSF is based on Ringer's solution [3] and includes NaCl, KCl, NaH_2PO_4, Na_2HPO_4, $NaHCO_3$, glucose, $MgCl_2$, and $CaCl_2$, however, the concentration of the ionic components of the aCSF varies slightly from lab to lab to meet different experimental purposes (or personal superstitions). As metabolic processes and the maintenance of the ionic gradient across the cell membrane constantly consumes ATP, and oxygen diffusion to the center of the slice will be impaired, glucose must be supplied in higher levels to overcome the constant demand to generate energy (allowing increased glycolysis). To reduce cellular damage resulting from the slicing procedure, a chemical approach can also be utilized, in which the aCSF is altered to provide transient protection to the tissue, as detailed in Table 1.

Prior to slicing, most investigators replace NaCl with high sucrose solution (to maintain osmolarity) or other cation, such as N-methyl D-glucamine (NMDG), or Choline, to prevent the membrane depolarization that leads to excitotoxicity. There is no consensus among investigators in regards to the best Na^+ substitution, however, Yasuyo and colleagues showed that NMDG based aCSF was most effective is maintaining GABAergic interneurons [28]. Hence, prior to slicing, the aCSF we are using contains (in Mm): 92 NMDG, 2.5 KCl, 1.25 NaH_2PO_4, 30 $NaHCO_3$, 20 HEPES, 25 glucose, 2 thiourea, 5 Na-ascorbate, 3 Na-pyruvate, 0.5 $CaCl_2$, 10 $MgSO_4$ (310 mOsm; pH 7.4), as previously reported by [30]. The aCSF should be ice-cold, to induce hypothermia and continuously carbogenated to avoid Ca^{2+} precipitation. As calcium can precipitate in pH above 7.8, do not add calcium before the aCSF is fully carbogenated.

The above NMDG-aCSF is used for the preslicing and slicing episodes; however, for the recovery period it should be replaced with aCSF that contains physiological concentrations of calcium and sodium, which are essential for functional activity of the neurons. Thus, during the recovery period, we use aCSF that contains (in Mm): 125 NaCl, 2.5 KCl, 1 $MgCl_2$, 1.25 NaH_2PO_4, 2 $CaCl_2$, 25 $NaHCO_3$, 2 thiourea, 5 Na-ascorbate, 3 Na-pyruvate, 20 HEPES, 25 dextrose, and saturated with carbogen (310 mOsm; pH 7.2–7.4).

2.2 Retinal External Solutions

Ames media is one of the most common external solutions used by retinal electrophysiologists. It was conceived as a solution that would closely mimic not only the ionic concentrations of CSF, but also the amino acid and protein content. Ames and Nesbett showed that Ames media, with 37 added amino acids, vitamins and supplements, prolonged the health of retinal tissue, by comparing light stimulated electrophysiological responses, and protein synthesis [56]. However, basic aCSF as described for brain slices above can also be used for retinal tissue recordings, especially in cases where components of the extracellular solution must be altered for the experimental paradigm. However, for long term incubation, Ames media would be the most appropriate incubation solution.

3 Methods to Reduce Cellular Damage

Slice viability can be improved by both chemical and technical means that reduce cellular damage. While the chemical approach is mainly through the addition of chemicals that reduce excitotoxicity and edema, the technical approach uses high impact machinery, such as the vibrotome that minimizes the damaged area and the Braincubator that provides a stable and optimal environment. The main methods to improve slice viability and extend their life span are detailed below.

3.1 Hypothermia

Hypothermia is divided into four sequential categories: mild (34–37 °C), moderate (28–34 °C), deep (17–28 °C) and profound (4–17 °C). Although not fully understood, over the years it has been shown that hypothermia provides neuroprotection via several mechanisms, including reduced metabolic rate and energy depletion, decrease in excitatory transmitter release, reduced alterations in ion flux, and reduction in edema [64]. The impact of temperature on metabolic rate is defined by the temperature coefficient, Q-10, which describes the rate of change of a biological reaction after increasing the temperature by 10 °C. For most biological reactions, Q-10 lies between 2 and 3, however, the Q-10 for the cerebellum is 2.2 during moderate hypothermia, and 4.5 during deep hypothermia [65], which illustrates how deep and profound hypothermia enhances tolerance for ischemic episodes.

Adenosine triphosphate (ATP) is the common molecule that serves as the energy "coin" that supports all forms of biological processes and thus essential for tissue viability [66]. To preserve neuronal functions, the physiological ATP concentration must be maintained. Decreasing the tissue temperature slows the rate of glucose metabolism (and thus ATP formation), but also decreases phosphocreatine and ATP breakdown [67], which helps to maintain energy homeostasis. Furthermore, it reduces the formation of lactate and inorganic phosphate [68], and thus improves the recovery of energetic parameters after slicing.

Ion permeability through the membrane is also temperature dependent as Na^+ permeability increases in high temperatures, [63] and calcium excretion is compromised in low temperatures, which directly impact the tissue excitability. Moreover, hypothermia is found to alter stress-induced protein expression [69] and to reduce the lethal response to bacteria released endotoxins by attenuating the violent host-inflammatory response that endotoxins initiate [68]. For these reasons, it has been concluded that lowering the tissue temperatures during global cerebral ischemia helps to protect and mitigate neuronal injury by preserving brain ATP supply [65], reducing oxidative damage, and maintaining mitochondrial function [70]. We induce profound hypothermia at the preslicing and slicing phases through perfusion of 4 °C NMDG-aCSF, as detailed below. Following slicing, and **short incubation** at 32 °C for 10–12 min, we reduce the tissue temperature to 15–16 °C until use.

To induce deep/profound hypothermia before the ischemic episode, experimenters often use cardiac perfusion of cold (4 °C) aCSF, prior to decapitation. This ensures the temperature decrease not only on the surface of the brain, but also in the core of the tissue [43]. The steps for a successful animal perfusion, protective cutting, and recovery incubation that further extends slice viability are as detailed below.

3.2 Animals Perfusion of aCSF

Tools: Insulin syringe with anesthetics, blunt scissors, fine scissors, blunt forceps, hemostat, butterfly needle, perfusion pump with pipes, polystyrene block, and needles to fix the limbs.

1. Deeply anesthetize the animal with an intraperitoneal injection of anesthetics (e.g., a mixture of ketamine (100 mg/kg) and xylazine (10 mg/kg) or sodium pentobarbital (10 mg/kg)) using an insulin syringe to minimize the stress of administration. Wait until the animal is unresponsive to external noxious stimuli (absence of reaction to pinch of the paws or the tail with forceps) before proceeding to the next step.
2. Place the animal in supine position and pin the limbs onto a polystyrene block with needles. Alternatively, tape the limbs using a masking tape.
3. Lift the skin from the xiphoid process using tweezers and cut through to the body cavity with blunt scissors. Make incisions to the lateral sides and cut through the diaphragm while holding the sternum. The ribcage can then be cut off to expose the heart. Pay careful attention to avoid damaging the organs and rupturing the circulation unnecessarily.
4. Stop the carbogenation of the NMDG-aCSF solution to avoid bubbles in the pipes. Start the perfusion pump to fill the pipes with the ice-cold solution. Gently hold the beating heart with blunt forceps and pierce the butterfly needle into the left ventricle. Secure the needle in place with a hemostat. Be careful not to push the needle through the heart.
5. Immediately cut the right atrium with fine scissors to allow the solution to flow through and drain out of the circulation. Clearing of the liver, as well as the solution at the atrium indicates the success of the perfusion.
6. Following the perfusion, animal can be decapitated and the brain can be removed as detailed below. The perfusion procedure will essentially provide slices with increased viability; however, it is main impact is on aged animals, as we have noticed that young animals (>14 days old) are less susceptible to ischemic processes and thus typically yield high quality slices even without perfusion.

3.3 The Protective Cutting Method

As the aCSF is important for tissue viability, its composition is often altered for slice preparation, storage, and recording. In their quest to preserve motorneuron health, Aghajanian and colleagues modified the "cutting aCSF" to reduce sell swelling and lysis which follows passive chloride entry [27]. They have shown that substitution of NaCl with high sucrose during slicing and recovery periods (>1 h), yielded a higher number of viable motorneurons. Following their work, Richardson and Messer showed that alteration of the aCSF to include high osmolarity, zero calcium, and kynurenic acid, significantly decreased cytotoxic edema and enhanced cellular viability [71]. These early studies led to the development of the "protective cutting method," reviewed by [30], in which the aCSF used for the slicing procedure contains sodium chloride substitutions, such as sucrose, choline, or NMDG to provide neuroprotection during the traumatic slicing phase. Through a systematic investigation of the protective cutting method, Ting and colleagues have optimized the key ingredients essential to improve the health of adult tissue slices [30]. They have shown that adding HEPES and ascorbate–thiourea combination reduces slice edema and slows slice deterioration, which is of special importance in experiments involving visual identification of neurons, as it improves the overall visibility of the slice. Moreover, Hong-Ye and colleagues showed that replacement of NaCl with glycerol during slicing is also neuroprotective as it prevents passive chloride entry, subsequent cell swelling and lysis [29].

The steps for successful protective cutting of neural tissue slices are as follows:

Reagents: ice-cold, continuously carbogenated NMDG aCSF in a beaker (for components and concentration, see previous section).

Tools: rodent guillotine, scalpel with blade, sharp scissors, fine forceps, spatula, plastic Pasteur pipette, beaker (~30–50 mL), plastic bag for the carcass.

1. Following perfusion decapitate the animal using a rodent guillotine or large scissors. The tool of choice would be in correlation to the size of the animal (use scissors for young and small animals, and guillotine for old and large rodents).
2. Remove the brain quickly and place it into ice-cold NMDG-aCSF saturated with carbogen (310 mOsm; pH 7.4), as detailed in **step 2.1**. In order to expose the bone, perform a median incision using sharp scalpel, starting from the nasal bone to the occipital surface. To get a clear, visible surface, clean the cervical part from remaining muscles. Once the bone is exposed, use sharp scissors to make an incision from the middle of the occipital bone to the nasal bone along the sagittal suture. For old animals, make additional incisions on the lateral sides of the skull, to help removing the bone in the next step.
3. Gently remove the interparietal and parietal bones with forceps from both sides; hence, the dorsal surface of the brain becomes visible. Take great care when removing the bone to not damage the tissue with the forceps.
4. Hold the skull above the beaker in a vertical position and gently scoop the brain with a spatula into the ice-cold carbogen saturated NMDG-aCSF solution. The brain must be removed carefully to avoid mechanical damage to the tissue, and as fast as

possible (<1 min) to minimize hypoxia. The aCSF solution in contact with the brain must be constantly saturated with a mixture of 95% O_2–5% CO_2, to provide oxygenation and balanced pH of 7.2–7.4.

5. After 30 seconds in the cold aCSF, remove the brain and prepare a flat surface that will be used to glue the brain to the cutting stage in the next step. *For neocortical slices (S1/M1)*, separate the brain into two hemispheres by a longitudinal cut along the midline. This will generate two hemispheres with a flat medial surface that can be later placed on the cutting stage.
6. Place the removed brain on an appropriate cutting stage and secure it with super-glue (preferably the gel form). Make sure the glue dries before submerging the preparation into ice-cold carbogen saturated NMDG aCSF solution. Pay attention to the orientation of the brain in reference to the vibrating knife, as this will determine the orientation of the slices and structures that will be maintained (coronal vs sagittal slices).
7. Use a vibrating microtome, previously calibrated with Vibrochek to minimize the Z-deflection vibration, to cut 300 μm thick brain slices in the preferred orientation that includes the region of interest. Adjust vibration amplitude, speed and frequency of the blade before slicing to ensure clean cutting. The entire slicing procedure should be completed in less than 10–15 min.
8. Transfer slices using a cutoff plastic Pasteur pipet into a custom built incubation system (Braincubator), containing probes for measuring pH levels (7.2–7.4), carbogen flow and temperature, as previously described ([35]; see also Figs. 3, 4, and 5). For the initial incubation period, slices can also be transferred into a chamber containing NMDG aCSF for 12 min at 32–34 °C, as previously described by [30], and then placed into the Braincubator, which includes an incubation aCSF.

3.4 Preparation of Retinal Whole Mount and Slices

In contrast to brain slices, the mammalian retina can either be isolated intact, or sliced (Fig. 6). The tissue is relatively thin (100–400 μm depending on species) and therefore remains oxygenated without being sliced. Avascular retina in particular (e.g., rabbit and guinea pig) survives removal from its host particularly well since oxygenation via aCSF is similar to in vivo [72]. However, dissection procedures for the retina vary considerably from that of brain slices. The main difference is that, depending on the purpose of the experiment, most retinal dissections are completed under far red light (>750 nm). This adds a significant level of complexity to the dissection and necessitates light tight dissection rooms and incubation apparatus. However, this inconvenience is certainly

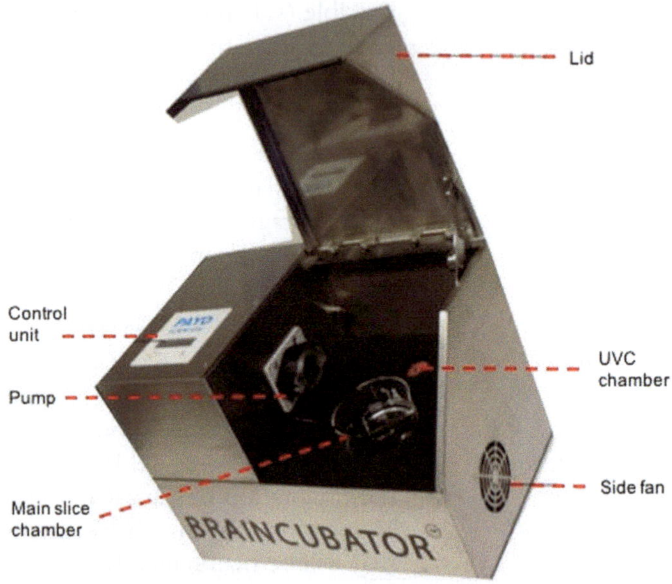

Fig. 3 Description of the Braincubator. (**a**) Photograph of the recovery incubation system made of a main chamber sitting on a Peltier plate, and a separate hidden UVC chamber. The circulation of aCSF between the chambers is mediated by a peristaltic pump

countered by the advantage of being able to activate retinal circuits with light. We present here ways to produce the most viable retinal tissue that can be maintained for >24 h:

3.4.1 Dissection

The majority of papers report that retinal dissections are completed at room temperature, as opposed to <4 °C for brain slices. Certainly, in our hands, retinal viability is greatly reduced when dissection is completed at <4 °C [73].

Whole Mount

Retinal neurons require high levels of oxygenation, especially when dark adapted [74]. Therefore, the eye must be quickly dissected following enucleation. Once the cornea, lens, and vitreous are removed, the eye-cup preparation can be stored for long periods of time without significant degradation. Furthermore, incubation in a custom-made chamber at 15–16 °C with tightly regulated pH (Braincubator) greatly increases the amount of time the tissue will remain viable [75].

Retinal Slices

To access interneurons (bipolar, horizontal, and amacrine cells) of the retina, slice preparations are frequently used. Vertical slices are most commonly prepared but horizontal slices can provide access to wide-field amacrine and horizontal cells without truncating their extensive dendritic fields [76]. In both instances, because of the delicate and thin nature of the retina, it must either be embedded in

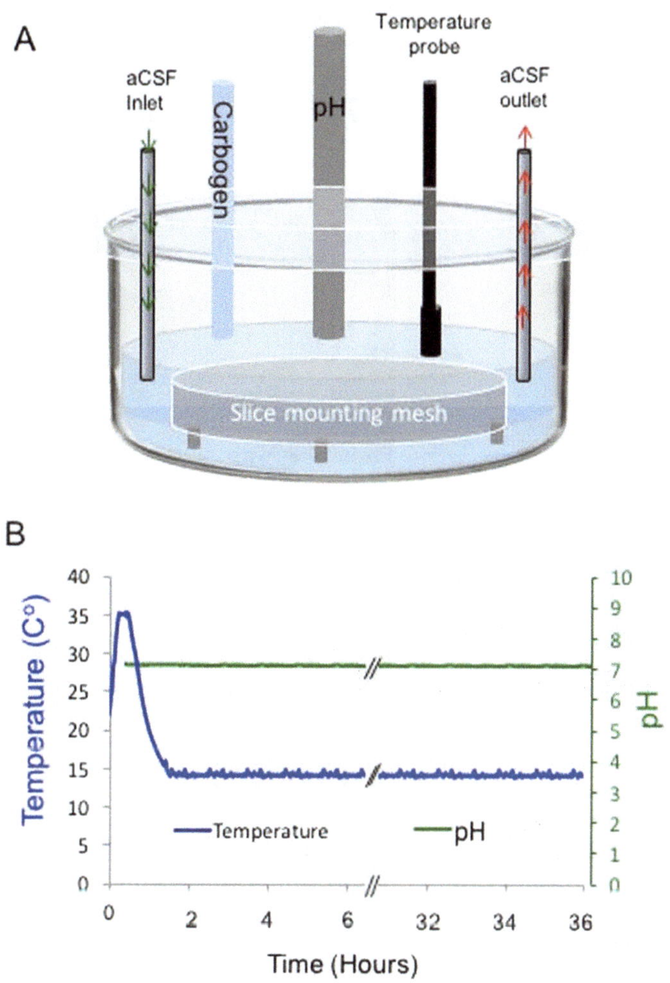

Fig. 4 Monitoring aCSF conditions. (**a**) Schematic diagram of the main slice chamber and the measuring probes in relation to the slice mounting mesh. (**b**) Plot of the aCSF parameters, illustrating the temperature and pH of the aCSF during incubation over a period of 36 h

low-gelling temperature agarose, made up in HEPES-buffered aCSF, prior to slicing; or attached to nitrocellulose filter paper and then sliced with a tissue chopper. The low-gelling temperature allows the retina to be embedded at <38 °C to avoid heat damage of the tissue. Once embedded, the agarose is rapidly cooled with ice-cold aCSF to solidify and suspend the retina for slicing (Fig. 6; [73]). Slicing can then be completed as described for brain slices.

Retinae attached to nitrocellulose is chopped with a McIlwain tissue chopper, and then orientated correctly by embedding the edges of the paper in vacuum grease in the recording chamber with the transverse section of the retina facing upward for easy access to bipolar, amacrine, and horizontal cells [77]. The advantage however, of suspending the retina in agarose is that the width of the

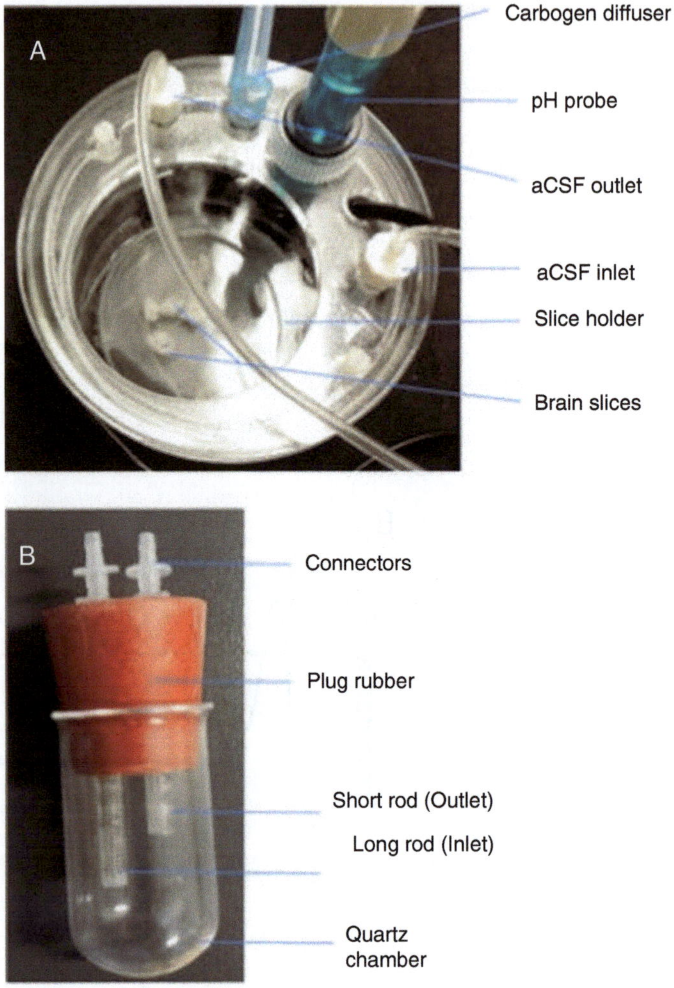

Fig. 5 Braincubator chambers. (**a**) Top view photograph of assembled slice chamber depicting the pH and temperature measuring probes along with the aCSF inlet and outlet. Note the brain slices mounted on a mesh in the bottom of the chamber. All measuring probes are attached to the lid to allow structural stability. (**b**) A side view photograph of the assembled UVC chamber depicting its organization. The chamber is made of Quartz tube to allow UVC light penetration to sterilize the aCSF flowing in the UVC chamber

tissue is extended allowing the retinal slice to lie flat and creating more stability in the slice. The researcher must choose the method appropriate to their experiments, however the tissue chopper is generally easier to use under dim red light when preparing dark adapted slices. Slices can also be incubated in the Braincubator under the same conditions as whole mount tissue, and remain viable for at least 24 h.

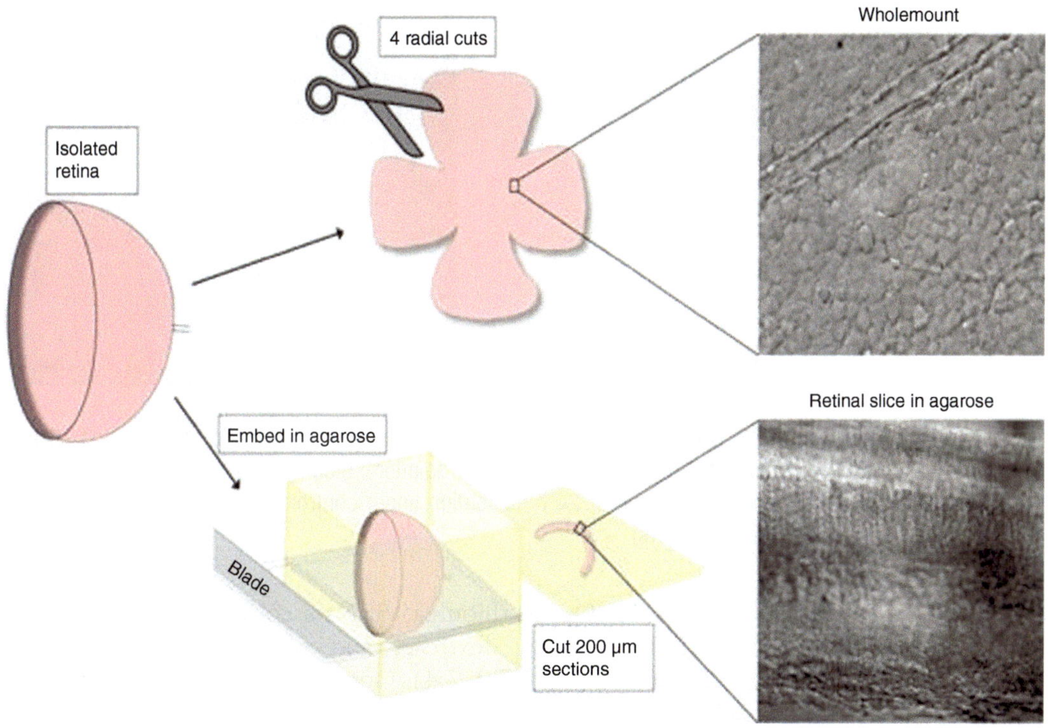

Fig. 6 Two options for retinal dissection. Once isolated, the retina can either be cut radially to allow it to be flattened (whole mount; top), or embedded in agarose and sliced with a vibrating microtome to produce retinal slices (bottom). Representative differential contrast images (DIC) of whole mount (top; ganglion cell up) and vertical retinal slice (bottom) show the products of these two approaches

3.5 Braincubator

To reduce cellular damage during incubation, we have designed an active recovery incubation chamber, which controls several parameters of the extracellular environment, such as temperature, pH, and bacteria levels (Fig. 5). As hypothermia has been shown to reduce metabolic rate and thus cellular damage, we have incubated brain slices and retinal whole mounts at 15–16 °C. The combination of hypothermia and UVC filtration to irradiate bacteria, proved to reduce bacteria levels and extend slice viability for >24 h (Fig. 7).

The design of this chamber, called the Braincubator, has been detailed previously [35, 38, 75, 78]. In this system, the aCSF is circulated through two chambers: the main chamber holds the slices and is constantly monitored for temperature and pH that are controlled through a closed loop system, and a secondary UVC chamber is exposed to UVC light to eradicate bacteria (Figs. 3–5).

3.5.1 Maintaining Neural Tissue in the Braincubator

1. Preset the temperature of the aCSF in the main chamber to 32 °C. The aCSF in the Braincubator should contain (in Mm): 125 NaCl, 2.5 KCl, 1 $MgCl_2$, 1.25 NaH_2PO_4, 2 $CaCl_2$,

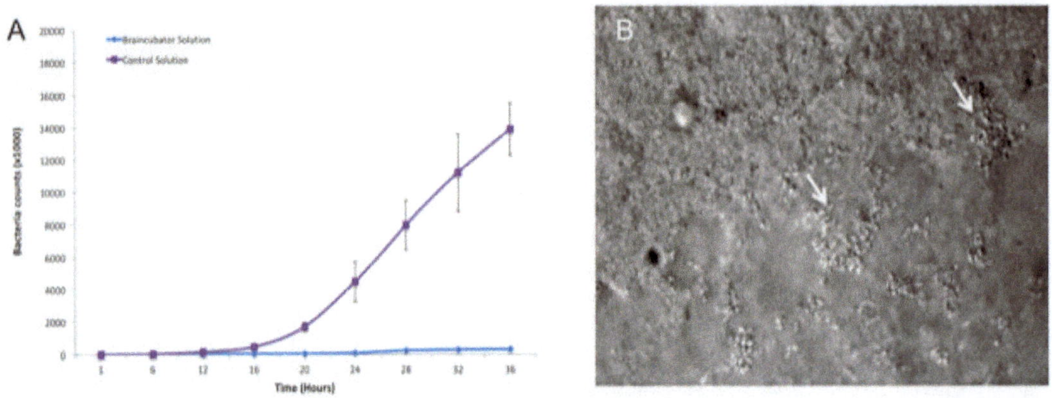

Fig. 7 Bacterial growth in aCSF. (**a**) Plot depicting the bacterial growth over a period of 36 h post-slicing, purple—under control conditions (room temperature, no UV irradiation); blue—at 15 °C with UVC irradiation. (**b**) A photograph of the surface of a slice after 24 h incubation under control conditions depicting bacteria colonies grow on the slice surface (arrows)

25 $NaHCO_3$, 2 thiourea, 5 Na-ascorbate, 3 Na-pyruvate, 20 HEPES, 25 dextrose, and saturated with carbogen (310 mOsm; pH 7.2–7.4). Once all slices are in the Braincubator (will take approximately 10–12 min), reduce temperature to room temperature for 10 min (22 °C), and then to 15–16 °C.

2. Using a peristaltic pump, circulate the aCSF through a second UVC chamber isolated from the main chamber and periodically exposed to 1.1 W UVC light (254 nm, 5 W/2P sterilizer UV lamp) to eradicate bacteria floating in the solution, as previously described [35, 38, 78]. Make sure the UVC chamber is "light tight," so no light is getting out of this chamber, as it can be hazardous to human tissue and can damage the slices. Control UVC light timing via a programmable timer using a random feature that turns ON at times varying between 15 and 26 min every 15–30 min to avoid excessive heating of the aCSF. For dark-adapted retinal tissue, cover the main chamber to exclude light. The main chamber can be mounted on a Peltier thermoelectric plate to either cool or heat the aCSF in the main chamber to the desired temperatures (0–50 °C). Set the flow rate in the UVC chamber to 10–12 mL/min.

3. Keep slices in the Braincubator at 15–16 °C until needed for imaging or electrophysiology experiments (up to 36 h, Fig. 5), as previously described [35, 79]. Make sure that no carbogen bubbles are on the mesh, as these can lead to lipid peroxidation and slice deterioration.

3.6 Antibiotics

Several protocols aiming to improve slice viability over time have adopted a different approach which involves the use of antibiotics, such as penicillin, to reduce bacteria levels [32]. However this approach introduces more complexity, as most antibiotics impact the biophysical properties of neurons [80], and therefore, should be examined carefully by the experimenter before their use, to avoid misinterpretation of the results.

4 Summary

Cellular deterioration is influenced by many functions, which we can be divided into two categories after slicing: (a) rapid and direct impact of the slicing procedure, and (b) late/delayed processes, that occur long after the slicing took place (>1 h), see Table 3. To reduce cellular damage during the short term incubation period, experimenters developed methods that focused on replenishing ATP levels (e.g., glucose/glutathione, sodium L-ascorbate, ethyl pyruvate, and myoinositol) [9], which proved to reduce energy failure and cell swelling.

The short-term vulnerability of acute neural tissue preparation is a result of several processes, including (1) the slicing procedure itself, which damages neuronal compartments and thus makes them leaky and exposed to excitotoxicity, (2) altered enzymatic activity, due to significant drop of energy molecules, such as ATP, (3) modification of cellular network, due to lack of innervation from isolated regions, (4) cell swelling due to increased permeability to chloride and further movement of water molecules to balance osmolarity, and (5) excitotoxicity caused by spillover of excitatory neurotransmitters such as glutamate followed by activation of NMDA receptors and increased intracellular calcium levels [42].

Most of the electrophysiological experiments involving imaging of neuronal tissue (e.g., patch clamp, photoactivation) are done in young animals. This is mainly due the fact that the viability and visibility of individual cells is much better, as the neuropil is less dense than in aged animals, allowing better diffusion of metabolites to the tissue. Moreover, tissue from young animals is more resistant to ischemia. However, tissue viability for in vitro experiments can be improved and extended, once an optimal environment is maintained as described in the methods section.

Development of the isolated tissue preparation has been pivotal to our understanding of the functions of neurons and neural networks. In this chapter, we have detailed methods currently employed to ensure, as far as is possible, that the excised tissue is viable and the experimental results obtained are meaningful. However, the experimenter should always keep in mind that the tissue has been removed from its normal environment, and care must be taken in the design of experiments and interpretation of the results.

References

1. Warburg O, Wind F, Negelein E (1926) The metabolism of tumors in the body. J Gen Physiol 8:519–530
2. Ashford CA, Dixon KC (1935) The effect of potassium on the glucolysis of brain tissue with reference to the Pasteur effect. Biochem J 29:157–168
3. Dickens F, Greville G (1935) The metabolism of normal and tumour tissue: neutral salt effects. Biochem J 29:1468–1483
4. Mcilwain H, Buchel L, Cheshire JD (1951) The inorganic phosphate and phosphocreatine of brain especially during metabolism in vitro. Biochem J 48:12–20
5. Li C-L, Mcilwain H (1957) Maintenance of resting membrane potentials in slices of mammalian cerebral cortex and other tissues in vitro. J Physiol 139:178–190
6. Hillman H, Mcilwain H (1961) Membrane potentials in mammalian cerebral tissues in vitro: dependence on ionic environment. J Physiol 157:263–278
7. Yamamoto C, McIlwain H (1966) Electrical activities in thin sections from the mammalian brain maintained in chemically-defined media in vitro. J Neurochem 13:1333–1343
8. Otsuka M, Konishi S (1974) Electrophysiology of the mammalian spinal curd in vitro. Nature 252:733–734. https://doi.org/10.1038/252497a0
9. Mitra P, Brownstone RM (2012) An in vitro spinal cord slice preparation for recording from lumbar motoneurons of the adult mouse. J Neurophysiol 107:728–741. https://doi.org/10.1152/jn.00558.2011
10. Kerkut G, Bagust J (1995) The isolated mammalian spinal cord. Prog Neurobiol 46:1–48
11. Takahashi T (1978) Intracellular recording from visually identified motoneurons in rat spinal cord slices. Proc R Soc London 202:417–421
12. Gibb A, Edwards F (1994) Patch clamp recording from cells in sliced tissues. Microelectrode Tech:255–274
13. Rekling JC, Funk GD, D a B et al (2000) Synaptic control of motoneuronal excitability. Physiol Rev 80:767–852. https://doi.org/10.1001/jama.2014.15298.Metformin
14. Carp JS, Tennissen AM, Mongeluzi DL et al (2008) An in vitro protocol for recording from spinal motoneurons of adult rats. J Neurophysiol 100:474–481. https://doi.org/10.1152/jn.90422.2008
15. Carlin KP, Jiang Z, Brownstone RM (2000) Characterization of calcium currents in functionally mature mouse spinal motoneurons. Eur J Neurosci 12:1624–1634
16. Bennett DJ, Li Y, Siu M (2001) Plateau potentials in sacrocaudal motoneurons of chronic spinal rats, recorded in vitro. J Neurophysiol 86:1955–1971
17. Genovese T, Esposito E, Mazzon E et al (2009) Beneficial effects of ethyl pyruvate in a mouse model of spinal cord injury. Shock 32:217–227. https://doi.org/10.1097/SHK.0b013e31818d4073
18. Ottoson D, Svaetichin G (1953) The electrical activity of the retinal receptor layer. Acta Physiol Scand 29:31–39. https://doi.org/10.1111/j.1748-1716.1953.tb00995.x
19. Tomita T (1965) Electrophysiological study of the mechanisms subserving color coding in the fish retina. Cold Spring Harb Symp Quant Biol 30:559–566. https://doi.org/10.1101/SQB.1965.030.01.054
20. Buskila Y, Farkash S, Hershfinkel M, Amitai Y (2005) Rapid and reactive nitric oxide production by astrocytes in mouse neocortical slices. Glia 52:169–176
21. Cameron MA, Al AA, Buskila Y et al (2017) Differential effect of brief electrical stimulation on voltage-gated potassium channels. J Neurophysiol. https://doi.org/10.1152/jn.00915.2016
22. Buskila Y, Crowe SE, Ellis-Davies GCR (2013) Synaptic deficits in layer 5 neurons precede overt structural decay in 5xFAD mice. Neuroscience 254:152–159
23. Flynn JR, Conn VL, Boyle KA et al (2017) Anatomical and molecular properties of long descending propriospinal neurons in mice. Front Neuroanat 11:1–13. https://doi.org/10.3389/fnana.2017.00005
24. Agmon A, Connors B (1991) Thalamocortical responses of mouse somatosensory (barrel) cortexin vitro. Neuroscience 41:365–379
25. Buskila Y, Morley JW, Tapson J, van Schaik A (2013) The adaptation of spike backpropagation delays in cortical neurons. Front Cell Neurosci 7:192. https://doi.org/10.3389/fncel.2013.00192
26. Kantevari S, Buskila Y, Ellis-Davies GCR (2012) Synthesis and characterization of cell-permeant 6-nitrodibenzofuranyl-caged IP3. Photochem Photobiol Sci 11:508–513

27. Aghajanian GK, Rasmussen K (1989) Intracellular studies in the facial nucleus illustrating a simple new method for obtaining viable motoneurons in adult rat brain slices. Synapse 3:331–338. https://doi.org/10.1002/syn.890030406
28. Tanaka Y, Tanaka Y, Furuta T et al (2008) The effects of cutting solutions on the viability of GABAergic interneurons in cerebral cortical slices of adult mice. J Neurosci Methods 171:118–125. https://doi.org/10.1016/j.jneumeth.2008.02.021
29. Hong J, Zhang J, Xiao C, Kong J (2006) Patch-clamp studies in the CNS illustrate a simple new method for obtaining viable neurons in rat brain slices: glycerol replacement of NaCl protects CNS neurons. J Neurosci Methods 158:251–259. https://doi.org/10.1016/j.jneumeth.2006.06.006
30. Ting J, Daigle T, Chen Q, Feng G (2014) Acute brain slice methods for adult and aging animals: application of targeted patch clamp analysis and Optogenetics. Methods Mol Biol 1183:221–242. https://doi.org/10.1007/978-1-4939-1096-0
31. Brahma B, Forman RE, Stewart EE et al (2000) Ascorbate inhibits edema in brain slices. J Neurochem 74:1263–1270. https://doi.org/10.1046/j.1471-4159.2000.741263.x
32. Connors BW, Gutnick MJ, Prince DA (1982) Electrophysiological properties of neocortical neurons in vitro. J Neurophysiol 48:1302–1320
33. Llinás R, Muhlethaler M (1988) An electrophysiological study of the in vitro, perfused brain stem-cerebellum of adult Guinea-pig. J Physiol 404:215–240
34. Geiger JR, Bischofberger J, Vida I et al (2002) Patch-clamp recording in brain slices with improved slicer technology. Eur J Phys 443:491–501. https://doi.org/10.1007/s00424-001-0735-3
35. Buskila Y, Breen PP, Tapson J et al (2014) Extending the viability of acute brain slices. Sci Rep 4:4–10. https://doi.org/10.1038/srep05309
36. Aitken PG, Dudek FE, Eskessen K et al (1995) Making the best of brain slices: comparing preparative methods. J Neurosci Methods 59:151–156
37. Cameron M, Kekesi O, Morley JW et al (2016) Calcium imaging of am dyes following prolonged incubation in acute neuronal tissue. PLoS One 11:e0155468. https://doi.org/10.1371/journal.pone.0155468
38. Breen PP, Buskila Y (2014) Braincubator: an incubation system to extend brain slice lifespan for use in neurophysiology. In: 2014 36th Annual International Conference of the IEEE Engineering in Medicine and Biology Society. IEEE, pp 4864–4867
39. Fujikawa DG (2015) The role of Excitotoxic programmed necrosis in acute brain injury. CSBJ 13:212–221. https://doi.org/10.1016/j.csbj.2015.03.004
40. Karnatovskaia LV, Wartenberg KE, Freeman WD (2014) Therapeutic hypothermia for neuroprotection: history, mechanisms, risks, and clinical applications. The Neurohospitalist 4:153–163. https://doi.org/10.1177/1941874413519802
41. Whittingham TS, Lust WD, Christakis DA, Passonneau JV (1984) Metabolic stability of hippocampal slice preparations during prolonged incubation. J Neurochem 43:689–696
42. Feig S, Lipton P (1990) N-methyl-D-aspartate receptor activation and Ca2+ account for poor pyramidal cell structure in hippocampal slices. J Neurochem 55:473–483
43. Davie JT, Kole MHP, Letzkus JJ et al (2006) Dendritic patch-clamp recording. Nat Protoc 1:1235–1247
44. Rice ME (1999) Use of ascorbate in the preparation and maintenance of brain slices. Methods 18:144–149. https://doi.org/10.1006/meth.1999.0767
45. Lee J, Grabb MC, Zipfel GJ, Choi DW (2000) Tissue responses to ischemia brain tissue responses to ischemia. J Clin Invest 106:723–731
46. Verkhratsky a KH (1996) Calcium signalling in glial cells. Trends Neurosci 19:346–352. https://doi.org/10.1016/0166-2236(96)10048-5
47. Won SJ, Kim DY, Gwag BJ (2002) Cellular and molecular pathways of ischemic neuronal death. J Biochem Mol Biol 35:67–86
48. Lee J, Grabb MC, Zipfel GJ, Choi DW (2000) Brain tissue responses to ischemia. J Clin Invest 106:723–731
49. Silver I, Erecinska M (1998) Oxygen and ion concentrations in normoxic and hypoxic brain cells. In: Oxygen transport to tissue XX. Springer, Boston, MA, pp 7–16
50. Moyer JR, Brown TH (2002) Patch-clamp techniques applied to brain slices. In: Patch-clamp anal. Humana Press, New Jersey, pp 135–193
51. Buskila Y, Amitai Y (2010) Astrocytic iNOS-dependent enhancement of synaptic release in mouse neocortex. J Neurophysiol 103:1322–1328. https://doi.org/10.1152/jn.00676.2009

52. Aitken PG, Breese GR, Dudek FF et al (1995) Preparative methods for brain slices: a discussion. J Neurosci Methods 59:139–149
53. Cohen P (1997) The search for physiological substrates of MAP and SAP kinases in mammalian cells. Trends Cell Biol 8924:353–361
54. Espanol MT (1996) Adult rat brain-slice preparation for NMR studies of hypoxia. Anesthesiology 84:201–210
55. Watson GB, Lopez OT, Charles VD, Lanthorn TH (1994) Assessment of long-term effects of transient anoxia on metabolic activity of rat hippocampal slices using triphenyltetrazolium chloride. J Neurosci Methods 53:203–208
56. Ames A, Nesbett FB (1981) In vitro retina as an experimental model of the central nervous system. J Neurochem 37:867–877. https://doi.org/10.1111/j.1471-4159.1981.tb04473.x
57. Hudson B, Uphol WB, Devinny J, Vinograd J (1969) The use of an ethidium analogue in the dye-buoyant density procedure for the isolation of closed circular DNA: the variation of the superhelix density of mitochondrial DNA. Biochemistry 62:813–820
58. Latt SA (1973) Microfluorometric detection of deoxyribonucleic acid replication in human metaphase chromosomes. Proc Natl Acad Sci 70:3395–3399
59. Williamson D, Fennell D (1975) The use of fluorescent DNA-binding agent for detecting and separating yeast mitochondrial DNA. Methods Cell Biol 12:335–352
60. Monette R, Small DL, Mealing G, Morley P (1998) A fluorescence confocal assay to assess neuronal viability in brain slices. Brain Res Brain Res Protoc 2:99–108
61. Khatri N, Man HY (2013) Synaptic activity and bioenergy homeostasis: implications in brain trauma and neurodegenerative diseases. Front Neurol 4:199. https://doi.org/10.3389/fneur.2013.00199
62. Malarkey EB, Parpura V (2008) Mechanisms of glutamate release from astrocytes. Neurochem Int 52:142–154
63. Reid KH, Edmonds HL, Schurr A et al (1988) Pitfalls in the use of brain slices. Prog Neurobiol 31:1–18
64. Barone FC, Raymond ZF (1997) Brain cooling during transient focal ischemia provides complete neuroprotection. Neurosci Biobehav Rev 21:31–44
65. Erecinska M, Thoresen M, Silver IA (2003) Effects of hypothermia on energy metabolism in mammalian central nervous system. J Cereb Blood Flow Metab 23:513–530. https://doi.org/10.1097/01.WCB.0000066287.21705.21
66. Attwell D, Laughlin SB (2001) An energy budget for signaling in the Grey matter of the brain. J Cereb Blood Flow Metab 21:1133–1145. https://doi.org/10.1097/00004647-200110000-00001
67. Waardes A, Van Thillart G, Van Den Erkelensy C et al (1990) Functional coupling of glycolysis and phosphocreatine utilization in anoxic fish muscle. J Biol Chem 265:914–923
68. Tisherman SA, Sterz F (2005) Therapeutic hypothermia. Springer
69. Hicks SD, DeFranco DB, Callaway CW (2000) Hypothermia during reperfusion after asphyxial cardiac arrest improves functional recovery and selectively alters stress-induced protein expression. J Cereb Blood Flow Metab 20:520–530. https://doi.org/10.1097/00004647-200003000-00011
70. Canevari L, Console A, Tendi EA et al (1999) Effect of postischaemic hypothermia on the mitochondrial damage induced by ischaemia and reperfusion in the gerbil. Brain Res 817:241–245. https://doi.org/10.1016/S0006-8993(98)01278-5
71. Richerson GB, Messer C (1995) Effect of composition of experimental solutions on neuronal survival during rat brain slicing. Exp Neurol 131:133–143. https://doi.org/10.1016/0014-4886(95)90015-2
72. Yu DY, Cringle SJ (2001) Oxygen distribution and consumption within the retina in vascularised and avascular retinas and in animal models of retinal disease. Prog Retin Eye Res 20:175–208. https://doi.org/10.1016/S1350-9462(00)00027-6
73. Cameron MA, Suaning GJ, Lovell NH, Morley JW (2013) Electrical stimulation of inner retinal neurons in wild-type and retinally degenerate (rd/rd) mice. PLoS One 8:e68882. https://doi.org/10.1371/journal.pone.0068882
74. Medrano CJ, Fox DA (1995) Oxygen consumption in the rat outer and inner retina: light- and pharmacologically-induced inhibition. Exp Eye Res 61:273–284. https://doi.org/10.1016/S0014-4835(05)80122-8
75. Cameron M, Kékesi O, Morley JW et al (2016) Calcium imaging of AM dyes following prolonged incubation in acute neuronal tissue. PLoS One 11:e0155468. https://doi.org/10.1371/journal.pone.0155468
76. Feigenspan A, Babai NZ (2017) Preparation of horizontal slices of adult mouse retina for electrophysiological studies. J Vis Exp:1–6. https://doi.org/10.3791/55173

77. Kulkarni M, Schubert T, Baden T et al (2015) Imaging Ca2+ dynamics in cone photoreceptor axon terminals of the mouse retina. J Vis Exp: e52588. https://doi.org/10.3791/52588
78. Buskila Y, Breen P, Wright J (2015) Device for storing a tissue sample WO2015021513
79. Cameron MA, Kekesi O, Morley JW et al (2017) Prolonged incubation of acute neuronal tissue for electrophysiology and. J Vis Exp 120:1–6. https://doi.org/10.3791/55396
80. Grøndahl TO, Langmoen IA (1993) Epileptogenic effect of antibiotic drugs. J Neurosurg 78:938–943. https://doi.org/10.3171/jns.1993.78.6.0938

Chapter 11

Confocal Imaging of Fast Flash Photolysis of Caged Compounds in Cultured Neurons

Eduard Korkotian and Menahem Segal

Abstract

The enhanced ability to visualize small neuronal compartments in live tissue, such as individual dendritic spine, is accompanied in recent years by a need for a precise, high temporal and spatial resolution ability to activate or suppress electrophysiological as well as biochemical properties within such compartments. Parallel rapid progress in molecular, cellular, and physics methodologies enabled the recruitment of novel technologies to the analysis of a wide spectrum of issues, from long lasting imaging of subcellular compartments in vitro as well as in vivo, to simultaneous recording of activities of networks of hundreds of neurons in behaving animals. In the present review, we will focus on a fast UV flash (4 ns) photolysis of caged molecules in a small sphere (<1 μm) near or within dendritic spines of cultured neurons. This method is faster and cheaper than the commonly used 2-photon uncaging, which requires a large investment in complex equipment. Our method can best be used with two-dimensional networks of neurons, grown in culture.

Key words Uncaging, Confocal microscope, Dendritic spines, Calcium

1 Introduction

Using high temporal and spatial resolution imaging of calcium variations in individual dendritic spines of living neurons, studies in the past two decades have demonstrated that the spine is an independent calcium compartment [1, 2]. Synaptic and chemical stimulation have been used in order to raise $[Ca^{2+}]i$ to higher levels in the spine head than in its parent dendrite. Consequently, dynamic changes in the shape, size and rate of formation of dendritic spines following exposure to plasticity-producing stimulation have been reported [2, 3]. The small size of the dendritic spine necessitated the development of methods for focal application of minute quantities of chemicals to its vicinity. The use of minimal electrical stimulation of afferent fibers has some advantages, but its main drawback is that the stimulation current can spread to adjacent presynaptic fibers and other spines and dendrites that are not

Nicholas J. D. Wright (ed.), *Basic Neurobiology Techniques*, Neuromethods, vol. 152,
https://doi.org/10.1007/978-1-4939-9944-6_11,

within the field of view could also be activated. Iontophoresis through a micropipette is another method for local drug delivery [4]. Using this method, the drug application pipette needs to be placed in close proximity to the studied element so as to maintain its spatial selectivity, which is not always easy to achieve and standardize. The concentration of the applied drug and the duration of its application do not warranty that the drug of interest will not spread to adjacent, unmonitored regions.

An alternative method of local drug application, which has been used successfully in recent years, involves the local flash photolysis of caged molecules [5–9]. It is based on the ability to chemically "cage" the molecule of interest, which makes it inert biologically, and breaking the caging bonds with UV light, which brings the activated atom to a higher energy level, which then causes a breakdown of the activated bond, to release the caging molecule during the subsequent "dark" interval and rapidly increase the concentration of the biologically active molecule. This method has been used for the past 30 years, initially in peripheral and model cells (Lester 87), and later with neurons, in vitro and in vivo, with growing efficacy. Currently, there are about 10–15 caged molecules that are being used extracellularly, including neurotransmitter agonists and antagonists (carbachol, norepinephrine, GABA, glutamate, etc.) and some intracellular second messengers, (e.g., ATP, cAMP, and calcium). These compounds vary in their caging molecule, the energy for release from the cage, and the effective concentration. Several variations of methods for flash photolysis have been described in the past [10–16], and they are aimed at producing a highly localized photolysis of caged molecules. Initially, the spatiotemporal resolution of uncaging was not so accurate, and a sphere of 30 μm was reported [17]. Likewise, the duration of photolysis was not critical [18] when it involved activation of caged molecules inside neurons with no specific request for a localized response. With the advent of more precise methods, these parameters have been attended to more closely.

Obviously, flash photolysis of caged molecules has many advantages over the more conventional methods for drug delivery, and the ability to uncage molecules inside a small neuronal compartment is just one of them [19]. With this method it is crucial to be able to precisely localize measure and try to minimize the sphere of photolysis so as to allow topical uncaging of the active substance and be able to study their kinetics. There are several methodological issues associated with flash photolysis of caged molecules, including the spatiotemporal resolution of the uncaging method as well as the diffusion of the active molecule from the original location, the delivery of the caged molecule to the region of interest, and the ability to have the reporter molecule in high enough concentrations so as to afford a high signal to noise ratio in the detecting system without affecting directly the ability of the neuron to respond to the stimulation (given that some detectors, e.g., those responding to change in ambient calcium concentration, are actually themselves calcium buffers, so that their high concentration within the recorded/imaged neuron will affect the parameters of the response to exposure to the uncaged molecule). Both chemical detectors that need to penetrate neuronal membrane, as well as biological detectors, that are incorporated in the genome (GCamp [20]) have been used extensively. Furthermore, artificial receptor molecules that can be activated by an artificial ligand, also present a challenge to the ability to activate these receptors locally [21]. Issues related to the presence of the detector, as well as the interpretation of changes in fluorescence intensity following uncaging (i.e., change in volume vs. a change in concentration of the detectors) and other methodological issues will be dealt with below. This review is not meant to provide a comprehensive summary of flash photolysis of caged molecules, which can be found elsewhere [22, 23] but to focus on the methods developed and used in our laboratory, specifically with cultured central neurons.

We have developed an effective system for fast UV laser-induced flash photolysis of caged molecules in a restricted sphere of less than 1 μm in diameter, which can be used for both extracellular and intracellular delivery of caged molecules aimed at studying calcium variations in individual dendritic spines of cultured hippocampal and cortical neurons.

2 Materials and Methods

2.1 Imaging Setup

The imaging platform includes an upright Pascal scanning confocal microscope (Zeiss Axioskop2) equipped with standard Argon and HeNe lasers for excitation at wavelength of 488 and 543 nm, respectively. Figure 1 is a schematic diagram of the configuration of lasers and optical devices, Figs. 2, 3, 4, 5, and 6 are detailed description of the different elements in the system. An air-cooled tripled ND:YAG laser (NewWave Research Inc.) [24] emitting 355 nm, 3.6 mJ, 4.0 ns single light pulses at a rate of 1 Hz is used for the photolysis of caged molecules. The UV laser light is coupled to the confocal microscope using a dichroic mirror with high reflection at 355 nm and high transmission of visible light (Locus, Optical Devices & Elements Ltd.) (Figs. 1b.8, 2, and 3) and is focused on the image plane through a 63× 0.9NA Achroplan water immersion objective (Figs. 1b.9 and 4). Due to chromatic aberrations of the objective (i.e., different focal distances for the visible and ultra violet spectrum), the focal plane for 355 nm UV light is about 20 μm lower than for blue/green light. To compensate for this chromatic aberration, a focus-correcting lens (KPX073 EFL-250 mm, Newport) (Figs. 1.7 and 5) is introduced at the entrance of the UV beam to the microscope. A second laser, Melles

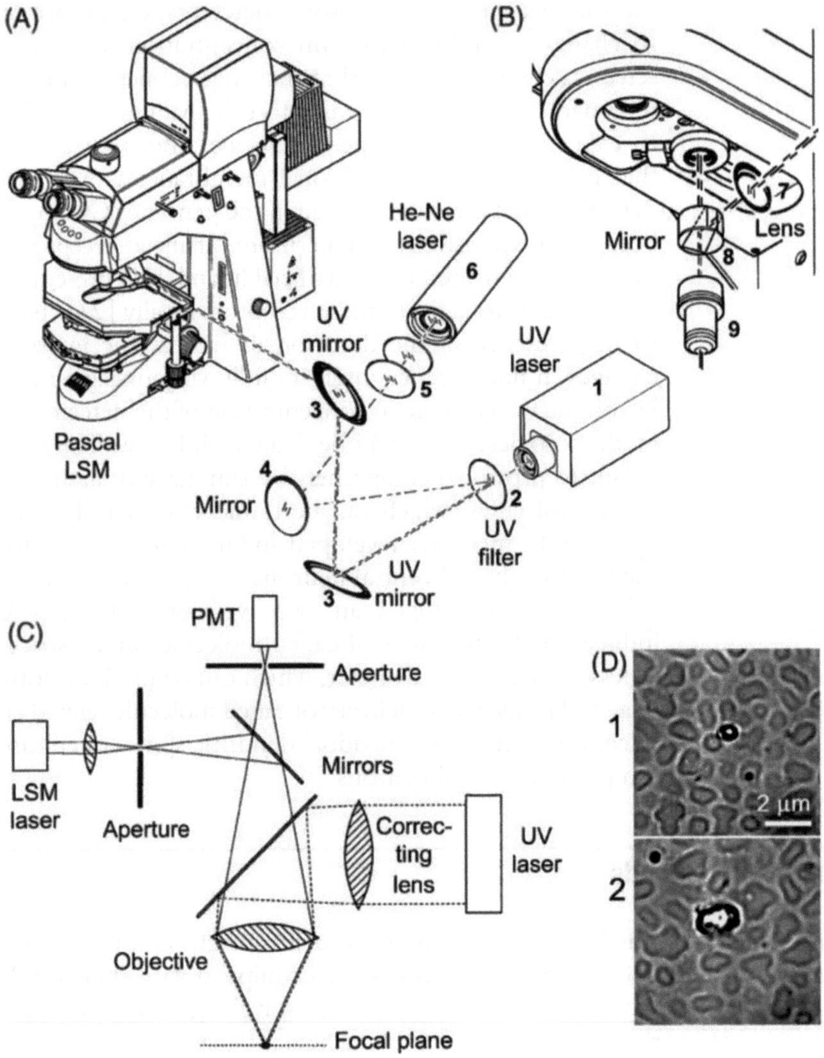

Fig. 1 Schematic diagram of the Pascal confocal microscope equipped for flash photolysis. (**a**) Low power image of the microscope, showing the two lasers, the UV laser, and the HeNe red laser used for alignment. The two light paths are merged with mirrors to enter the microscope through a mirror located above the objective. Numerals are addressed in the text. (**b**) A higher resolution image showing the light path deflected via a dichroic mirror above the objective lens. (**c**) Schematic diagram of the light paths, including the LSM lasers used for the imaging, and the UV laser for photolysis. A correction lens is placed in the light path to adjust the focal plane of the two lasers. (**d**) Two images showing the impact of different intensities of UV laser on a plane of focus, a glass slide coated with blue marker, visualized in phase microscopy. The light pulse creates a hole in the plane of the dye. The smaller hole, obtained with a full array of UV filters, above, creates a hole that is less than 1um in diameter. The larger hole was created by a higher power laser light pulse without filters (modified from [24])

Griot red helium–neon (1 mW, 632 nm) (Figs. 1.6 and 3) is used to align the UV laser in the field of view. The red beam is steered using Newport 10D20 ER2 dielectric mirrors (Figs. 1.4 and 2) and the two beams are combined before coupling to the microscope using

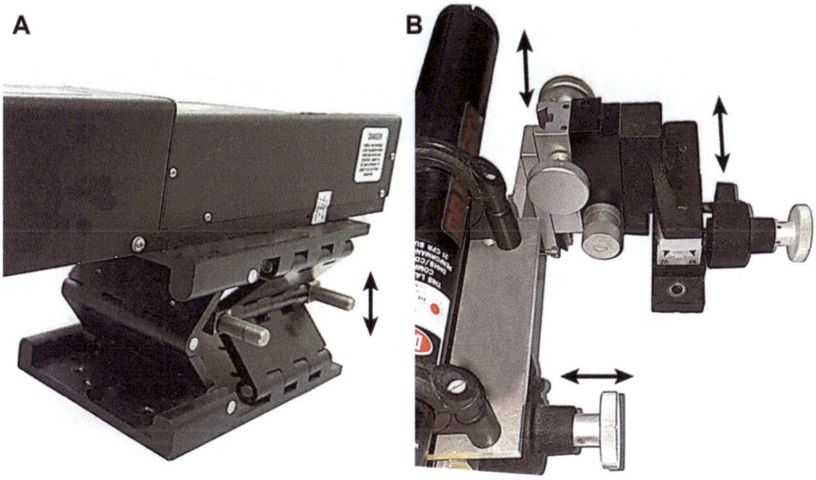

Fig. 2 The UV laser (**a**) and the red laser (**b**) placed on adjustable stages. xyz controls are shown for the red laser, that can be controlled at 3D (arrows mark the three dimensions)

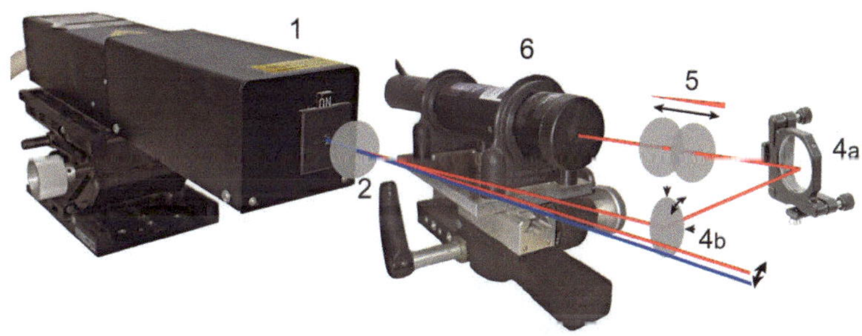

Fig. 3 The two lasers are aligned so as to produce a single combined beam (*see also* **Note 1**). The UV laser [1] emits blue light, the red one [6] emits red light, and the latter is adjusted with sets of mirrors and correction lenses [2, 4, 5]. Numerals in here and Figs. 4–6 are at the same locations as shown in Fig. 1

one of the UV attenuating filters (Figs. 1.2 and 3) as a dichroic mirror. The focal depth of the red beam is further corrected to match that of the UV beam using a telescopic lens system (Fig. 1.5). Both 355 nm UV light and 632 nm red light are directed into the confocal microscope using two high energy UV mirrors Y3-1025-45-P (from CVI Laser corporation) (Figs. 1.3 and 4). Prior to the experiment, the size of the focused UV laser beam is verified to be of less than 1 μm^2 using blue marker paint on glass, as illustrated in Fig. 1. Various high-energy UV filters (Fig. 1d) adjusted for 355 nm light (optical densities of 0.5, 1.0, or 2.0, corresponding to a reduction of 3, 10, or 100, respectively, all from CVI Laser Corporation) are used separately or in combination to control the intensity of the beam. When all three UV filters are used in series, the light intensity is reduced to 1/3000 of nominal intensity, about 1 nJ. Final adjustment of the UV laser power is

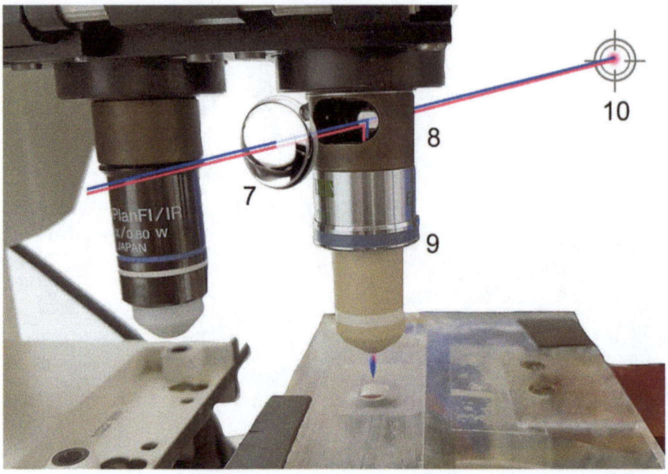

Fig. 4 A view of the turret which is adjusted to allow the combined beam into the lens and the specimen. Same numerals as in Fig. 1. The combined red/blue light beam enters the turret; the beam is split so that it is focused at a distance [22] to allow verification of the combined beam, and is directed through the lens into the specimen [9]

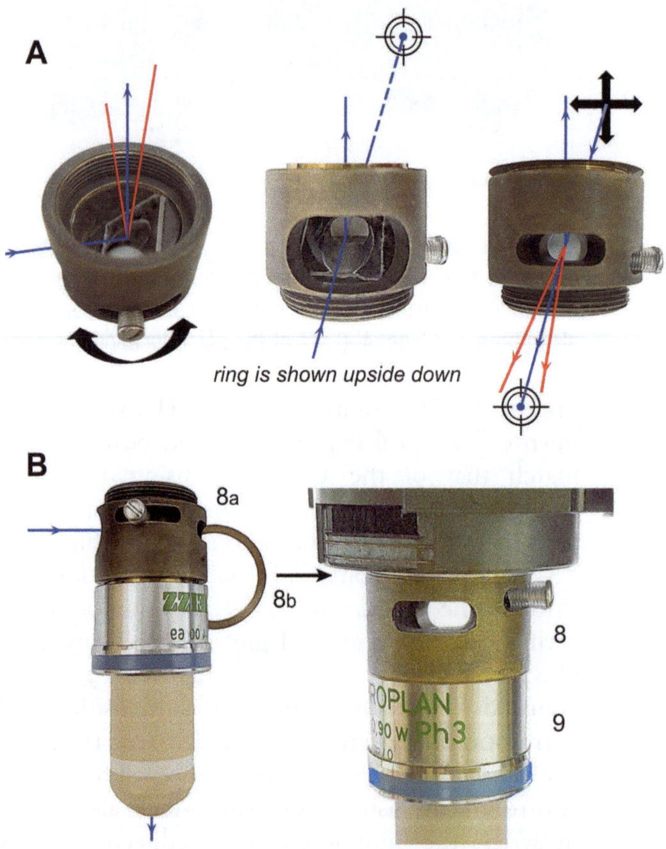

Fig. 5 Zoomed in view of the cubicle that contains the mirror (**a**) and its position above the water immersion lens (**b**) that is used for the imaging. The combined beams can be finally adjusted to be focused in the center of the field of view. Numerals are from Fig. 1

done by inserting one to four UV-grade uncoated glass windows into the beam path. Each window reduces the UV light intensity by about 10%. When the additional reduction in light intensity due to filters, mirrors, glass lenses and passage through water is taken into account, the calculated UV light intensity amounts to 200–300 nJ, which produce a flux of 4–6 × 10^{11} photons per pulse, focused on a less than femtoliter of medium containing from 1000 to 100,000 caged molecules. Fine positioning of the combined beam is produced by microscale adjustments (Fig. 6).

An important facet of the uncaging system assembly and alignment is the type of microscope used which can be either an inverted or upright microscope. In the case of the upright system, a water immersion objective is used, and the light path includes only one compartment. An upright microscope better fits the requirement of light stability, as the uncaging and reference laser light beams converge into the microscope lens, and the precise positioning of objective in *z* axis is critical. Therefore, either stage must move up and down with respect to the objective for focusing or the accurate level of focal plane must be achieved each time and verified prior the

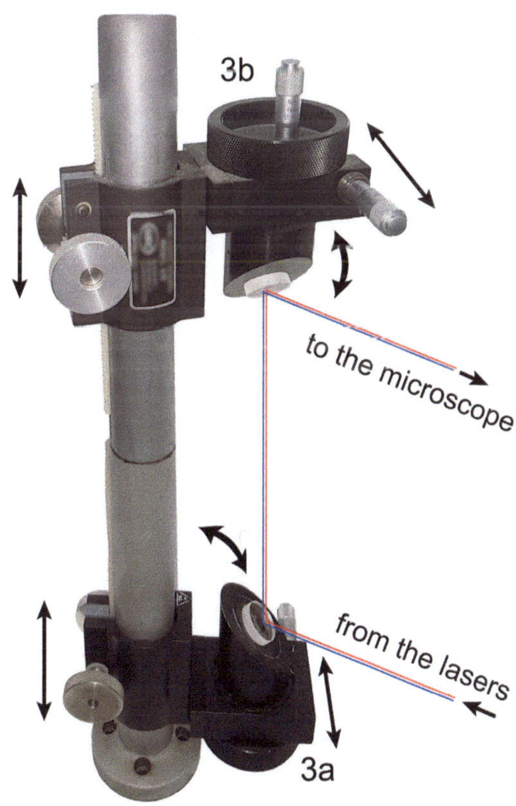

Fig. 6 The pole that is secured to the air table (bottom) and where two mirrors are used to align the combined light beam from the lasers to the microscope. The position and direction of the beam can be adjusted (arrows)

experiment. In contrast, an inverted microscope, even though it allows easier access to the tissue, the light beam has to pass through two media, air/oil and the cover glass, resulting in less stable light and more complex alignment.

Another important issue is the insertion of the uncaging lights into the microscope. In a case of regular non-scanning fluorescent microscope, any appropriate optical port (such as camera or illumination ports) can be used to direct the light into the imaging system. The advantage of the current system is that uncaging light beams does not interfere with regular fluorescence illumination. This is not feasible in the case of scanning imaging systems as they continuously shift the excitation light beams along the specimen. Considering typical pixel time to last several microseconds and about 4 ps for UV flash duration, uncaging will happen at any random place, if not additionally controlled. A reasonable solution is to insert the "uncaging lights" after the scanner. One option is the use of thin optical fibers driven by external manipulator and positioned right near the place of interest. Our careful tests however demonstrated that 1–3 μm thin fiber is unable to supply sufficient amount of UV light (data not shown). By that we concluded that placing a transit dichroic mirror between the microscope and the objective (Fig. 10 elements 8 and 9 in Fig. 1) was optimal.

The uncaging system requires use of optical table, large enough to accommodate all its optical elements. A square of 40 × 40 cm may be sufficient. The system will be aligned according to position of UV light source and therefore its position should be carefully preselected. The height of UV laser location is an important issue as it will determine the setting of other optical elements. We recommend setting about 12 cm from the base to the light beam output. The laser stand must be flexible in z dimension with particular stability to avoid possible slow drifts (Fig. 2a, arrow). Reference light source is adjusted according the UV laser. It is equipped with 3D-flexible stand with course and fine tuning of the height and lateral positions (Fig. 2b, arrows).

Figure 3 shows the set of optical elements used to consolidate the UV and reference light beams. The red light beam is slightly extended using adjustable telescopic optical element 5 (arrow) which helps proper farther focusing after its transition through the UV correcting lens (see Fig. 4, element 7). Mirrors 4a and 4b on Fig. 3 play a role of near field and far field adjustment elements. Particularly, 4a is used to consolidate the UV and red light beams on the filter/mirror optical element 2, while 4b helps to parallelize the two beams in their farthest joint axial path (curved arrow).

Two carefully parallelized beams then reach the elevator system (Fig. 6) consisting of two identical UV-reflecting mirrors placed on two opposite tunable platforms (3a and 3b) allowing to adjust both lateral position of the mirror and its angle regarding the horizontal and vertical planes (arrows on the right). The height of each platform can be set independently using two knobs equipped with mechanical retainers to avoid unexpected shifts (arrows on the left). The output of this optical system is the coupled light beams directed toward the microscope and strictly parallel to the horizontal surface.

The internal construction of the cylinder, containing dichroic mirror and situated between the body of the microscope and the objective is shown on Fig. 5a. Note the upside down orientation, chosen here for convenience. The inner part of cylinder can be rotated around the central axis and fixed using the adjusting screw (left panel, arrow). This will affect the final precise position of the light spot under the microscope, marked as blue line (correct) and red lines (incorrect). This will happen only if the approaching light is not ideally centered on the surface of dichroic mirror. Light beams enter the cylinder through the front window (Fig. 5a, middle). Correct positioning of the front window relative to the microscope body can be regulated using the set of thin washers (Figs. 5b and 8a, b), between cylinder and the microscope. Most of the light is reflected down, toward the objective (upside on the figure) by dichroic mirror while the regular fluorescent illumination as well as emission will successfully pass through it. Some minor fraction of "uncaging lights" will penetrate the mirror and leave the cylinder through the backside window (right panel). This fraction of the light is used to hit the target, situated about 1 m apart. Matching the center of the target is provided by upper mirror platform (see Figs. 3b and 6), schematically illustrated here as crossed arrows. Matching the target (Fig. 4, element 10) guarantees the accurate central location of uncaging light spot in the field of view of the microscope (Fig. 4, elements 8 and 9).

2.2 Chemicals

Caged glutamate (*N*-(-carboxy-2-nitrobenzyl)-l-glutamic acid (*N*-(CNB-caged) l-glutamic acid) (molecular probes)) is prepared in the dark from stocks, and added to the recording medium at a final concentration of 0.1–0.5 mM. Caged fluorescein (CMNB-caged fluorescein dipotassium salt (2 mM) and NP-EGTA tetrapotassium salt (6–10 mM), molecular probes (*see* **Note 2**)) are back-filled into a micropipette tip, which is then filled with 3 M K-acetate, and injected into selected neurons. Alternatively, chemicals were loaded into cells using the acetomethyl-ester (AM) form. Ambient light is reduced to minimum throughout the recording session. The solutions are used within 2–3 h of preparation, to avoid the effects of spontaneous uncaging over time.

2.3 High-Resolution Imaging of Individual Dendrites and Spines

A coverslip is transferred from the 24 well plate (**Protocol 1**) where neurons grow for 2–4 weeks [25, 26] into the recording chamber, placed in an automated X–Y stage (Luigs & Neuman, Germany) of the confocal microscope, where it is bathed with recording medium containing (in mM, **Protocol 2**) NaCl 129, KCl 4, $MgCl_2$ 1, $CaCl_2$ 2, glucose 4.2, and HEPES 10, pH is adjusted to 7.4 with NaOH, and osmolality to 320 mOsm with sucrose, at room temperature. Voltage gated sodium currents are blocked by TTX. In some experiments cells are transfected beforehand with DsRed or GFP to allow detailed visualization of cellular morphology (**Protocol 3**). Individual cells are impaled with a micropipette containing Alexa 555 (5 mM) for viewing cell morphology and Fluo-4 (5 mM), or 10 mM Oregon Green-1 for imaging calcium. The dyes are rapidly injected into the cell, and are allowed to equilibrate in the cell for at least 20 min before experiments are commenced (*see* **Note 3**). Alternatively, recording is made with patch pipettes having axial resistance of 4–6 MOhm, containing (in mM) K-acetate 120, Mg ATP 2, Hepes 20, KCl 10, Na_2 phosphocreatine 10, and Alexa 555 and/or Fluo-4, 0.02–0.1. Images of 512 × 512 pixels are taken with the 63× water immersion objective. 3D reconstruction of the dendrites is made from successive 0.2–0.5 μm optical sections. A line is scanned through the center of a dendrite/spine pair (0.7 ms per line) to detect fast changes in fluorescence during a response to drug application. The imaging laser lights (488 and 543 nm) are reduced to 3–5% of nominal intensity. Using this setting, we are able to continuously stimulate the same spine–dendrite pair with no significant loss of reactivity due to dye bleaching or photodynamic damage. Also, baseline fluorescence does not change across the 2–3 h observation time. The acquisition program of line scans in the confocal microscope triggers the UV laser flash using a Master 8 stimulator (AMPI, Israel). This also triggers the electrical stimulation and acquisition of membrane currents in the recorded cells. Signals are amplified with Axopatch 200 (Axon Instruments, California), stored and analyzed using PClamp, MiniAnalysis (Synaptosoft Inc.) and Origin software. Data are analyzed using Matlab based programs. Curve fitting is done with KaleidaGraph software. Background noise is subtracted in all linescan recordings before calculation of DF/F.

2.4 Determination of the Location and Size of the UV Spot

Two methods are used to estimate the location and size of the UV laser light spot. In the first (Fig. 1d) a glass coverslip is painted with a blue marker pen, allowed to dry, immersed in water and placed in the microscope light path. Following focusing of the red light, the UV light pulse is applied, causing the formation of a hole in the paint. Fine adjustment is sometime needed to verify that the UV laser overlaps with the center of the red laser spot (Figs. 7 and 8) (*see* **Note 1**). The size of the hole varies as a function of the glass filter used, with minimum effective size being less than 1 μm. This method is used routinely to produce small adjustments in the alignment of the UV laser with the red laser spot in the light path. The second method employed caged fluorescein (Figs. 7 and 9). A thin layer of caged fluorescein is painted on a glass slide,

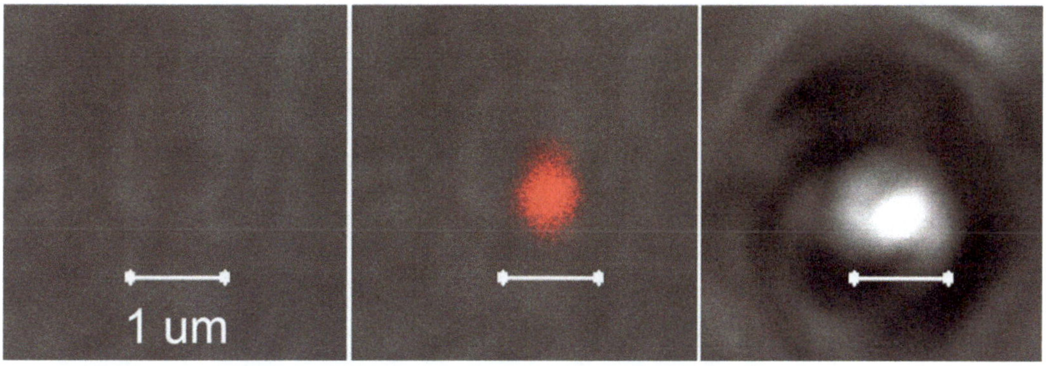

Fig. 7 Alignment of the red laser with the UV laser. Three images of the same location, showing (from left to right) the substrate in phase contrast (image a bit blurred) with a calibration bar of 1 μm, the red marking spot about 0.5 μm in diameter, and the position of the UV beam, creating a hole in the substrate, at exactly the same location as the red marking spot

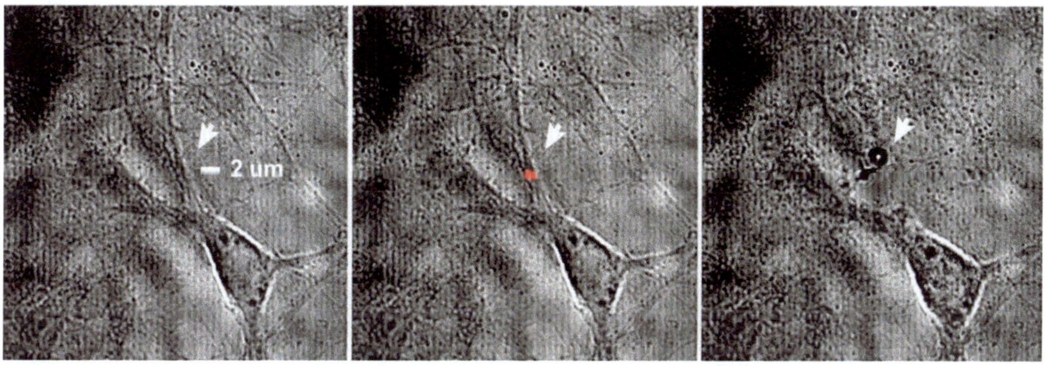

Fig. 8 The UV laser, when applied at high power, can be used as a micro-knife, to severe dendrites of cultured neurons. The images are in phase contrast, showing a cultured hippocampal neuron, with a pyramidal shape. (bar = 2 μm). In the center image the red marker is aimed at the apical dendrite, and following a flash, (right), the top dendrite is severed from the soma. The soma can keep on living after the dissection of its dendrite

allowed to dry, covered with a glass coverslip and sealed with varnish, to avoid exposure of the dye to water. The glass coverslip is then immersed in water and placed in the microscope. The size of the fluorescent spot can be measured through the microscope image using fast line scan. This method is used to measure directly the size of the UV laser spot. The width of the fluorescent spot is less than 1 μm, and the lateral diffusion of the fluorescent dye is minimal.

An interesting application of the UV laser flash can be to severe dendrites of live neurons in culture (Fig. 8). When the UV laser is used at high intensity, with few if any reduction filters, it can be aimed at a dendrite, and following a flash, it can severe it, leaving the mother neuron intact. This method can be used to study the role of dendrites in electrical and synaptic properties of neurons, but will not be dealt with herein.

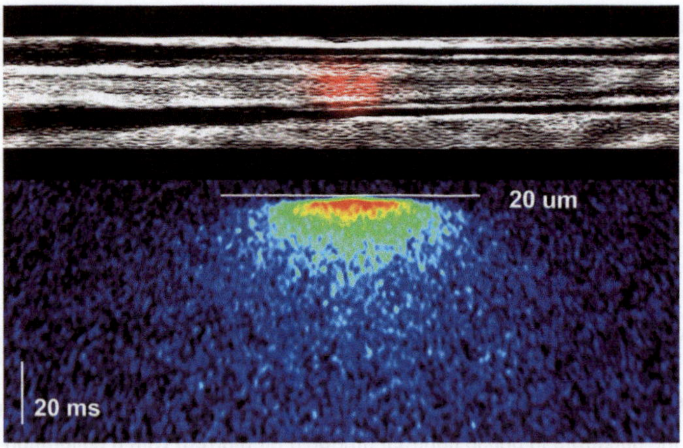

Fig. 9 Flash photolysis of caged fluorescein. A glass tube (top, phase image) was filled with caged fluorescein and a UV flash was focused in the center of the tube. A line scan (0.7 ms per line) along the glass tube shows the increase in fluorescence following the flash, and the diffusion of the fluorescence within the glass tube laterally. While the shortest latency of increased fluorescence is in the middle of the tube, about 2um from the center, fluorescence spread to both sides, and decayed slowly, as the only factor involved in the decay is diffusion to both ends of the tube (compare to the following figure). (unpublished observations)

Glutamate is one of the more popular caged molecules, as it is being used for mapping of glutamate responses in culture and brain slices [14, 27–29]. The use of caged glutamate (e.g., Figs. 10 and 11) highlights another major issue, that is, the diffusion of the uncaged molecule, to activate high affinity membrane receptors remote from the site of photolysis. Obviously, the effective sphere is dependent on a number of factors, including the concentration of the photo-released compound, the efficacy of light to "uncage" bonds, the affinity of the studied ligand to the receptor, and the size, duration and intensity of the light flash. We have examined different caged molecules, in terms of their stability and size of the uncaging sphere, as well as their impact on the sensing molecule (Fluo-4 AM, for the case of intracellular calcium concentrations). Systematic experiments were conducted with caged EGTA ([19, 30]; Figs. 11, 12, and 13), caged fluorescein (Fig. 9) and caged glutamate [31]. The system produces a reliable, restricted and reproducible local photolysis of caged molecules, which are used successfully for the analysis of dendritic spines, as demonstrated below.

2.5 Flash Photolysis of Caged EGTA

We used caged EGTA to study spine–dendrite coupling in cultured hippocampal neurons (Fig. 14). Two to three-week-old cultures, in which cells (5–10 cells in the standard conditions) were transfected 2–5 days earlier with DsRed (Clonetech) for imaging of cell morphology, were used. Cultures were incubated with 6 μM

NP-EGTA-AM, and the reporter molecule, Fluo-4 AM (2 μM) at room temperature for 1 h before use. The compounds penetrate the cell membrane where they are converted to the active form. Alternatively, cells were injected on the recording stage under visual guidance with CMNB-caged fluorescein dipotassium salt (2 mM) or NP-EGTA tetrapotassium salt (6–10 mM), Alexa Fluor555 (5 mM) and Fluo-4 (5 mM) using a sharp micropipette containing 1 M K-Acetate. Glasses were placed in the recording chamber, controlled by an automated X-Y stage (Luigs and Neumann, Ratingen, Germany) of the confocal microscope, where it was bathed in the standard recording medium at room temperature. In some experiments cells were recorded with a 1.5 mm OD patch pipette, where 136 mM CsCl replaced the K-gluconate and 0.1 mM Alexa Fluor 555, and 0.1 mM Fluo-4 are added.

Images of 512 × 512 pixels are captured with the 63X water immersion objective. 3D reconstruction of the dendrites was composed of successive 0.2–0.5 μm optical sections. A line was scanned through the center of a spine–dendrite segment (0.7 ms per line) to detect fast changes in fluorescence following a flash of the UV laser. The imaging laser lights (488 nm and 543 nm) were reduced to 3–5% of nominal intensity. Using this setting, we are able to continuously stimulate the same spine–dendrite segment with no significant loss of reactivity due to dye bleaching or photodynamic damage. Also, baseline fluorescence is not reduced significantly, and transient fluorescence signals were reduced by up to. Twenty percent across the 1.5–2 h observation times, indicating that the cells were viable for the duration of the experiments. Another indication to that fact involved the assessment of the morphology of the transfected neurons; healthy cells had smooth dendrites but they blabbed when the cell began to deteriorate following extensive photolysis. Data were not collected from such cells. The dendritic segments selected for the analysis in the present study had medium size spine heads (1.05 ± 0.12 μm) and varied in length from being stubby, less than 0.5 μm, to long ones, over 2 μm. We avoided measuring from headless spines, or from abnormally long ones (>4 μm). Images of the spines were taken at high resolution several times during the experiment, using the red wavelength. Rest periods of 5–10 min were taken between sessions of UV pulses, to allow cells to recover. Images of Fluo-4 fluorescence were not taken, because of low resting fluorescence levels.

A transient rise of $[Ca^{2+}]i$ in individual spine heads was recorded in Fluo-4 loaded neurons following flash photolysis of caged EGTA (Fig. 10). The elevated $[Ca^{2+}]i$ decayed rapidly, with a half-life of less than 9 ms. $[Ca^{2+}]i$ rise advanced along the spine neck in a nonlinear fashion, and $[Ca^{2+}]i$ variations in the parent dendrites were inversely correlated with spine neck length. Elevated fluorescence in the dendrite was also seen with a membrane-bound fluorophore. Also, fluorescence recovery from bleaching of a calcium-

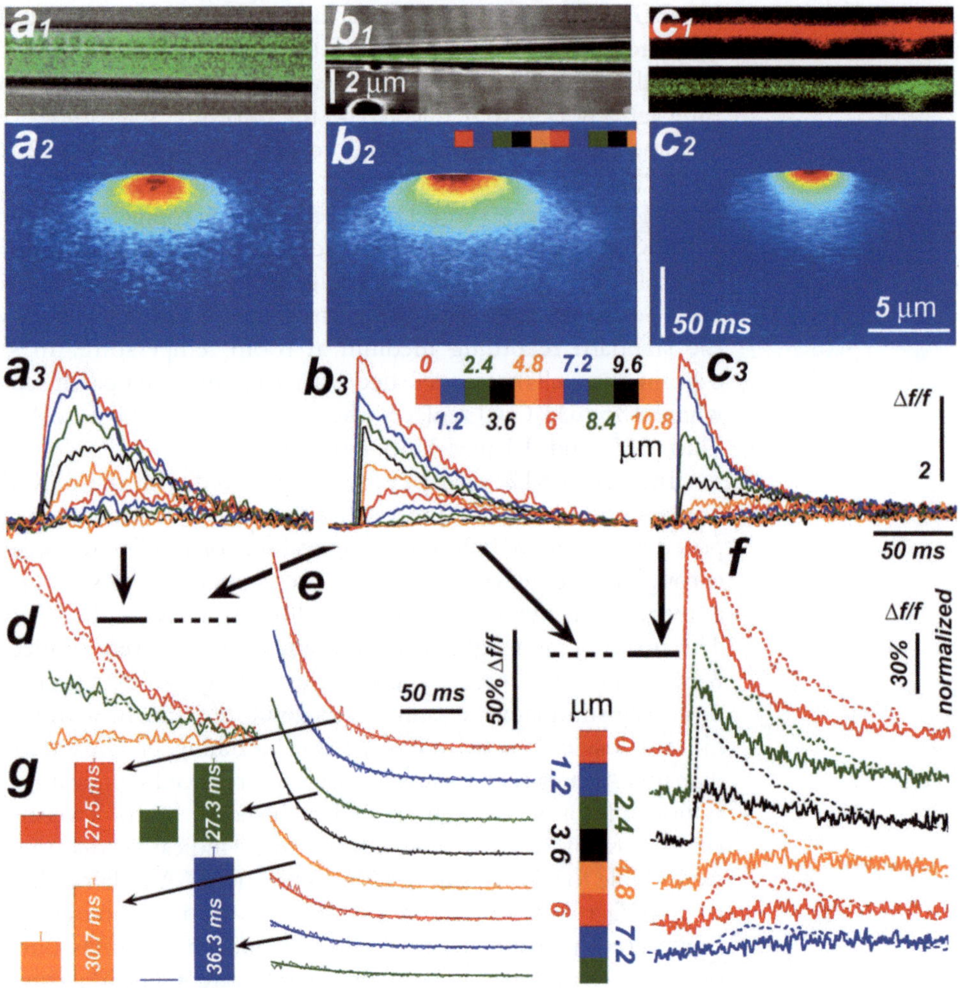

Fig. 10 Diffusion of flash photolysed calcium in glass tubes and thin dendrites. Comparisons were made between thick (about 5 µm, **a**) and thin (about 1.2 µm, **b**) glass tubes, illustrated in (*a1* and *b1*), respectively and a thin dendrite (**c**). (*a2*, *a3*, *b2*, *b3*, *c2* and *c3*) Line scans, from top to bottom, and appropriate graph presentations illustrating the rise of [Ca^{2+}]i measured by Fluo-4. (**d**) Comparison of decays from (*a3*) (solid line, thick tube) and (*b3*) (dotted line, thin tube). (**e**) Single exponent decay curve fits for (*b3*) (thin tube). Note colors, corresponding to distance color bars. (**f**) Comparison of (*b3*) (thin tube, dotted lines) and (*c3*) (dendrite, solid lines) responses at sequential distances from the focus of uncaging. There are obvious differences between thin glass tube and the dendrite of the same size in the distance of calcium spread and decay time shown in (**g**) (left bars for dendrite and right bars for the thin glass tube for same distances from the center of uncaging marked by appropriate colors) (modified from [20])

bound fluorophore was much slower than expected from changes in fluorescence seen upon release of [Ca^{2+}]i, indicating that the dendritic fluorescence change is not due to the diffusion of the calcium-sensitive fluorophore but reflects a genuine diffusion of free [Ca^{2+}]i. This conclusion is supported also by a mathematical analysis of calcium diffusion through spine necks. Flash-photolysed fluorescein decayed much slower than $[Ca^{2+}]i$ transient (Fig. 9), indicating that powerful mechanisms are responsible for the fast removal of elevated $[Ca^{2+}]i$. Both calcium stores and Na/Ca exchanger are involved in the fast removal of excess $[Ca^{2+}]i$.

Finally, spine length varied with the recent history of the $[Ca^{2+}]i$ variations in the spine. Thus, with fewer than 20 flashes, spines extended, whereas more flashes induced their shrinkage. Consequently, elongation of 0.4 μm resulted in complete isolation of the spine from the dendrite, while shrinkage enhanced coupling with the parent dendrite (Fig. 14). These studies highlight a dynamically regulated coupling between a dendritic spine head and its parent dendrite.

2.6 Flash Photolysis of Caged Glutamate: Interaction with Back Propagating Action Potentials

The functional significance of the great heterogeneity among dendritic spines of central neurons has been the subject of much speculation until recent years, due to the lack of methods for their quantitative analysis in live tissue. We have used confocal line scan imaging of rapid changes in dendritic spines and parent dendrites of cultured hippocampal neurons following a focal flash photolysis of caged glutamate (Fig. 11 [26]). Two-week-old cultured neurons were patch clamped with a pipette containing Alexa 555 for imaging morphology of the cells, and Fluo-4 (0.1 mM) for imaging of changes in $[Ca^{2+}]i$. Current responses were recorded from the soma, and back-propagating action currents were produced by depolarizing commands. Glutamate was uncaged using a fast UV pulse which produced a sphere of about 1 μm, which was localized near spine heads through the use of a red laser guide. Photolysis of glutamate caused a rapid and transient change in spine $[Ca^{2+}]i$ and a smaller and slower change in the parent dendrite (Fig. 11). As we have seen with photolysis of calcium in the dendritic spine head, the spine–dendrite disparity changed as a function of spine length, such that larger $[Ca^{2+}]i$ transients were seen in parent dendrites of shorter spines than of longer ones. Interestingly, the current response recorded at the soma, about 15pA, the size of individual spontaneous postsynaptic current, was not correlated with spine length. Back-propagating action currents facilitated $[Ca^{2+}]i$ responses to glutamate in the parent dendrites more than their effect on spine $[Ca^{2+}]i$ transients. The somatic current response but not the local $[Ca^{2+}]i$ transient was affected by the distance of the spine from the soma, such that smaller current responses were recorded from the more distal spines (unpublished observations). In general, there was no correlation between the magnitude of the spine $[Ca^{2+}]i$ transient and the current response recorded at the soma.

2.7 Spine–Dendrite Interaction

We explored the effects of spine length on calcium changes in the parent dendrite (Fig. 14). Assuming that the spine–dendrite interaction is bidirectional, and that changes in spine length may affect signals coming from the dendrite to the spine, we caused a local rise

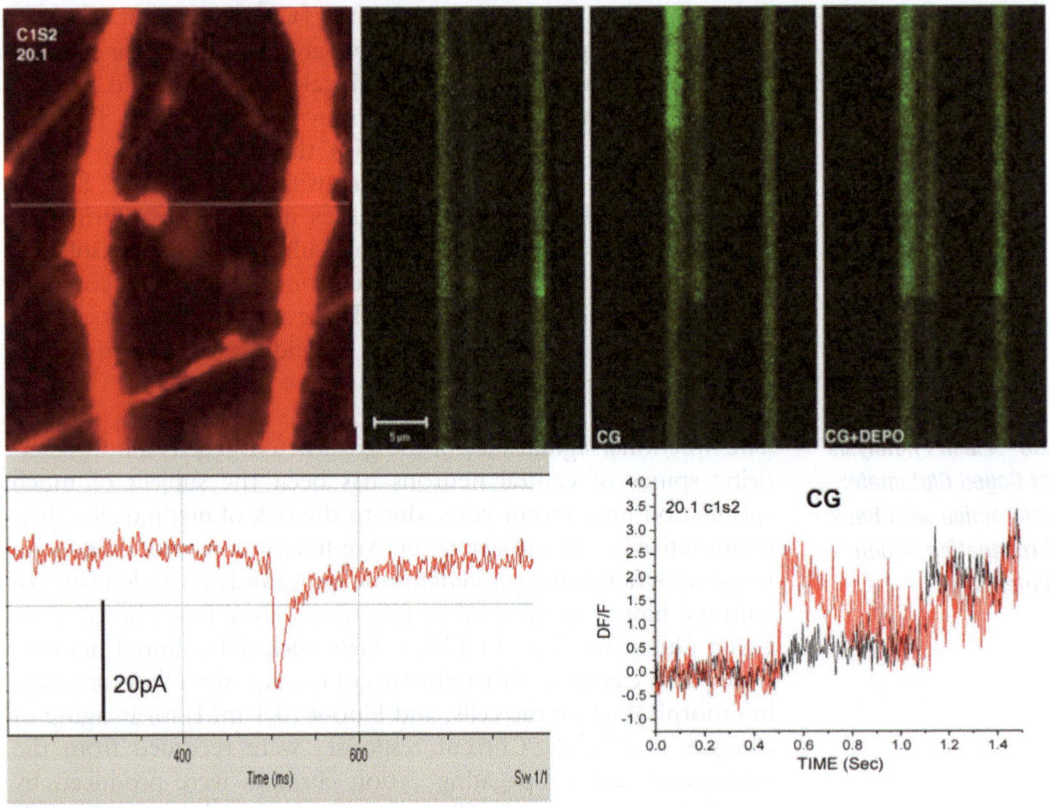

Fig. 11 Simultaneous recording and imaging of calcium responses to back propagating action potentials and flash photolysis of caged glutamate Top left, Alexa Fluor and Fluo-4 filled neuron used to image the morphology (red) and calcium variations (green) of the dendrites and spines. A thin line through the spine head, parent dendrite on the left and a sister dendrite on the right, is drawn. The green images, from left to right, are line scan responses to a back propagating action potential, response to caged glutamate, and the combination of both. Note the large response in both dendrite and a minor response in the spine head, to the action potential, and the selective response in the spine head to the uncaging of glutamate. The combined stimulation enhances the response of both dendrites. Bottom left, averaged electrical responses recorded in the soma, to the uncaging of glutamate. Bottom right, analysis of the line scans of the response to glutamate, illustrating the response in the spine head, red, and the parent dendrite, green (unpublished observations)

of intracellular calcium in the dendrite, using caged EGTA, and explored the magnitude of spread of free calcium from the dendrite to the adjacent spine. More interestingly, we explored the role of spines in the spread of calcium signals along the dendrite. In that respect, we choose dendritic segments of the same neurons that are either endowed with spines or devoid of them, and compare the spread of calcium signal along the dendritic shaft. It is important to make these comparisons in the same neurons, as different neurons may have different buffering capacity for the elevated calcium. Finally, we compared the spread of the calcium signals in the dendrite to the spread of such signals in a glass tube, where no extrusion mechanisms or internal calcium buffers exist, with the natural situation.

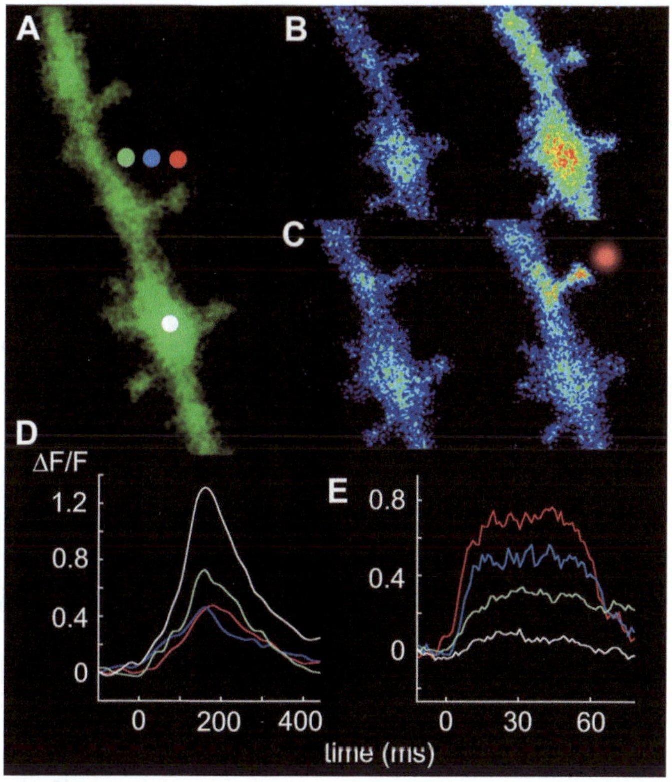

Fig. 12 Uncaging of glutamate compared to back-propagating action potential. (**a**) A dendrite of a cell injected with Fluo-4. (**b**) Two frames, demonstrating the calcium levels before and at the peak of the back- propagating action potential. (**c**) Two frames demonstrating the calcium levels before and at the peak of spine response to flash photolysis of caged glutamate. (**d**, **e**) are two graphs, representing calcium dynamics during the back- propagating action potential (**d**, as in **b**) and caged glutamate photolysis (**e**, as in **c**). In both cases, the color codes are representing: red for spine head, blue for spine neck, green for the shaft, and white for a dendrite remote from the selected spine. Note different time scales between D&E (unpublished observations)

Another aspect of the spine /dendrite communication addresses the mobility of synaptic constituents between the dendrite to the spine head. The relevant question is- does the spine neck constitute a barrier for the transfer of receptor molecules from the dendrite to their final location on the spine head. To study this, we use transfected neurons, where glutamate receptors tagged with green fluorescent protein (GFP (**Protocol 3**)) is followed along the dendrite, and examine their inclusion into the spine head following the conditioning protocol, in long and short spines. These experiments examine the most fundamental issue of the role of spines in synaptic communication.

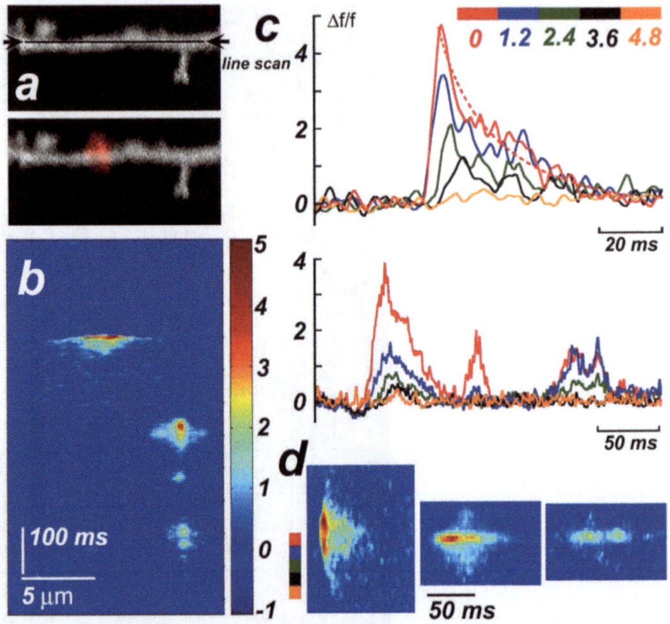

Fig. 13 Line scan of flash photolysed calcium, recorded in a short dendritic segment. (**a**) Images of the dendrite, where two spines, on right and left are seen, using the DsRed fluorescence. Line scan track is seen on a top and photolysis flash reference point on the bottom panels. (**b**) Line scan, from top to bottom, illustrating the change in [Ca^{2+}]i seen following the flash. Further down on the right are local spontaneous events. (**c**) Analyzed regions of interest (as seen in (**d**)), representing the focal plane of the flash/center of spontaneous event, and successive regions to its right and left, in 1.2μm intervals marked by red, blue, green, black, and orange successive colors. Dotted red line on c top represents single-exponent decay curve fit. The f/f was performed in this and all other raster images and analyzed scans. (**d**) The evoked and spontaneous events aligned to compare amplitudes and the extent of spread of the calcium rise (modified from [19])

3 Notes

1. The alignment of the two lasers, the UV and the red lasers are of critical importance for the localization of the photolysis spot. Naturally, different experiments are using different widths of the carrier glass chamber, and what is aligned for one glass is not necessarily proper for a second glass. Thus, since both laser beams must be focused in the same plane, care has to be taken to align them using the blue marker pen technique, so as to assure that they overlap. This cannot be made once the culture glass is inserted into the chamber.
2. For all photoactivatable compounds, care has to be taken to prepare the stock solution in the dark, and to keep the aliquots stored at −80 °C until use. It is advisable to paint the micropipette used to load the cells with the dye with black ink (e.g., permanent marker).

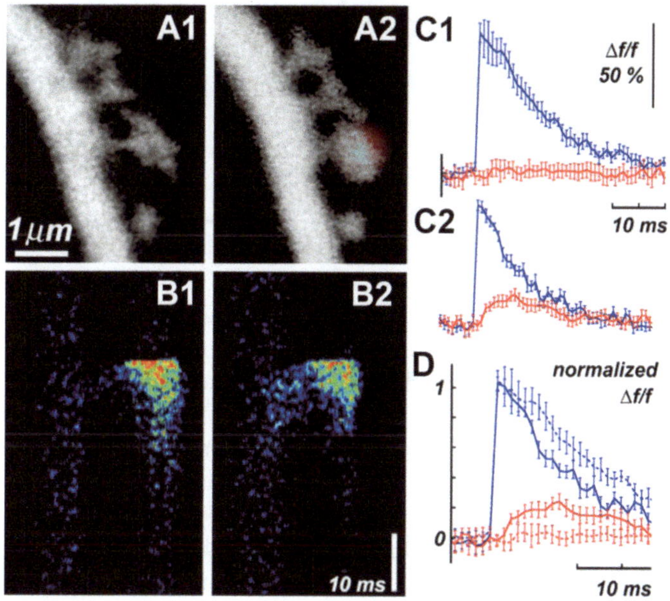

Fig. 14 Spine neck regulates calcium communication between spine head and parent dendrite. The cell, initially transfected with DsRed for imaging morphology, was loaded with Fluo-4 as calcium indicator, and NB-EGTA, both in cell permeant, AM form. The UV flash was aimed at the spine head, indicated by a bright spot on the red-marked spine. **A1** and **A2** are two images of the same set of spines, taken 30 min apart. Note that the neck of the "red" spine shrunk in length. This was accompanied by an apparent increase in spine neck diameter. **B1** and **B2** are two sample line scans of the same spine at the corresponding time as in **A1** and **A1**. Note that in **B1**, the rise in $[Ca^{2+}]$i is restricted to the spine head, whereas in **B2** the $[Ca^{2+}]$i rise spread into the parent dendrite, to the left of the spine. **C1** and **C2** line scan analysis of the transient rises of $[Ca^{2+}]$i, averaging 14 flashes each. In both cases, the blue trace is the summary of the $[Ca^{2+}]$i changes in the spine head, and the red trace is of the parent dendrite. Note that in **C1** there is no change in the parent dendrite $[Ca^{2+}]$i level, whereas in **C2** there is a small rise in $[Ca^{2+}]$i in the parent dendrite following a larger, shorter latency and faster rise in the spine head. The two sets of experiments are normalized and plotted together in **D** for comparison. Similar concentrations of NP-EGTA in the same spine–dendrite pair at its longer (**A1**) and shorter (**A2**) states (averaged responses to 14 repetitive flash stimulations for every time point at 0 (**C1**) and 30 min (**C2**)) (Reproduced from [30])

3. Micropipettes (1.2 mm o.d.) are pulled on a horizontal Sutter Instruments puller to produce tip resistance (in 1 M K-gluconate medium) of 50–80 MOhm. Pipettes are advanced to the proximity of the selected neuron for recording, lowered gently so as to produce a dent in the membrane of the cell soma, visualized in light microscope. A short "buzz" is then applied to the Axoclamp 2B amplifier (Axon Instruments Inc.) and the entry of the content of the pipette into the cell can be visualized with the fluorescence light beam. Following the

filling up of the cell, which may require several short pulses (0.5 s each), the pipette is gently removed, and the cell is monitored with light microscope to verify that the operation did not kill the cell. This will be indicated by maintaining a smooth membrane, and having low [Ca^{2+}]i level.

4 Discussion

The present uncaging method is ideal for a simple, two-dimensional neuronal culture, in that it allows a precise, fast and reproducible placing of a flash photolysis spot near a structure of interest, such as a dendritic spine. Other methods employ 2-photon uncaging method, which may be more precise in the z dimension, where the two photons converge, but photoactivation requiring a long exposure to the light, up to several milliseconds, was reported [15]. This prolonged uncaging may result in lateral spread of the caged molecule, to affect adjacent spines that are not in the plane of focus of the imaging frame. Given that the density of spines in vivo is about a spine per um, it is highly likely that adjacent spines, may also respond to the uncaged molecule. On the other hand, our method is inferior to the 2-P method as it is not suitable for a three-dimensional structure, as is the case with a brain slice or in-vivo brain. Furthermore, the ability to redirect the laser beam within milliseconds from one location to another, so as to activate nearly simultaneously several targets, is a feature not currently available in our system. Nonetheless, the ability to directly activate small compartments in relative isolation is a feature unparalleled by the more complex 2-P uncaging system.

In recent years we have witnessed an increase in the use of high resolution microscopy, linked to the use of flashes of light to activate specific molecules in small spheres. Such is the case of flash photolysis of caged EGTA in presynaptic terminals, linked to TIRF (total internal reflection fluorescence), which allows sub-microscopic resolution of individual synaptic vesicles that are near the surface of the imaged field of view [32]. This method is useful with two dimensional cultured neurons, and it can employ flash photolysis to control the timing of a rise of intracellular calcium needed to cause release of neurotransmitters.

Another recent method that can employ high resolution optical control of neuronal activity uses light-operated potassium channel opener. In this method, the channel is present in the neuronal membrane, and light activates it, to open the G-protein coupled inward rectifying potassium channel (GIRK) which hyperpolarizes the cell and prevents it from action potential discharges [21]. While this method has only recently been described, it can be used in cultured hippocampal neurons, to explore the location of the channels most critical for their function in regulating neuronal excitability.

Another exciting development involves the combined use of super-resolution microscopy with single molecule tracking, using quantum dot technology [33, 34]. This method allows the tracking of single receptor molecules to examine the dynamics of lateral movement on the membrane of molecules related to neuronal plasticity. It allows the study of dynamic exchange of molecules inside and on the surface of neurons. The ability to use super-resolution microscopy allows a sub-micron resolution of spine morphology in living tissue. This combined methodology is expected to produce exciting insight into the dynamics of neuronal development and plasticity.

These and similar applications employing highly focused UV flashes as fundamental tools will undoubtedly continue to unravel the mechanisms of neuronal growth, synapse formation, plasticity and longevity at the subcellular level of analysis.

5 Specific Protocols

5.1 Protocol 1: Preparation of the Cultures: Media and Protocols

5.1.1 Materials

Rat P0 pups.

Horse serum (Cat.# 16,050, Gibco).

Fetal calf serum (Gibco).

HEPES (Cat # 83264, Sigma).

Enriched MEM (Minimal Essential Medium, enriched with glucose (0.6%), GlutaMAX (2 mM), and gentamicin (15 μg/mL) (Gibco #32360-026)).

24 well plates (Nunc).

B-27 (Cat. # 17504, Gibco).

L-15 (Cat# 21083, Gibco).

5-Fluoro-2-deoxyuridine (Cat # F0503, Sigma).

Trypsin (Cat# 15090, Gibco).

5.1.2 Procedure

Wistar Rat pups are decapitated on day of birth (P0), their brains removed and placed in a chilled (4 °C), oxygenated (by bubbling with a pipette tip) Leibovitz L15 medium (Gibco) enriched with 0.6% glucose and gentamicin (Sigma, 20 μg/mL) under a dissecting microscope. The hippocampus is removed and mechanically dissociated by trituration after incubation with trypsin (0.25%) and DNase (50 μg/mL), for 20 min at 37 °C 5% CO_2, in a humidified incubator. The tissue is passed to the plating medium. This consists of 1 mL minimum essential medium (MEM, Earl salts, Gibco)) containing 5% heat-inactivated horse serum (HS), 5% fetal calf serum, and B-27 (1 μL/1 mL), enriched with 0.6% glucose, gentamicin (20 μg/mL), and 2 mM GlutaMAX (enriched MEM). About 10^5 cells in 1 mL medium are plated in each well of a

24-well plate, onto a hippocampal glial feeder layer which is grown on the glass for 2 weeks prior to the plating of the neurons [25]. Cells are left to grow in the humidified incubator at 37 °C, 5% CO_2 for 4 days, at which time the medium is changed to 10% HS in enriched MEM, plus a mixture of 5′-fluoro-2-deoxyuridine/uridine (FUDR) (Sigma, 20 μg and 50 μg/mL, respectively), to block glial proliferation. Four days later the medium is replaced by 1 mL 10% HS in MEM which does not contain FUDR, and no further changes are made.

5.2 Protocol 2: Recording Medium (All Prepared in Double Distilled Water)

NaCl 129 mM.

KCl 4 mM.

Glucose 4.2 mM.

HEPES 10 mM.

$MgCl_2$ 1 mM.

$CaCl_2$ 2 mM.

pH is adjusted to 7.4 with NaOH.

Osmolality set to 320 mOsm with sucrose (Advanced Instruments, Inc. MicroOsmometer).

Note: Recording medium (without Ca and Mg) can be stored at +4 °C for weeks. $CaCl_2$ and $MgCl_2$ are added on day of use. Recording medium is then warmed to room temperature. Tetrodotoxin is added at 0.5 μM before use. All media are filtered with 0.22 μm filters before use.

5.3 Protocol 3: Transfection

Transfection is routinely conducted with 8–10 DIV cells. A Lipofectamine 2000™ (Invitrogen) mix is prepared at 1 μL/well with 50 μL/well Opti-MEM™ (Invitrogen), and incubated for 5′ at room temperature. This is mixed with 1.5 μg/well total DNA in 50 μL/well Opti-MEM™, and incubated for 15′ at room temperature. The mix is then added to the transfected culture wells, and allowed to incubate for 4–6 h before a change of medium. In most cases, at least several neurons are transfected on a single dish. Routinely, GluR1-eGFP with a ubiquitous CMV promoter is used. Cotransfection efficiency for several plasmids using this method is nearly 100%. Experiments are conducted routinely at 3–7 days after transfection.

References

1. Harris KM, Kater SB (1994) Dendritic spines: cellular specializations imparting both stability and flexibility to synaptic function. Ann Rev Neurosci 17:341–371
2. Segal M (2016) Dendritic spines: morphological building blocks of memory. Neurobiol Learn Mem. pii: S1074–7427
3. Sala C, Segal M (2014) Dendritic spines: the locus of structural and functional plasticity. Physiol Rev 94:141–188
4. Murnick GJ, Dube G, Krupa B, Liu G (2002) High resolution iontophoresis for single synapse stimulation. J Neurosci Methods 116:65–75

5. Adams SR, Tsien RY (1993) Controlling cell chemistry with caged compounds. Annu Rev Physiol 55:755–784
6. Augustine GJ (2001) Illuminating the location of brain glutamate receptors. Nat Neurosci 4:1051–1052
7. Canepari M, Nelson L, Papageorgiou G, Corrie JE, Ogden D (2001) Photo chemicaland pharmacological evaluation of 7-nitroindolinyl- and4-methoxy-7-nitroindolinyl-amino acids as novel, fast caged neurotransmitters. J Neurosci Methods 112:29–42
8. Delaney KR, Zucker RS (1990) Calcium released by photolysis of DM-nitrophen stimulates transmitter release at squid giant synapse. J Physiol 426:473–498
9. Zucker RS (1992 Jan) Effects of photolabile calcium chelators on fluorescent calcium indicators. Cell Calcium 13(1):29–40
10. Brown EB, Shear JB, Adams SR, Tsien RY, Webb WW (1999) Photolysis of caged calcium in femtoliter volumes using two photon excitation. Biophys J 76:489–499
11. Callaway EM, Katz LC (1993) Photo stimulation using caged glutamate reveals functional circuitry in living brain slices. Proc Natl Acad Sci USA 88:7661–7665
12. Dodt HU, Eder M, Schierloh A, Zieglgansberger W (2002) Infrared-guided laser stimulation of neurons in brain slices. Sci STKE 120 (2):1–11
13. Parpura V, Haydon PG (1999) 'Uncaging' using optical fibers to deliver UV light directly to the sample. Croat Med J 40:340–345
14. Pettit DL, Wang SS, Gee KR, Augustine GJ (1997) Chemical two-photon uncaging: a novel approach to mapping glutamate receptors. Neuron 19:465–471
15. Svoboda K, Tank DW, Denk W (1996) Direct measurement of coupling between dendritic spines and shafts. Science 272:716–719
16. Wang SS, Khiroug L, Augustine GJ (2000) Quantification of spread of cerebellar long-term depression with chemical two-photon uncaging of glutamate. Proc Natl Acad Sci U S A 97:8635–8640
17. Engert F, Paulus GG, Bonhoeffer T (1996 May) A low-cost UV laser for flash photolysis of caged compounds. J Neurosci Methods 66(1):47–54
18. Markram H, Segal M (1992) The inositol 1,4,5-trisphosphate pathway mediates cholinergic potentiation of rat hippocampal neuronal responses to NMDA. J Physiol 447:513–533
19. Korkotian E, Segal M (2006) Spatially confined diffusion of calcium in dendrites of hippocampal neurons revealed by flash photolysis of caged calcium. Cell Calcium 40 (5–6):441–449
20. Berlin S, Carroll EC, Newman ZL, Okada HO, Quinn CM, Kallman B, Rockwell NC, Martin SS, Lagarias JC, Isacoff EY (2015) Photoactivatable genetically encoded calcium indicators for targeted neuronal imaging. Nat Methods 12(9):852–858. https://doi.org/10.1038/nmeth.3480
21. Barber DM, Schönberger M, Burgstaller J, Levitz J, Weaver CD, Isacoff EY, Baier H, Trauner D (2016) Optical control of neuronal activity using a light-operated GIRK channel opener (LOGO). Chem Sci 7(3):2347–2352. https://doi.org/10.1039/C5SC04084A
22. Gurney AM, Lester HA (1987) Light-flash physiology with synthetic photosensitive compounds. Physiol Rev 67(2):583–617
23. Kaplan JH, Somlyo AP (1989) Flash photolysis of caged compounds: new tools for cellular physiology. Trends Neurosci 12(2):54–59
24. Korkotian E, Oron D, Silberberg Y, Segal M (2004) Confocal microscopic imaging of fast UV-laser photolysis of caged compounds. J Neurosci Methods 133(1–2):153–159
25. Korkotian E, Segal M (2011) Synaptopodin regulates release of calcium from stores in dendritic spines of cultured hippocampal neurons. J Physiol Lond 589(Pt 24):5987–5995
26. Korkotian E, Holcman D, Segal M (2004) Dynamic regulation of spine-dendrite coupling in cultured hippocampal neurons. Eur J Neurosci 20:2649–2663
27. Kandler K, Katz LC, Kauer JA (1998) Focal photolysis of caged glutamate produces long term depression of hippocampal glutamate receptors. Nat Neurosci 1:119–123
28. Matsuzaki M, Ellis-Davies GCR, Nemoto T, Miyashita Y, Iino M, Kasai H (2001) Dendritic spine geometry is critical for AMPA receptor expression in hippocampal CA1 pyramidal neurons. Nat Neurosci 4:1086–1092
29. Rossi FM, Margulis M, Tang CM, Kao JP (1997) N-Nmoc-l-glutamate, a new caged glutamate with high chemical stability and low pre-photolysis activity. J Biol Chem 272:32933–32939
30. Korkotian E, Segal M (2007) Morphological constraints on calcium dependent glutamate receptor trafficking into individual dendritic spine. Cell Calcium 42(1):41–57
31. Vlachos A, Korkotian E, Schonfeld E, Copanaki E, Deller T, Segal M (2009) Synaptopodin regulates plasticity of dendritic spines in hippocampal neurons. J Neurosci 29:1017–1033

32. Midorikawa M, Sakaba T (2015) Imaging exocytosis of single synaptic vesicles at a fast CNS presynaptic terminal. Neuron 88(3):492–498. https://doi.org/10.1016/j.neuron.2015.09.047
33. MacGillavry HD, Blanpied TA (2013) Single-molecule tracking photoactivated localization microscopy to map nano-scale structure and dynamics in living spines. Curr Protoc Neurosci 65:2.20.1–2.2019. https://doi.org/10.1002/0471142301
34. Ehlers MD, Heine M, Groc L, Lee MC, Choquet D (2007) Diffusional trapping of GluR1 AMPA receptors by input-specific synaptic activity. Neuron 54(3):447–460

Chapter 12

Analysis of the Functional States of an Astrocyte Syncytium

Yixing Du, Conrad M. Kiyoshi, David Terman, and Min Zhou

Abstract

In the central nervous system (CNS), astrocytes are cytoplasmically interconnected through specialized conduit proteins, called gap junctions. This establishes a syncytial network that runs across the entire brain. Because of this unique anatomic characteristic, the still mysterious function of astrocytes, particularly in the adult brain, is likely hidden within this syncytial network. Recent studies show that a strong electrical coupling is a system-wide feature of astrocyte networks, which enables astrocytes to constantly equalize their membrane potentials so that a syncytial isopotentiality can be achieved. Functionally, syncytial isopotentiality coordinates the individual astrocytes into a functional system in brain homeostasis. Dysregulated extracellular K^+ concentration impairs neuronal excitability and synaptic transmission. Under syncytial isopotentiality, the homeostatic regulation of extracellular K^+ can be performed with far greater efficacy than previously thought. In this chapter, this newly appreciated mechanism that governs the operation of astrocyte network will be introduced. A new functional readout concerning the functional state of an astrocyte network and its application will be described. Second, a new model system, freshly dissociated astrocyte miniature syncytium, will be introduced for the quantitative analysis of astrocyte syncytial network function. Third, a computational model will be introduced. A combination of these can be powerfully applied to decipher the dynamic change in the coupling strength of an astrocyte syncytium as well as astrocyte-neuron crosstalk at system levels. This new method also has a great potential to be used to identify a causal relationship between an impaired astrocyte network function and the disease under investigation. Overall, a novel approach described in this chapter allows examination of astrocyte function at the network level in the healthy and diseased brain.

Key words Astrocytes, Gap junctions, K^+ channels, Astrocyte syncytium, Membrane potential, Syncytial isopotentiality, Goldman–Hodgkin–Katz equation, Patch clamp, Computational modeling

1 Introduction

1.1 Astrocytes

Astrocytes are the most populous cell type in the brain. Based on their morphology and anatomic locations, astrocytes can be broadly divided into protoplasmic astrocytes in the grey matter and fibrous astrocytes in the white matter. Additionally, the radial-shaped Bergmann glia in cerebellum and Müller glia in the retina are also considered as astrocytes. In contrast to the traditional view of passive supporting cells in the brain, emerging evidence reveals the multifaceted roles of astrocytes in developing and adult brain

Nicholas J. D. Wright (ed.), *Basic Neurobiology Techniques*, Neuromethods, vol. 152,
https://doi.org/10.1007/978-1-4939-9944-6_12, © Springer Science+Business Media, LLC, part of Springer Nature 2020

function. In the developing brain, astrocytes actively regulate synaptogenesis and maturation. In adult brain, astrocytes provide neurotrophic support, maintain a homeostatic extracellular ion and neurotransmitter concentrations, ferry of energy substrates from blood flow into the brain, and regulate synaptic transmission and plasticity [1–5]. Despite increasing research attention on astrocytes, the physiological function of astrocytes, particularly in the adult brain, remains largely unknown. Additionally, neuronal excitability, synaptic potentials, and synaptic plasticity can be examined with well-established electrophysiological, Ca^{2+}-imaging, optogenetics, and chemogenetics methods; however, lack of a functional readout remains as a major obstacle to discern the abnormality from normality of an astrocyte network in various forms of neurological disorders [6]. It is our hope that the described method here will facilitate and lead your research toward this completely unexplored area.

1.2 Syncytial Coupling of Astrocytes Is Necessary for Normal Brain Function

Astrocytes are coupled to each other to form the largest syncytial networks through gap junction coupling in the brain [7–10]. While this suggests that astrocyte function may go beyond the single cell level as a functional system [2, 11, 12], the mechanisms dictating the operation of these prominent glial networks is currently under intensive investigation [13, 14]. For instance, the oscillation of intracellular Ca^{2+} has been indicated as a glial-specific signaling mechanism [4]. Also, syncytial coupling mediates the exchange of ions, metabolites, and calcium that play critical roles in brain homeostatic and signaling functions [8, 15–22]. The importance of this network in brain function has been further underscored through genetic deletion of two major astrocytic gap junctions, connexin 43 (Cx43) and 30 (Cx30). In these mice, disruption of syncytial coupling results in a disturbed K^+ and glutamate homeostasis interrupted synaptic long-term potentiation (LTP) and impaired sensorimotor and spatial memory tasks [9, 23–25]. This observation unequivocally shows that the disruption of syncytial coupling is causal to that observed dysfunction of basic and advanced brain function [26–29].

1.3 Astrocyte Syncytial Isopotentiality

Evidence that neurons and astrocytes can be intercellularly connected was unearthed over 50 years ago [30, 31]. For the former, the role of neuronal gap junction was immediately determined as "electrical synapse" that mediates a fast form of synaptic transmission for synchronized neuronal firing. Despite the speculation that the role of astrocyte gap junction coupling is to create an isopotential syncytium [32], definitive evidence that syncytial isopotentiality does exist came only recently [33]. A strong electrical coupling indeed equalizes membrane potentials among astrocytes (Fig. 1), and the resultant syncytial isopotentiality indeed coordinates astrocytes into a functional system. As a demonstrated

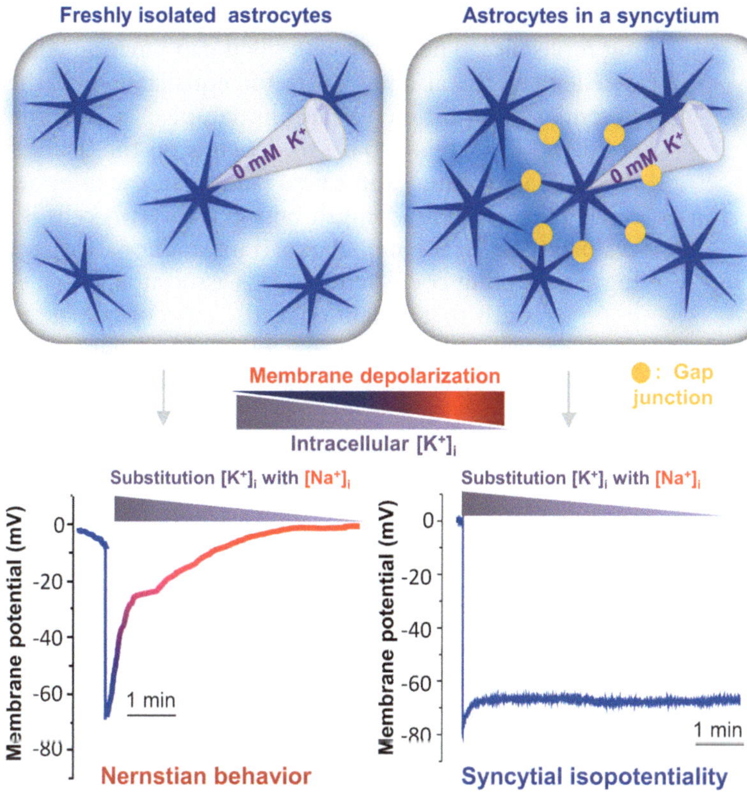

Fig. 1 Syncytial isopotentiality. Substitution of endogenous K^+ by Na^+ in recording solution ($[Na^+]_P$, 140 mM Na^+/0 mM K^+) depolarizes single freshly dissociated astrocyte V_M to ~0 mV following to GHK prediction (left), called Nernstian behavior. However, in a synctial coupled astrocyte, this depolarization can be compensated for to a quasi-physiological V_M level by the physiological V_M of the neighboring astrocytes (right)

example, syncytial isopotentiality maintains a sustained K^+ uptake driving force for high efficient uptake of K^+ ions focally applied to an individual astrocyte inside a syncytium [33]. Likewise, the syncytial isopotentiality should also facilitate the exchange of K^+ ions between the interior and exterior of a syncytium through astrocyte K^+ channels. This has provided strong support to the frequently discussed "K^+ spatial buffering" hypothesis [18, 33–35].

Innovative use of a combination of patch-clamp method, astrocyte syncytia in acute brain slices or after fresh dissociation into "miniature syncytia", and computational modeling has made the discovery of syncytial isopotentiality possible [36, 37]. In this chapter, we will begin with the biophysical rationale that led to the new experimental methods. A major focus, however, is to introduce hands-on experimental protocols that can be used by researchers with access to brain slice preparation, whole-cell patch recording system for analysis of the dynamic change in gap junctional coupling of an astrocyte network in healthy and diseased brain.

1.4 Physiological and Biophysical Principles Underlying the Methodology

Gap junctions permit the exchange of ions, metabolites and signaling molecules. The electrolyte-filled gap junction channels are also excellent electrical conductors for equalization of the membrane potential (V_M) difference among coupled cells, termed electrical coupling. The ability to do so depends on the quantity of the gap junction aggregates, also called plaque, in the interface of coupled cells. The coupling strength between the two cells is traditionally evaluated by dual patch-clamp transjunctional voltage analysis. A higher coupling coefficient (CC) is associated with a stronger coupling strength that is defined as the ratio of voltage transferred to the recipient cell ($\Delta V_{recipient}$) over the voltage induced by the delivered current (ΔI) in the stimulated cell ($\Delta V_{stimulated}$), or $CC = \Delta V_{recipient}/\Delta V_{stimulated}$ (Fig. 9). However, due to a significant shunting of injected current through the low membrane resistance, the CC measurement is of limited value for informing the actual coupling strength in neighboring astrocytes [10, 38]. To make this worse, each astrocyte directly couples to 7–9 astrocytes [39, 40]. This long-standing technical delemma has been recently resolved in our recent study [33]. Experimentally, this was achieved through elimination of membrane K^+ conductance by substitution of intracellular K^+ with Na^+ ion in dual patch recording. Under these conditions, a strongly coupling CC of 94% was revealed [33]. This innovative use of conventional patch-clamp recording method is another focus of this chapter.

Physiologically, our method is designed following a basic feature of astrocytes. Due to a high-density expression of K^+ conductance and minimal Na^+ and Ca^{2+} conductances, astrocyte behaves as a perfect K^+ electrode [41]. Consequently, one can experimentally lower the intracellular K^+ concentration ($[K^+]_i$) through dialysis of the recorded cell with reduced or K^+-free pipette solutions. This, in turn, alters the K^+ equilibrium potential (E_K) of the recorded cell to depolarized levels according to the prediction from Goldman–Hodgkin–Katz (GHK) equation. However, the associated syncytium should compensate for the loss of physiological membrane potential (V_M) in the recorded cell, and the level of the compensation should be determined by the coupling strength and the quantity of directly coupled nearest neighbors [33, 39]. Overall, the variation of the steady-state quasi-physiological V_M ($V_{M,SS}$) is a readout of the coupling strength among astrocytes in a syncytium (Fig. 4). A computational model is now made available for quantitative analysis the variation of coupling strength in the same brain region, for instance, during neuronal activation, or from brain region to region [39, 40]. We will guide you through the use of the modeling database in your experiments.

We developed a biophysically based, mathematical model of a network of gap junction coupled astrocytes in the hippocampal CA1 region. In this model, the coupling strength of a syncytium (S) is defined as the ratio of gap junction conductance (G_g) over the

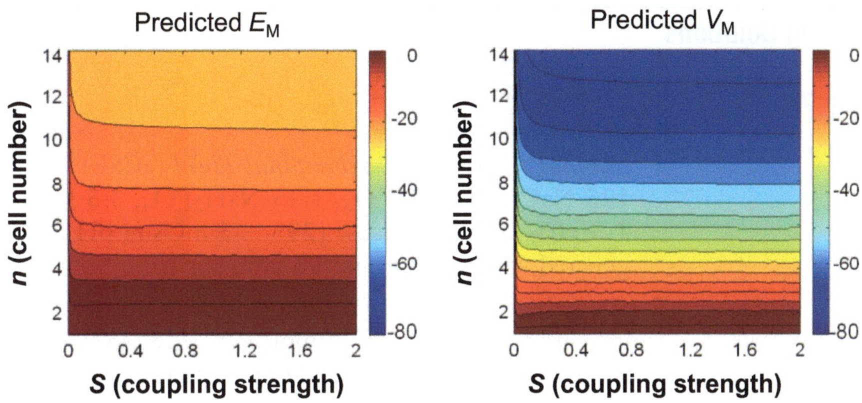

Fig. 2 The aggregate effect of syncytial control over the membrane potential of an individual astrocyte. A model astrocyte, recorded with a 140 mM Na^+ ($[Na^+]_p$) solution, is directly coupled within a syncytium of a variable number of astrocytes ($1 \leq n \leq 14$) and at variable levels of gap junction coupling strength ($0 \leq S \leq 2$). Left: In the $[Na^+]_p$ recorded cell, $[K^+]_i$ increases with more cells coupled to it, and with stronger *S*. These changes result in a progressive shift in E_M from 0 mV (0 mM $[K^+]_i$) to more negative values as predicted by the GHK equation (see color scale). Right*:* The measured membrane potential (V_M) in the $[Na^+]_p$ recorded astrocyte, which was determined by simulating the model for 2000 seconds. Both increasing the number of astrocytes coupled to the recorded cell and strengthening *S* deviates the recorded V_M from the GHK predicted E_M toward the physiological V_M (blue in color scale) in neighboring astrocytes (modified from Ma et al., 2016)

membrane conductance (G_M) in the form of $S = G_g/G_M$. The model predicts that the increasing syncytium size has an aggregate effect on the final resting potential of the recorded cell with K^+ free/Na^+- containing electrode solution ($[Na^+]_p$) (Fig. 2). The model also predicts that a syncytium of size 7–9 is sufficient to hold the target cell at the quasi-physiologically functional level of ~−70 mV. The modeling details are available in our publication [33], and a database that can be directly used for quantitative analysis of your experimental data is available as "supporting information" in Kiyoshi et al. 2018 [39].

2 Animals

In our study and all the results presented, the experimental procedures were performed in accordance with a protocol approved by the Animal Care and Use Committees of The Ohio State University. Animals of wild-type C57BL/6N, BAC-ALDH1L1-eGFP transgenic mice [42], and other mice with genetic manipulations can be used. Depending on the purpose of the study, the animals can be divided into the following age groups: postnatal day (P) 1–3 (neonatal) [43], P4–P20 (developing), and P21–28 (young adults) [44]. Animals of both sexes were used. This description of animal usage is applicable throughout this chapter.

3 Recording Solutions

Bath Solutions

1. *Standard artificial Cerebrospinal Fluid* (aCSF) (in mmol/L): 125 NaCl, 25 $NaHCO_3$, 1.25 NaH_2PO_4, 3.5 KCl, 2 $CaCl_2$, 1 $MgCl_2$ and 10 Glucose (295 ± 5 mOsm; pH 7.3–7.4; oxygenated with 95% O_2 and 5% CO_2).
2. *Acute brain slice cutting solution* (aCSF with reduced Ca^{2+} and increased Mg^{2+}) (in mmol/L): 125 NaCl, 3.5 KCl, 25 $NaHCO_3$, 1.25 NaH_2PO4, 0.1 $CaCl_2$, 3 $MgCl_2$ and 10 Glucose. Ice-cold and oxygenated with 95% O_2 and 5% CO_2.
3. *K^+-free aCSF* (in mmol/L): 128.5 NaCl, 25 $NaHCO_3$, 1.25 NaH_2PO4, 2 $CaCl_2$, 1 $MgCl_2$ and 10 Glucose (295 ± 5 mOsm; pH 7.3–7.4; oxygenated with 95% O_2 and 5% CO_2).
4. *Oxygen–glucose deprivation (OGD) aCSF* (in mmol/L): To simulate ischemic conditions in situ and in vitro, glucose was replaced with equimolar sucrose in aCSF described above, and O_2 was depleted by bubbling the solution with 95% N_2 and 5% CO_2 (pH 7.3–7.4) for 30 min prior to and throughout OGD treatment.

Electrode Recording Solutions

5. *K^+-based electrode solution* (in mmol/L): 140 K-gluconate, 13.4 Na-gluconate, 0.5 $CaCl_2$, 1.0 $MgCl_2$, 5 EGTA, 10 HEPES, 3 Mg-ATP, and 0.3 Na-GTP (280 ± 5 mOsm, pH 7.25–7.35).
6. *K^+-free/Na^+-based electrode solution* (in mmol/L): The 140 K-gluconate in the above K^+-based electrode solution is substituted by 140 Na^+-gluconate or 140 NaCl. When there is a need in experimental design, the K^+ can also be particularly substituted with Na^+ to create a partially "K^+-deficient astrocyte" as a reporter of syncytial coupling strength.
7. *Other monovalent cation-based K^+-free electrode solution*: In electrode K^+ can also be substituted with Li^+, Cs^+, or $NMDG^+$. The Li^+ substitution can create a "K^+-deficient astrocyte" where the Na^+-K^+ ATPase is also inhibited [45]. The Cs^+ substitution can create a "K^+-deficient astrocyte" with fully inhibited K^+ channel activity [46].

4 Equipment and Apparatus

1. *An electrophysiology setup* for submerged slices equipped with ×4 and ×40 infrared differential interference contrast (IR-DIC) visualization, a patch-clamp amplifier capable of current and voltage clamp and providing two-channel recording (e.g., MultiClamp 700A or 700B, Molecular Devices), an interface for converting digital-analog signals between the amplifier and a computer (e.g., Digidata 1322A, Molecular Devices).
2. *A fluorescent imaging system* (e.g., Polychrome V system from Till Photonics, Germany) is advantageous for high-resolution visualization of small glial soma and placing of dual patch electrodes on it. This system can also be used for identifying astrocyte based on SR101 positive staining or eGFP-expressing astrocytes in ALDH1L1-eGFP and MrgA1 transgenic mice.
3. *Two micromanipulators* for automatic loading of two electrodes to the top of the brain slice can shorten the loading time (e.g., PatchStar from Scientifica, UK).
4. *Glass electrodes pullers* were pulled from a Micropipette Puller (e.g., Model P-97 and P-87, Sutter Instrument).
5. *Vibratome* for brain slices cutting (e.g., Pelco 1500).
6. *Borosilicate glass capillaries* are available from Warner Instruments (Cat. no. 64-0772), worked well in this experiment. An open electrode resistance in the range of 2.5–5 MΩ, when filled with a KCl-based electrode solution ($[K^+]_P$) or NaCl-based electrode solution ($[Na^+]_P$).
7. *Slice anchors for Series 20 chamber* in our experiments are purchased from Warner Instruments, LLC (Hamden, CT, USA).

5 Acute Brain Slice and Miniature Syncytia Preparation

1. *Acute brain slices preparation.* After anesthesia with 8% chloral hydrate in 0.9% NaCl, the mouse was decapitated and the brain was rapidly removed from skulls and placed into ice-cold oxygenated cutting solution in a culture dish. Mouse brain was transferred onto an ice-cold mouse brain slicer matrix and was trimmed to coronal tissue block with the interested brain region (hippocampus). The coronal tissue block was mounted into the tray of the Vibratome with glue and then the tray was filled with ice-cold cutting solution. Coronal hippocampal slices (250 μm thickness) were sectioned at 4 °C with the Vibratome and then transferred into a small basket in a

container within oxygenated standard aCSF, incubated at 25 °C for 60 min for recovery from preparation damage.

2. *Sulforhodamine 101 (SR101) staining*. Slices were transferred from standard aCSF into another container with oxygenated aCSF supplemented 0.6 μM astrocytic marker sulforhodamine 101 (SR101) and incubated at 34 °C for 30 min. When BAC-ALDH1L1-eGFP transgenic mice or other transgenic mice with an eGFP-tag reporter, such as hGFAP:Gq-MrgA-1-eGFP transgenic mice [47], are used, this step can be eliminated.

3. *Freshly dissociated astrocytes and astrocyte miniature syncytia*.
   (a) After SR101 incubation or after slices recovery (for ALDH1L1-eGFP mice), the CA1 regions in our study were dissected out from slices and carefully cut into small pieces (1 mm^2) using two hypodermic needles, then transferred into a 1.5 mL Eppendorf tube containing oxygenated aCSF supplemented with 24 U/mL papain and 0.8 mg/mL L-cysteine for 7 min incubation at 25 °C.
   (b) After papain digestion, the loosened tissues were gently triturated with a fire-polished glass pipette 5–7 times into a cell suspension and transferred into the recording chamber.
   (c) In each cell suspension, more than 40 small tissue blocks were yielded that contained single to up to 20 astrocytes. The number of astrocytes and their domain territories in each tissue block can be readily visualized based on SR101 staining or eGFP fluorescence (Fig. 3, D2–3; Fig. 11, A). An important criterion in cell selection is to choose those astrocytes with their domain territories largely intact.

6 Methods

6.1 Basic Electrophysiological Properties of Astrocytes in Brain Slices

6.1.1 Introduction

In the mature rodent brain, astrocytes are nonexcitable cells in the brain. The soma size of astrocytes is typically <10 μm, which imposes a difficulty for the electrophysiological study of this type of brain cells. However, advancement in microscopic technique and high-quality camera in the imaging system has made astrocytes more readily accessible for electrophysiological study. This subtype of glial cell is characterized by exhibiting a more hyperpolarized membrane potential (V_M), ~−80 mV, and expression of a linear current-to-voltage (I–V) relationship, passive membrane K^+ conductance [48]. These basic features are most likely evolutionally conserved from the low species, like the invertebrate leech, to the human brain [31, 49]. In the developing brain, however, astrocytes exhibit a variable rectifying I–V, resulting from a complex

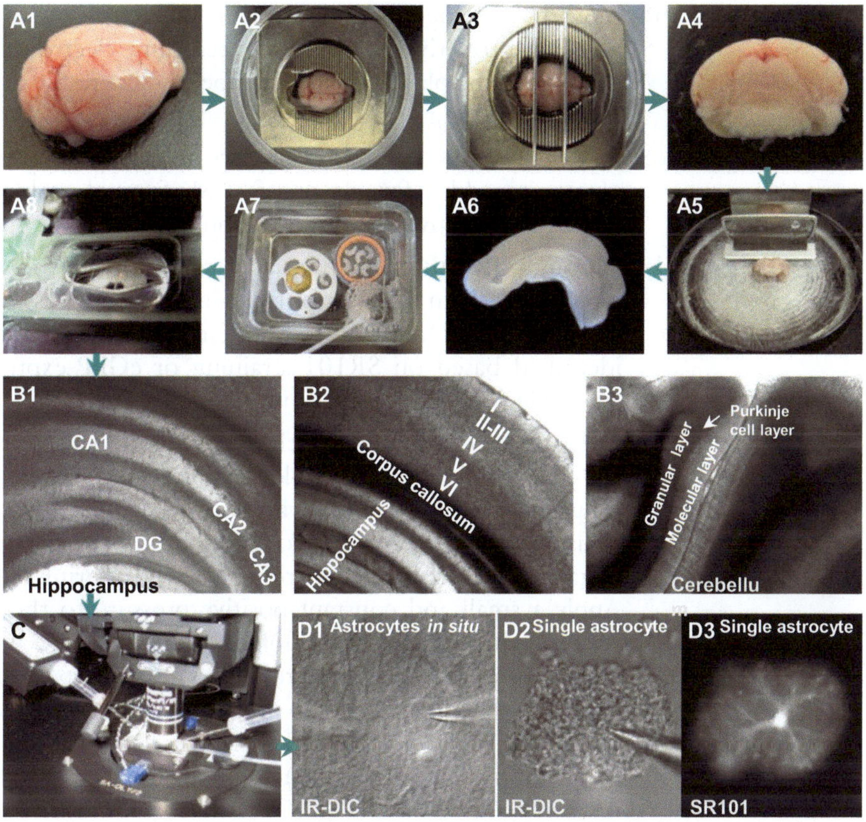

Fig. 3 Astrocytes in acute brain slice or after fresh dissociation. **(A1–8)** Illustration of the steps for acute brain slice preparation. **(B1–3)** Different brain regions visualized through a x4 objective under infrared differential interference contrast (IR-DIC). **(C)** Electrophysiology setup for submerged slices. **(D1)** IR-DIC image of dual electrodes patch clamp on two neighboring astrocytes in brain slice. **(D2)** Single electrode recording of a freshly dissociated astrocyte with its largely intact processes extended from the soma. **(D3)** The intact astrocyte morphology in **D2** can be further validated by SR101 staining using the fluorescent imaging system

expression of several voltage-gated outward K^+ conductances [44, 50]. Developmentally, the immature in the expression of membrane K^+ channels is associated with a progressive integration of newborn astrocytes into a syncytium [50, 51]. In contrast, the passive conductance is associated with an exceptionally low membrane resistance (R_M), indicating abundant expression of voltage-independent leak type K^+ channels [36, 52, 53]. The species and quantity of membrane K^+ channels is another important determinant for the optimal function of astrocyte syncytium. Specifically, it determines the efficacy of exchange of K^+ ions between the interior and exterior of a syncytium, or K^+ buffering.

6.1.2 Experimental Procedure

(a) After the perfusion rate at 2 mL/min is stably established for the recording chamber, an acute brain slice will be transferred to the chamber (Fig. 3, C).

(b) As shown in Fig. 3, C, the region of interesting should be adjusted properly for access by recording electrode. A slice anchor will be placed to the chamber to secure the slice from free moving (Fig. 3, A6).

(c) Lower the 10×, or 40× objective to the slice, then use a pipette to add a few drops of aCSF from the objective to make the objective water-immersion.

(d) Through IR-DIC, identify an astrocyte from soma shape, typically <10 μm with one or two primary processes branching out (Fig. 3, D1). Astrocyte can be more cell type specifically identified based on SR101 straining or eGFP-expression in ALDH1L1-eGFP reporter mouse.

(e) Fill patch electrode with K^+ electrode solution and install it to the electrode holder. This holder is connected through a ~1 ft. long silicone tubing to a 3-way stopcock. And a 1 mL syringe is also attached to the stopcock for adding either positive or negative pressure to the recording electrode during recording. Apply a small and constant positive pressure to the internal solution through electrode holder. Under IR-DIC with ×40 objective, advance the electrode to the surface of the brain slice first. Then move the tip of the electrode to the cell body of the identified astrocyte. In voltage clamp mode, apply a recurrent 5 mV square voltage pulse to monitor the electrode resistance.

(f) Approach the cell body slowly until a pressure caused "shadow" appears on the cell soma, then release the positive pressure. This is typically followed by a rapid increase in seal resistance. Apply a −80 mV negative potential through the commander panel to facilitate seal formation. This manipulation typically results in the rapid formation of a GΩ seal.

(g) Rupture the patch of membrane isolated inside the electrode tip by negative pressure pulses to establish the whole-cell configurations. A successful transition from cell-attach to whole-cell mode can be determined by a progressive decrease of seal resistance to <30 MΩ without an obvious negative shift in holding currents. After the whole-cell mode is established, wait at least 5 min to allow ion equilibration between electrode solution and intracellular cytoplasm.

(h) Set astrocyte first in the current clamp mode (I-C) from the Commander panel, use the "View" function in Clampex software to observe the resting V_M, then use the "Record" function to record resting V_M for the record. Then switch to voltage-clamp mode (V-C) from the Commander. Load a preprogrammed voltage comment protocol: holding potential, −80 mV, voltages steps from −180 mV to +20 mV,

10 mV increments and 50 ms duration. One can use the "View" function for a preview, or directly use "Record" function to record the current profile of an astrocyte.

(i) Clampfit software is used for post hoc analysis of the recorded data.

6.1.3 Methodological Application/Expected Results

(a) This section provides a basic introduction of whole-cell astrocyte recording from acute brain slices.

(b) Due to a high leak K^+ conductance, a critical reminder for quality control is how to distinguish K^+ leak conductance from the nonspecific leak conductance. After rupture into whole-cell recording, both conductances are linear in the *I–V* plot, which can be rather difficult for inexperienced researchers. One can rely on the following parameters to make a collective decision. First, when rupture the cell-attach to whole-cell in "seal test" mode with the cell held at −80 mV, the baseline of holding current should remain at close to 0 pA. A downward shift in holding current is a sign of loss of initially established GΩ seal. In some cases, the holding current can return to ~0 pA, but the overall quality of the recording is in question. Often time, the experiment should be terminated. Second, in the $I = 0$ mode, the V_M reading should be around −80 mV. Third, the whole-cell *I–V* plot should also have a reversal potential at ~ −80 mV.

(c) Due to an exceptionally low R_M, the "Membrane test" test available from the Clampex becomes dysfunctional. The biophysical rationale underlying this failure is described in our previous report [52]. In short, the passive properties of astrocytes, that is, access resistance (R_a), R_M, and C_M generated from "Membrane test" are all erroneous. Thus, the use of "Resistance test" from Clampex to have an estimate of the R_{in} from astrocytes. The passive properties can be better estimated from freshly dissociated single astrocytes [33, 36]. This will be introduced in Subheading 6.4.

6.2 Analysis of Astrocyte Syncytial Isopotentiality in Brain Slice

6.2.1 Introduction

Astrocytes are strongly electrically coupled through gap junction. Consequently, the V_M of an individual astrocyte cannot deviate substantially and independently away from the shared syncytial isopotentiality [33] (Fig. 1). Experimentally, this can be detected by the substitution of endogenous K^+ ion by a 140 mM Na^+/K^+ free electrode solution ($[Na^+]_P$). The purpose of the following section is to guide you through the experimental evaluation of the functional state of an astrocyte network. Additionally, examples are given to illustrate how this method can be used in combination with syncytial anatomy study that correlates the form to the function. Although this method was only recently established, it is foreseeable that this diagnostic tool will be extended to the study of astrocyte syncytium in healthy and diseased conditions very soon.

6.2.2 Experimental Procedure

(a) The experimental procedure is the same as described in Subheading 6.1 except for the Na^+ based $/K^+$ free electrode solution was used to fill the micropipette.

(b) The cell was ruptured under current clamp recording mode. The breakthrough of the membrane patch shifted the V_M immediately to a resting membrane potential of −78 mV (Fig. 4B). Over time, the V_M reaches a stationary level of −73 mV. This is in sharp contrast to the anticipated depolarization to 0 mV as predicted by the GHK equation for K^+ – free electrode solution. This indicates the existence of strong electrical coupling in a syncytium that prevents the $[Na^+]_P$-induced depolarization in the recorded cell.

(c) The initial V_M recording ($V_{M,I}$) reflects the true resting V_M of an astrocyte that is independent of the recording solutions used, that is, $[K^+]_P$ or $[Na^+]_P$. (Fig. 4B). The steady-state V_M ($V_{M,SS}$) does not differ in $[K^+]_P$ recording from $V_{M,I}$ (Fig. 4B). However, the $V_{M,SS}$ in $[Na^+]_P$ recording is ~5 mV more depolarized then $V_{M,I}$ in hippocampal astrocyte syncytium.

(d) During the V_M recording, it is essential that the input resistance (R_{in}) is periodically monitored by "Resistance test" (a 63 pA/600 ms pulse) available from Clampex software (Fig. 4B2). Increase in R_{in} reduces the accessibility of $[Na^+]_P$ to the interior, which in turn decreases the Na^+ concentration in the recorded cell. Consequently, this disqualifies the recorded astrocyte to be a Na^+-saturated "reporter" of the coupling strength of an astrocyte syncytium.

6.2.3 Methodological Application/Expected Results

1. In hippocampal astrocyte syncytium, the quasi-physiological $V_{M,SS}$ detected with $[Na^+]_P$, −73 mV, is 5 mV depolarized compared to the physiological V_M, −78 mV, in the coupled neighboring astrocytes. Our computational model predicts that variation of $V_{M,SS}$ in the same syncytium or syncytia of other parts of brain regions reflects the difference in syncytial coupling strength (S) (Figs. 2, and 4C). Specifically, as shown in Fig. 4C, the more the hyperpolarizing shifts in $V_{M,SS}$, the stronger the strength of syncytial coupling (S).
2. Due to the space limit, a complete excel database could not be included in this chapter, but you can consult the "supporting information" in Kiyoshi et al. (2018) Glia [39]. The relative change in coupling strength (S) is established based on the data generated from hippocampal CA1 syncytium. A further strengthening or weakening of coupling strength (S) can be

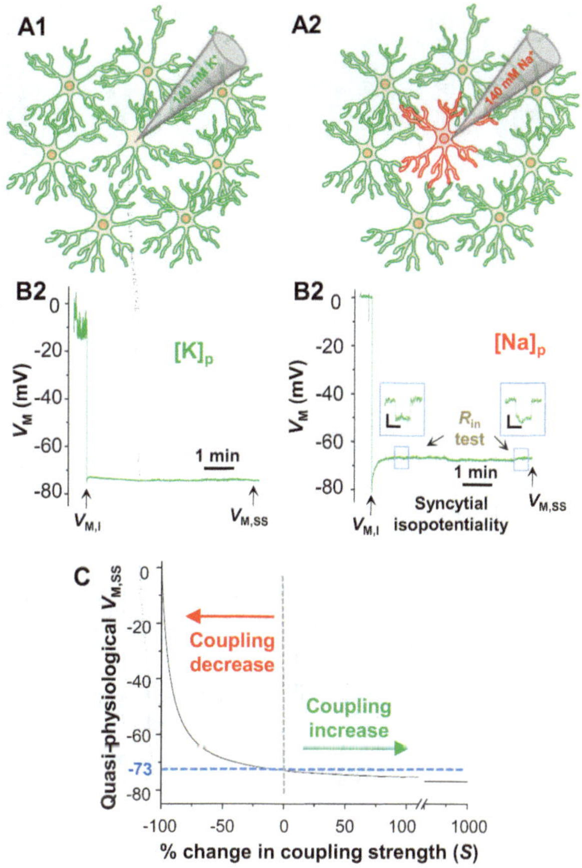

Fig. 4 Astrocyte syncytial isopotentiality. Illustration of recording from a syncytial coupled astrocyte with $[K^+]_p$ (**A1**, left) and $[Na^+]_p$ (**A2,** right). **B1** and **B2** are V_M recordings made with $[K^+]_p$ and $[Na^+]_p$, respectively. The V_M with $[Na^+]_P$ disobeys the Nernstian prediction. Instead, the V_M maintains at a quasi-physiological level. Also in **B2** shows the initial V_M ($V_{M,I}$), and the quasi-physiological V_M at the steady-state ($V_{M,SS}$). Input resistance (R_{in}) was periodically monitored during recording. **(C)** The relationship between $V_{M,SS}$ and coupling strength (S) predicted by computational modeling. The $V_{M,SS}$ from hippocampal CA1 syncytium, −73 mV, is used as the baseline in the model to predict the relative difference in coupling strength (S%) between CA1 syncytium and a syncytium of another brain region

identified from this database and compared with CA1 syncytium.

3. To illustrate how this model to be used for quantitative analysis of the difference in coupling strength (S), the correlation of the syncytial anatomy and coupling strength (S) is illustrated from visual cortex and cerebellum (Figs. 5 and 6).

4. For astrocyte syncytial anatomy study, a powerful approach is the combinative use of ALDH1L1-eGFP reporter mice with

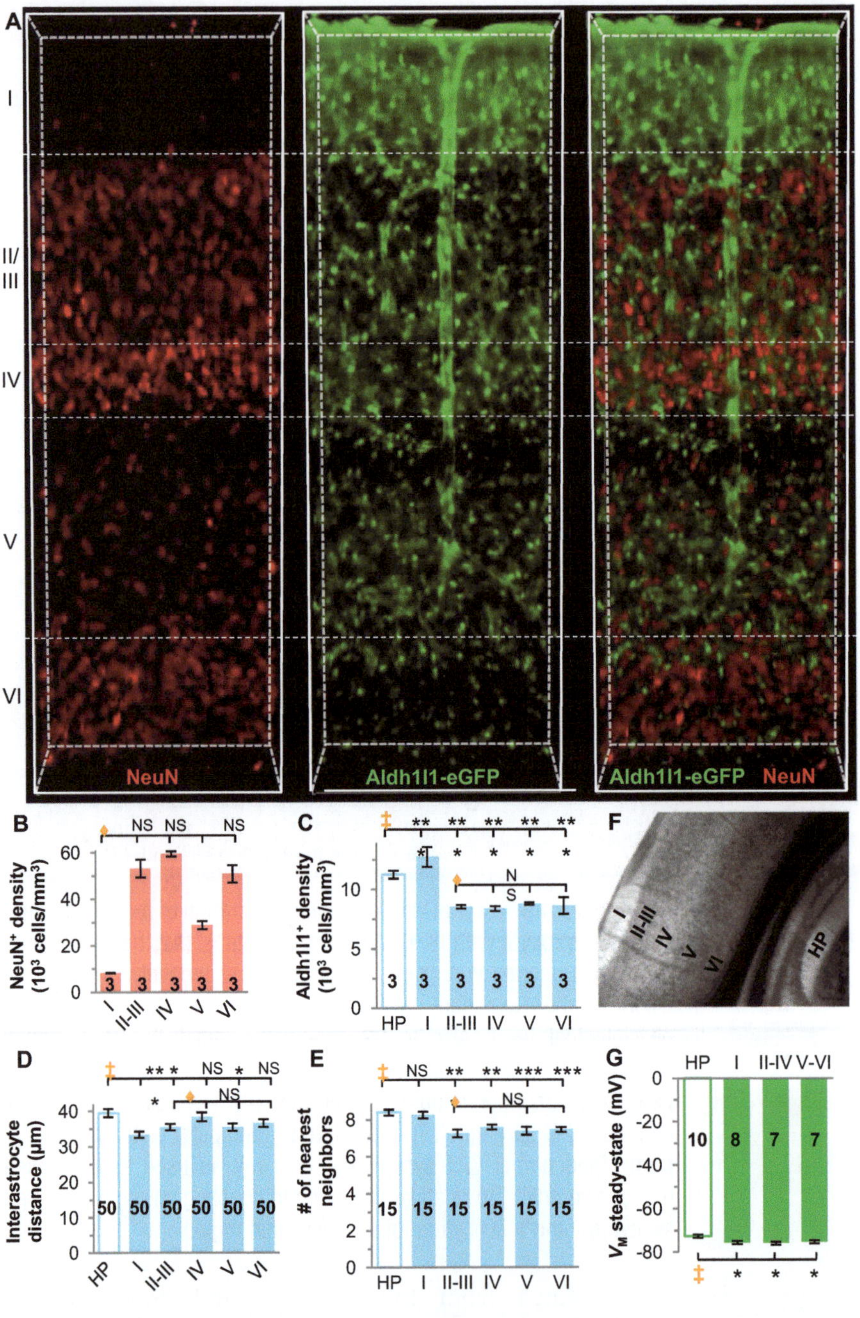

Fig. 5 Syncytial isopotentiality is present in all cortical layers of the visual cortex. **(A)** CUBIC optically clear visual cortex with NeuN immunostaining in Aldh1l1-eGFP transgenic mice. **(B)** $NeuN^+$ cell density varies across layers. **(C)** Aldh1l1-$eGFP^+$ astrocytes are denser in layer I, but less dense in Layers II–VI when compared to CA1 hippocampus. **(D)** Interastrocytic distance varies across cortical layers. **(E)** The number of nearest neighbors is significantly less in Layers II–VI. **(F)** IR-DIC imaging of the visual cortex layers as indicated. **(G)** $V_{M,SS}$ in the cortical layers was significantly more negative. **(B–E)** One-way ANOVA, Tukey's post hoc (); (C–F) One-way ANOVA, Dunnett's post hoc (‡). $*p < 0.05$, $**p < 0.01$, $***p < 0.005$, NS: $p > 0.05$

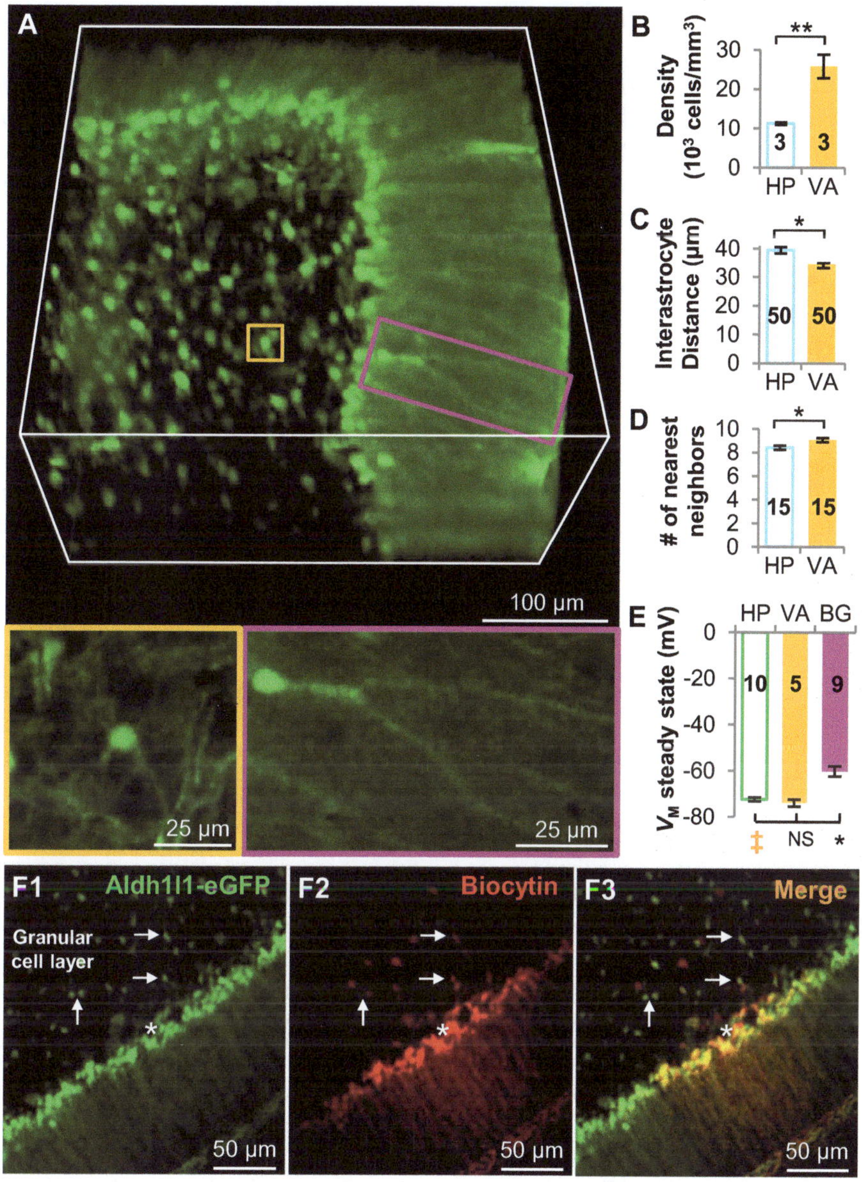

Fig. 6 Syncytia established by Bergmann glia and velate astrocytes vary in coupling strength. **(A)** CUBIC optically clear cerebellum reveals the morphological and spatial organizational pattern heterogeneity of velate astrocytes (gold) and Bergmann glia (magenta). Velate astrocytes are higher in density **(B)**, shorter in the interastrocytic distance **(C)**, and adjacent with one more nearest neighbors **(D)** than hippocampal CA1astrocytes. **(E)** The quasi-physiological $V_{M,SS}$ in Bergmann glia was significantly more depolarized than velate and hippocampal astrocytes. **(F)** Diffusion of pipette-loaded biocytin among Bergmann glia and to adjacent velate astrocytes. The recorded cell is marked by white asterisk. Arrows point to biocytin-positive velate astrocytes in the granular cell layer. **(B–D)** Unpaired *t*-test; **(E)** One-way ANOVA, Dunnett's post hoc (‡). $*p < 0.05$, $**p < 0.01$, NS: $p > 0.05$

the tissue-clearing method. In our hand, we used the latest CUBIC method for tissue clearing [54–56]. This facilitates the visualization of astrocyte syncytium (Figs. 5 and 6). In the different layers of the visual cortex, the anatomy, in terms of cell density, interastrocytic distance and the number of the nearest neighbors vary in morphometric analysis. The coupling strength (*S*), however, does not differ between layers. However, the coupling strength (*S*) of visual cortex is stronger than hippocampal CA1 syncytium (Fig. 5).

5. Another example we give here is the observation from the cerebellum. Cerebellar Bergmann glia and velate astrocytes are derived from the same progenitor pool but strikingly different in their morphology [57]. Bergmann glia are characterized by having their cell bodies arranged in rows alongside with the soma of Purkinje neurons and extending of their processes along the Purkinje cell layer toward the pia in the cerebellum. Velate astrocytes are cerebellar protoplasmic astrocytes that exhibit characteristic velate sheath processes and are more dispersed in arrangement [58]. In the dye coupling analysis, these two subtypes of astrocytes are coupled (Fig. 6F). Despite a striking difference in syncytial anatomy, the syncytial isopotentiality occurs to syncytial networks established by both subtypes of astrocytes. However, the coupling strength (*S*) is significantly weaker in Bergmann glia network than those established by valet astrocytes as well as hippocampal astrocytes (Fig. 6).

6.3 Transjunctional Voltage/Current Analysis of the Electrical Coupling of Astrocytes

6.3.1 Introduction

Traditionally, the coupling strength (*S*) between a pair coupled cells is examined by dual patch whole-cell recording. The currents/command voltages are delivered from one cell, termed stimulated cell ($Cell_S$), to induce voltages or currents in the same cell. The delivered currents/voltages also travel across the gap junction to induce voltages/currents in the coupled cell, termed recipient cell ($Cell_R$). The ratio of the induced voltages in the $Cell_R$ over the $Cell_S$ is the coupling coefficient (CC), and this can also be expressed by the ratio of induced currents in $Cell_R$ over the $Cell_S$ [10, 33, 39, 50, 59]. For the reasons described earlier in the Introduction that transjunctional voltages or currents measurement does not accurately inform the coupling strength of a syncytium. However, it does provide the coupling information at the cellular levels. Additionally, dye coupling has been widely used to gain insight into astrocyte syncytial coupling. However, evidence from others and us indicates that transjunctional measurement has higher sensitivity than dye coupling analysis [10, 60–62]. Thus transjunctional voltage/current analysis remains as a powerful method for deciphering the coupling information from an astrocyte syncytium. In this section, we will introduce the procedure of conducting dual patch recording in brain slices. The same procedure is also applicable for carrying out dual patch transjunctional study from pairs of freshly dissociated astrocytes. Example of application will also be provided to determine where this method is mostly applicable in your research.

6.3.2 Experimental Procedure

(a) Chamber perfusion, brain slice loading and securing are described above.

(b) Identify two neighboring astrocytes based on cell morphology, and that should be further confirmed by astrocyte-specific marker expression. In Fig. 7, SR101 staining was used as a marker for a pair of neighboring astrocytes identification.

(c) Fill two patch electrodes with K^+ electrode solution and install them to the electrode holders. Apply constant positive pressure to the internal solution for both electrodes. Under IR-DIC with ×4 objective, advance the electrodes to the surface of the brain slice. Switch to the water immersion ×40 objectives. Apply a few drops of bath solution to the objective

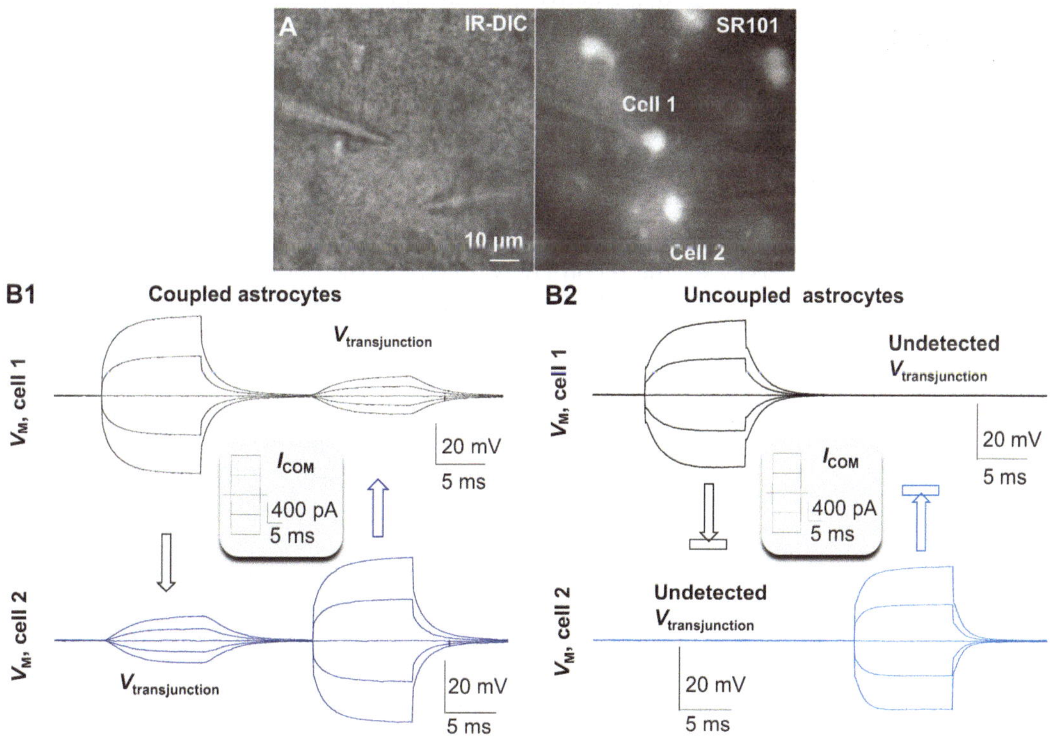

Fig. 7 Dual patch recording made from neighboring astrocytes. **(A)** Dual electrodes patch recording from a pair of neonatal astrocytes in CA1 stratum radiatum. The recorded cells (DIC, left) were selected based on SR101 staining (right). **(B)** A pair of electrically coupled **(B1)**, and uncoupled astrocytes **(B2)**. Both electrodes were set on current clamp (I-C mode). The current steps (I_{COM}), shown in the inset, were alternately delivered to the stimulated cell ($C_{stim.}$) that induced transjunctional voltage ($V_{transjunction}$) only in the recipient cell ($C_{reci\cdot}$) of the coupled pair **(B1)**

that runs down to form a fluid bridge between the lens and the slice. Move the tips of electrodes to the two identified astrocyte. In voltage clamp mode, apply a recurrent 5 mV square voltage pulse to monitor the electrode resistance as described in Subheading 6.1.

(d) Similar to the description above, approach the two cell soma slowly with one electrode at a time. Once on-cell, release the electrode pressure. Apply a −80 mV holding the potential to facilitate the formation of a GΩ seal. Repeat this procedure with the second electrode. Check and keep the position of the first electrode relative to the cell. Finally, ensure both electrodes to form GΩ seal on two astrocytes. Because both cells reside close to one another, it is critical to carefully coordinate the movement of the two electrodes and readjustment of the relative position between cell soma and electrodes. Be sure to approach the cell gently. Avoid larger movement of the electrodes, including approaching, going up or down.

(e) Rupture the patch of membrane isolated inside the electrode tip by negative pressure pulses to establish the whole-cell configurations. Again, a steady holding current during the decline of seal resistance to <30 MΩ is an indicator of a successful transition from cell-attach to the whole-cell mode. Repeat this for the second electrode. After both whole-cell configurations are established, wait at least 5 min to allow ion equilibration between the electrode solution and intracellular cytoplasm of the recorded cells.

(f) Set the recorded astrocytes in the desired mode. As shown in Fig. 7B1, both cells are in current clamp (I-C) mode under the control of Clampex 9.2. Program both channels in I-C mode, the command current steps are delivered to the $Cell_S$ that induced voltages in the $Cell_S$ and transjunctional voltages in the coupled $Cell_R$. As shown in Fig. 7B1, the $Cell_S$, and $Cell_R$ have alternately defined for injection of command current steps.

6.3.3 Methodological Application/Expected Results

1. The first application of transjunctional analysis is to determine if the transjunctional voltages/currents are bidirectional. Evidence shows that different gap junction subunits exhibit different permeability that can generate preferential unilateral conductivity of transjunctional voltages [63]. This occurs to retinal glial cells [64], but not to hippocampal protoplasmic astrocytes [10].
2. The second application of transjunctional analysis is to determine the voltage-dependence of the transjunctional voltages. In the heterologous system, both connexin 43 (Cx43) and 30 exhibit deactivations at transjunctional voltages >20 mV

[65–68]. However, the transjunctional voltages or currents of astrocytes do not exhibit the same properties in situ [10, 38, 50, 69].

3. The third application of transjunctional analysis is to determine whether the electrical coupling occurs to the same or different brain cell types. One such example is illustrated in Fig. 7, where some of the neonatal astrocytes already established their gap junctional coupling that can be detected by transjunctional voltage analysis (Fig. 7B1), but others remain as "solitary" individuals [50]. The absence of electrical coupling was also determined by this method between different brain cell types, that is, astrocyte–neuron, astrocyte–NG2 glia, NG2 glia–neurons, and neuron–neuron [59].

4. As mentioned before, transjunctional voltage analysis is more sensitive in detecting the gap junctional coupling than dye coupling analysis. One such example is illustrated in Fig. 8. A

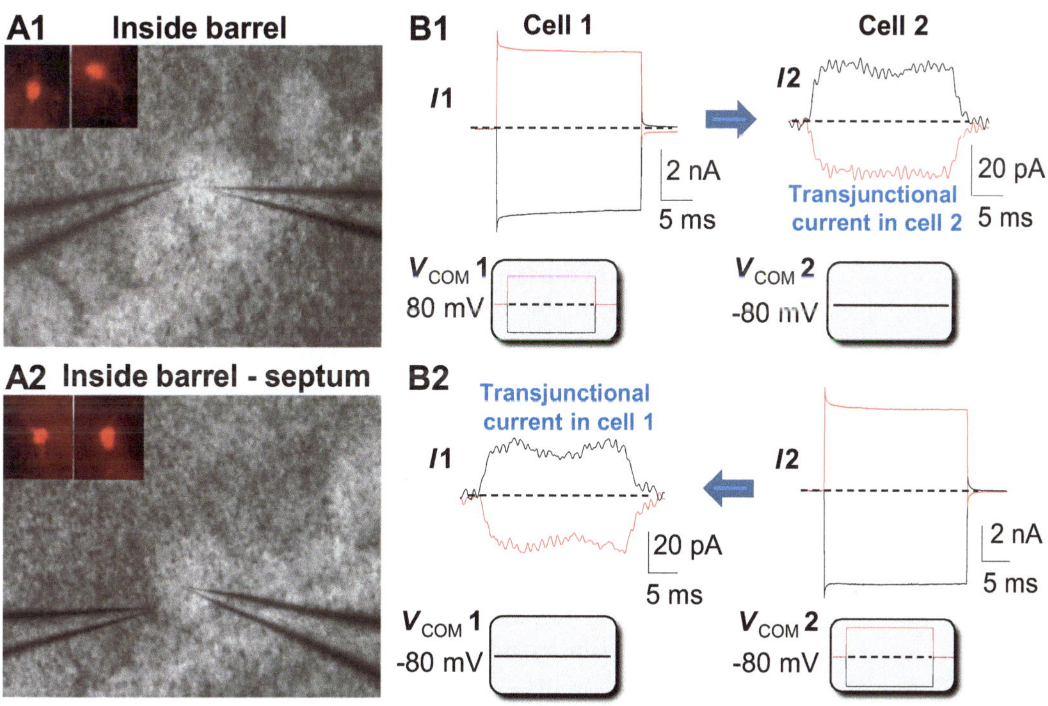

Fig. 8 Dual electrodes voltage clamp on neighboring astrocytes to study gap junctional coupling in astrocyte syncytium. **(A)** Dual electrodes patch clamp on a pair of neighboring astrocytes in somatosensory barrel cortex that identified by the IR-DIC image and SR101 fluorescence. The pair was selected either with two astrocytes within the same barrel (inside barrel, **A1**), or one in the barrel the other in the adjacent septum (inside barrel—septum, **A2**). **(B)** Both electrodes were set on voltage clamp (VC mode) and held the membrane potential at −80 mV. The command voltages (V_{COM}), ±100 mV/50 ms, were alternately delivered to cell1 and cell2 as indicated below the recording traces, and the induced transjunctional current in the adjacent astrocyte is also indicated. Note that the transjunctional current is bidirectional

published study showed that dye coupling could not or barely detectable between astrocytes inside the layer VI somatosensory barrels. This is associated with an enriched expression of Cx43 inside the barrels, but weak in the septa [70]. To determine if the barrel organization indeed indicate a restricted gap junctional communication with the barrels, the transjunctional analysis can be performed. As shown in Fig. 8B, the transjunctional currents can be induced from pairs of astrocytes resided either both inside barrels or one inside and another outside of the barrels.

6.4 Freshly Dissociated Single, Doublet Astrocytes and "Miniature Syncytia"

6.4.1 Introduction

In the previous sections, we introduced the rationale and methods to examine the astrocyte syncytial function. However, syncytial coupling also creates a formidable barrier to examine the basic functional properties, such as cellular expression of ion channels, receptors, and transporters [71]. Additionally, syncytial coupling also prevents an accurate analysis of the strength of gap junction coupling between astrocytes and the aggregate effect of syncytial coupling to an astrocyte network [33]. To get to the heart of these questions, freshly dissociated astrocytes has now been demonstrated as a highly valuable model for the physiological and pathological study of astrocytes [33, 36, 37]. We will start by introducing the dissociation method that yield single, doublet astrocytes, and multiple gap junction coupling astrocytes in their intact state, we called it miniature syncytia. Importantly, the morphology of individual astrocytes, the syncytial anatomy and the function of astrocytes, in terms of the ion channels and gap junction coupling, are all well-preserved in this reduced bona fide model that is ideal for astrocyte study.

6.4.2 Experimental Procedure

(a) Details in fresh tissue dissociation are described in 5.3.
(b) In each cell suspension, more than 40 small tissue blocks were yielded that contained single to up to 20 astrocytes. The number of astrocytes and their domain territories in each tissue block can be readily visualized based on SR101 staining or eGFP fluorescence (Fig. 3, D2–3; Fig. 11A). An important criterion in cell selection is to choose those astrocytes with their domain territories largely intact.
(c) The dissociated cells were allowed 1–2 min to touch down to the bottom of the chamber before switching on a constant aCSF perfusion at a flow rate of 1.5 mL/min.
(d) Following this tissue dissociation protocol, the hippocampal tissues from neonatal to young adult animals contained single to multiple astrocytes that could be readily identified by SR101 staining, and viable astrocytes inside the tissues could be readily identified based on the somatic shape and the

well-preserved spatial domain that matched closely to astrocytes in situ [72].

(e) The whole-cell recordings were made at least 5 min after the settlement of cells in the recording chamber. Morphologically intact single dissociated astrocytes were identified by SR101 or eGFP fluorescence.

(f) To examine the physiological properties of individual astrocytes or miniature syncytia, standard K^+ electrode solution is used. The V_M is recorded in current clamp and current profile evoked by voltage steps is recorded in voltage clamp. The input resistance (R_{in}) is tested by using "Membrane test" in voltage clamp.

(g) To examine the passive properties of individual astrocytes, that is, membrane capacitance (C_M), the coupling strength between a pair of coupled astrocytes, or a coupling strength of a miniature syncytium, K^+ free/140 mM Na^+ electrode solution is used.

(h) To simulate ischemic condition, glucose was replaced with equimolar sucrose in aCSF, and oxygen was depleted by deoxygenating bath solution with 95% N_2 and 5% CO_2 (pH 7.3–7.4) for 30 min prior to and throughout the 30 min oxygen–glucose deprivation (OGD) in experiments.

(i) The dissociated cell suspensions were discarded 2 h after papain digestion, and the above procedure was repeated to obtain a new dissociated cell suspension in the same day's experiment.

6.4.3 Methodological Application/Expected Results

1. The first application of this model is for the study of the functional properties of individual astrocytes. Astrocytic processes are the primary cellular locations where ion channels, receptors, and transporters perform cellular functions [71]. An example of this study is shown in (Fig. 9). A single uncoupled astrocyte can be identified by its positive staining for SR101. Whole-cell patch clamp recording can be performed as described in Subheading 6.1. In this case, single astrocytes exhibit the same resting V_M as their counterparts in situ, and the command voltages (V_{COM}) induced passive conductance does not differ from in situ astrocyte recording. An important clarification revealed from this single astrocyte study is that the low R_M is an intrinsic feature of astrocytes. By combining with a fast solution perfusion system, the expression of ionotropic GABA, AMPA and transporter currents can be definitively studied without the influence of neuronal activity [73, 74]. Additionally, the cytoplasm of the individual astrocytes or astrocyte after recording can be harvested for gene profiling by RT-PCR or qRT-PCR. In our hands, this has

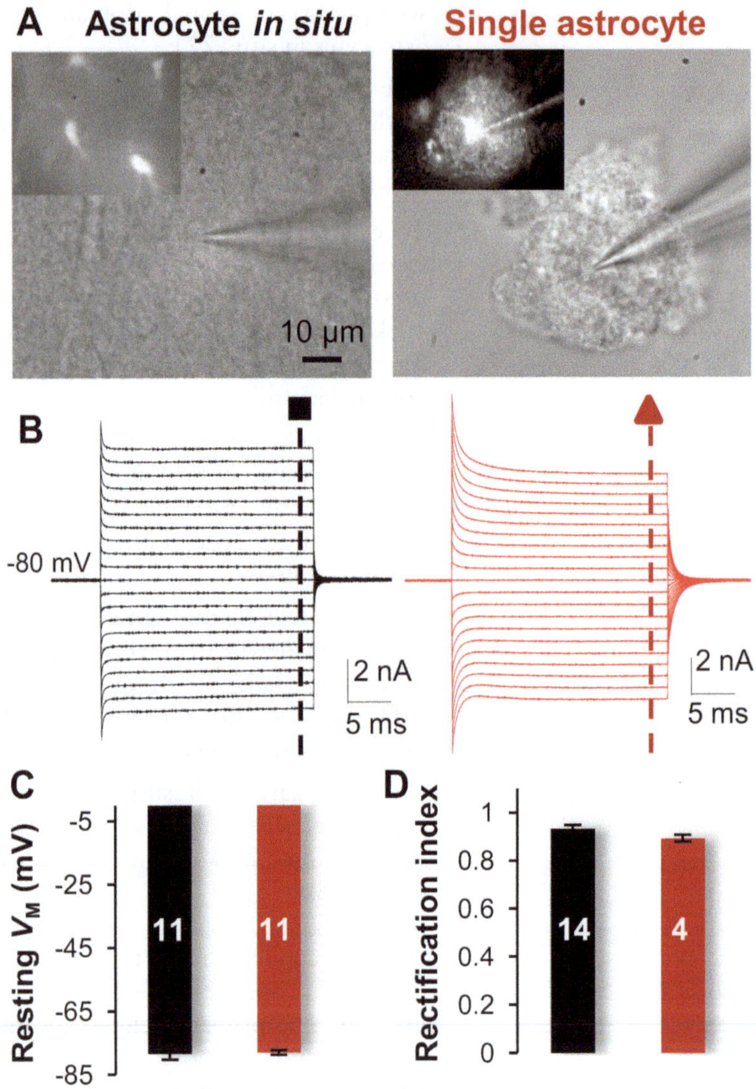

Fig. 9 (A) The IR-DIC images were obtained during patch recording from a syncytial coupled astrocyte (left), and a single astrocyte in freshly dissociated tissue preparation (right). Insets showed the SR101 staining. **(B)** Both astrocytes showed a similar passive membrane conductance. **(C)** The single astrocytes showed comparable resting membrane potentials as astrocytes in situ. **(D)** The rectification index (RI) values were comparable between single and syncytial coupled astrocytes

been successfully used to identify the genes encoding astrocytic marker GFAP, and various astrocyte K^+ channels [73, 75–77].

2. Use of pairs of freshly dissociated astrocytes is another advantageous application of this model. As shown in Fig. 10, in the SR101 staining, the spatial domains of a pair of coupled astrocytes can be inspected for quality before going on for dual

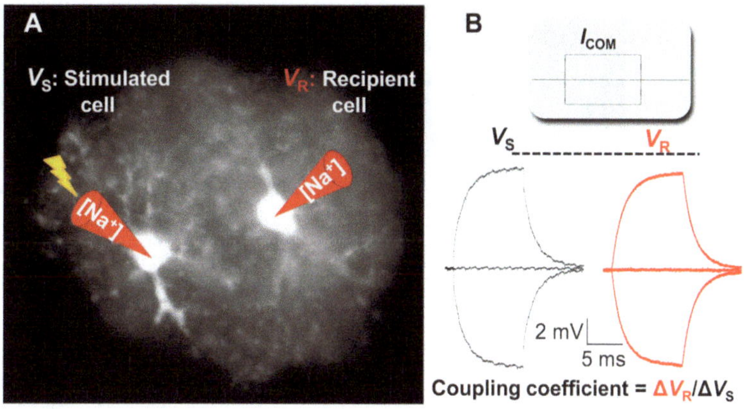

Fig. 10 **(A)** Illustration of the use of K^+ free/Na^+-based electrode solution to maximally reduce the membrane K^+ conductance for accurate transjunctional voltage (V_{tr}) measurement and calculation of coupling coefficient (CC). **(B)** The current injection induced V_M depolarization in the stimulated cell (V_S) and V_{tr} recorded in coupled recipient astrocyte (V_R). The calculated coupling coefficient (CC) is 94% in this astrocyte doublet

patch recording. The protocol for this recording mode is described in Subheading 6.3. For the transjunctional voltage study, the shunting of injected currents is already substantially reduced in two coupled astrocytes. Additionally, K^+-free/ Na^+ electrode solution can be used to eliminate the K^+ conductance. The example is shown in Fig. 10 used Na^+ electrode recording solution for transjunctional voltage study. Remarkably, a coupling coefficient (CC) of 94% was unveiled. Such a strong electrical coupling suggests that the electrical barrier between astrocytes is nearly absent, therefore a pair of coupled astrocytes electrically behaves as one cell. This at the cellular levels explains how syncytial isopotentiality is achieved. Yet no similar study has been done in other parts of the brain, nor under the diseased conditions. Thus, once designed properly, transjunctional voltage study from dissociated astrocyte doublets can provide important insights into the coupling state in healthy and diseased conditions.

3. A powerful application of miniature syncytia for stroke mechanism study is illustrated in Fig. 11. The rationale for this experimental design is similar to the Na^+ electrode recording to identify the existence of syncytial isopotentiality. However, the Na^+ loading now occurs to the neighboring astrocytes around the astrocyte recorded with K^+-based solution. The Na^+-loading and intracellular K^+ ($[K^+]_i$) dissipation are caused by energy failure due to oxygen–glucose deprivation (OGD), and this is anticipated to depolarize astrocytes V_M. In a single astrocyte shown in Fig. 11B, the OGD-caused energy failure

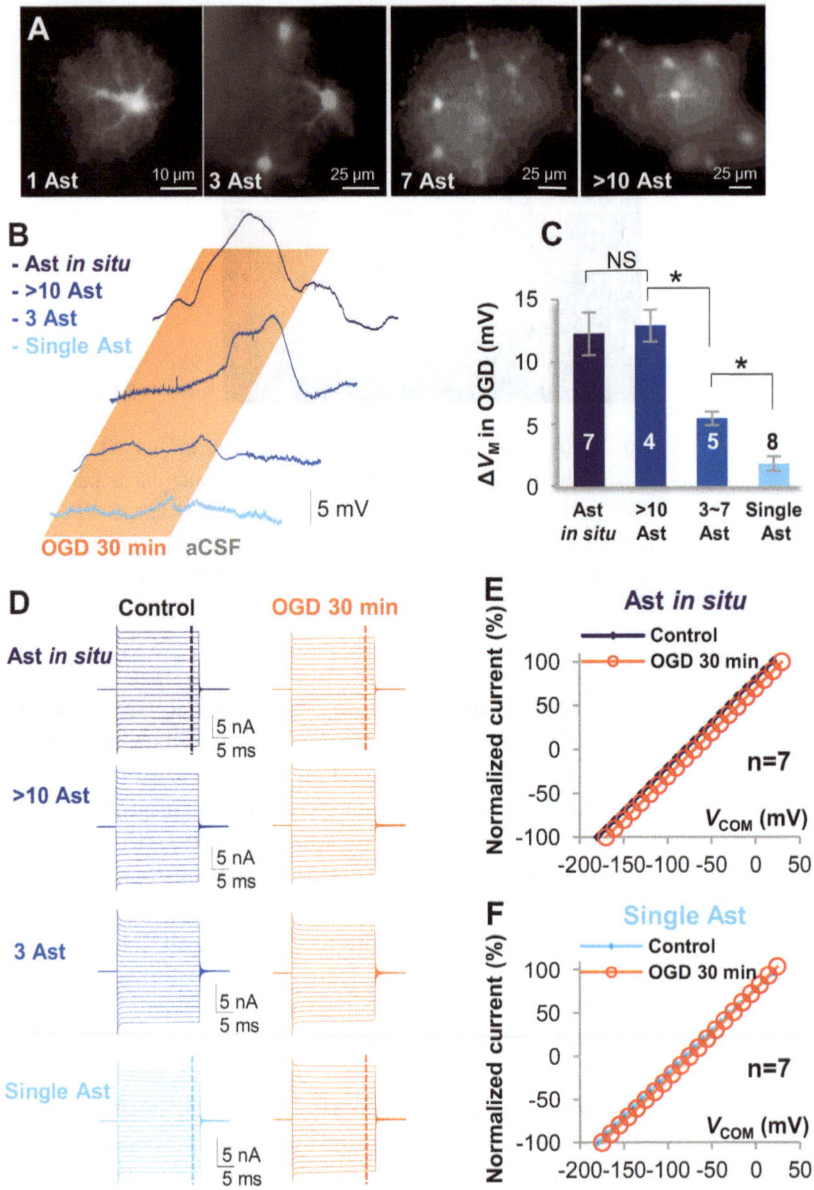

Fig. 11 Functional study of the reduced-sized syncytium in OGD condition. **(A)** SR101 staining of a single freshly dissociated astrocyte, and three "miniature syncytia" with a varied number of astrocytes as indicated. Ast: astrocyte. **(B)** OGD-induced V_M depolarization increases with the number of coupled astrocytes. **(C)** Quantification of OGD-induced ΔV_M in (b) (NS, $p > 0.05$; $*p < 0.05$, ANOVA). **(D)** Representative astrocyte whole-cell currents recorded before and at the end of 30 min OGD under uncoupled and varied coupling syncytial sizes. **(E)** Normalized *I–V* plots show an OGD-induced positive shift in *I–V* plot (depolarization), but no suppression of passive conductance change in astrocyte in situ. **(F)** OGD induced neither a shift in *I–V* plot nor reduction in passive conductance in single astrocytes

could not induce V_M depolarization due to a strong concentration-clamping of the recorded cell by the physiological $[K^+]_i$ in the electrode. However, the ability of this concentration-clamping control progressively declines as the size of the miniature syncytia expanse. Ultimately, the OGD-induced depolarization appears in miniature syncytia containing >10 astrocytes. This size of depolarization becomes comparable to OGD induced astrocyte depolarization in situ [37]. In the model, potential OGD induced impairment of membrane K^+ channel conductance can be concomitantly examined (Fig. 11D, E). In this case, a 30 min acute OGD does not alter the functional K^+ conductance in astrocytes. Through the use of new astrocyte model, for the first time, a major mechanism underlying cerebral ischemia-induced astrocyte depolarization was revealed. Accordingly, the miniature syncytia model has a great potential to be used in the future for astrocyte physiology and pathology study.

Acknowledgments

This work is sponsored by a grant from the National Institute of Neurological Disorders and Stroke RO1NS062784 and R56NS097972 (MZ) and National Science Foundation DMS1410935 (DT).

References

1. Barres BA (2008) The mystery and magic of glia: a perspective on their roles in health and disease. Neuron 60(3):430–440.. Epub 2008/11/11. https://doi.org/10.1016/j.neuron.2008.10.013
2. Verkhratsky A, Nedergaard M (2018) Physiology of astroglia. Physiol Rev 98(1):239–389.. Epub 2018/01/20. https://doi.org/10.1152/physrev.00042.2016
3. Kimelberg HK (2010) Functions of mature mammalian astrocytes: a current view. Neuroscientist 16(1):79–106.. Epub 2010/03/20. https://doi.org/10.1177/1073858409342593
4. Khakh BS, McCarthy KD (2015) Astrocyte calcium signaling: from observations to functions and the challenges therein. Cold Spring Harb Perspect Biol 7(4):a020404. Epub 2015/01/22. https://doi.org/10.1101/cshperspect.a020404
5. Khakh BS, Sofroniew MV (2015) Diversity of astrocyte functions and phenotypes in neural circuits. Nat Neurosci 18(7):942–952.. Epub 2015/06/26. https://doi.org/10.1038/nn.4043
6. Kiyoshi CM, Zhou M (2019) Astrocyte syncytium: a functional reticular system in the brain. Neural Regen Res 14(4):595–596. Epub 2019/01/12. https://doi.org/10.4103/1673-5374.247462
7. D'Ambrosio R, Wenzel J, Schwartzkroin PA, McKhann GM 2nd, Janigro D (1998) Functional specialization and topographic segregation of hippocampal astrocytes. J Neurosci 18(12):4425–4438. Epub 1998/06/10
8. Rouach N, Koulakoff A, Abudara V, Willecke K, Giaume C (2008) Astroglial metabolic networks sustain hippocampal synaptic transmission. Science 322(5907):1551–1555.. Epub 2008/12/06. https://doi.org/10.1126/science.1164022
9. Wallraff A, Kohling R, Heinemann U, Theis M, Willecke K, Steinhauser C (2006) The impact of astrocytic gap junctional coupling on potassium buffering in the hippocampus. J Neurosci 26(20):5438–5447.. Epub 2006/05/19. https://doi.org/10.1523/JNEUROSCI.0037-06.2006
10. Xu G, Wang W, Kimelberg HK, Zhou M (2010) Electrical coupling of astrocytes in rat

hippocampal slices under physiological and simulated ischemic conditions. Glia 58 (4):481–493. Epub 2009/10/02. https://doi.org/10.1002/glia.20939

11. Kimelberg HK, Nedergaard M (2010) Functions of astrocytes and their potential as therapeutic targets. Neurotherapeutics 7 (4):338–353. Epub 2010/10/01. https://doi.org/10.1016/j.nurt.2010.07.006
12. Ransom BR (1996) Do glial gap junctions play a role in extracellular ion homeostasis? In: Spray DC, Austin DR (eds) Gap junctions in the nervous system Editors. RG Landes Company, TX, pp 159–173
13. Nimmerjahn A, Bergles DE (2015) Large-scale recording of astrocyte activity. Curr Opin Neurobiol 32:95–106. Epub 2015/02/11. https://doi.org/10.1016/j.conb.2015.01.015
14. Giaume C, Koulakoff A, Roux L, Holcman D, Rouach N (2010) Astroglial networks: a step further in neuroglial and gliovascular interactions. Nat Rev Neurosci 11(2):87–99.. Epub 2010/01/21. https://doi.org/10.1038/nrn2757
15. Kuga N, Sasaki T, Takahara Y, Matsuki N, Ikegaya Y (2011) Large-scale calcium waves traveling through astrocytic networks in vivo. J Neurosci 31(7):2607–2614.. Epub 2011/02/18. https://doi.org/10.1523/JNEUROSCI.5319-10.2011
16. Langer J, Stephan J, Theis M, Rose CR (2012) Gap junctions mediate intercellular spread of sodium between hippocampal astrocytes in situ. Glia 60(2):239–252.. Epub 2011/10/26. https://doi.org/10.1002/glia.21259
17. Lin JH, Weigel H, Cotrina ML, Liu S, Bueno E, Hansen AJ et al (1998) Gap-junction-mediated propagation and amplification of cell injury. Nat Neurosci. 1 (6):494–500. Epub 1999/04/10. https://doi.org/10.1038/2210
18. Orkand RK, Nicholls JG, Kuffler SW (1966) Effect of nerve impulses on the membrane potential of glial cells in the central nervous system of amphibia. J Neurophysiol 29 (4):788–806. Epub 1966/07/01
19. Rose CR, Ransom BR (1997) Gap junctions equalize intracellular Na+ concentration in astrocytes. Glia 20(4):299–307. Epub 1997/08/01
20. Simard M, Arcuino G, Takano T, Liu QS, Nedergaard M (2003) Signaling at the gliovascular interface. J Neurosci 23(27):9254–9262. Epub 2003/10/10
21. Wang F, Smith NA, Xu Q, Fujita T, Baba A, Matsuda T et al (2012) Astrocytes modulate neural network activity by Ca(2)+−dependent uptake of extracellular K+. Sci Signal 5(218): ra26. Epub 2012/04/05. https://doi.org/10.1126/scisignal.2002334
22. Nagy JI, Rash JE (2000) Connexins and gap junctions of astrocytes and oligodendrocytes in the CNS. Brain Res Brain Res Rev 32 (1):29–44. Epub 2000/04/07
23. Pannasch U, Vargova L, Reingruber J, Ezan P, Holcman D, Giaume C et al (2011) Astroglial networks scale synaptic activity and plasticity. Proc Natl Acad Sci U S A 108(20):8467–8472. Epub 2011/05/04. https://doi.org/10.1073/pnas.1016650108
24. Lutz SE, Zhao Y, Gulinello M, Lee SC, Raine CS, Brosnan CF (2009) Deletion of astrocyte connexins 43 and 30 leads to a dysmyelinating phenotype and hippocampal CA1 vacuolation. J Neurosci 29(24):7743–7752. Epub 2009/06/19. https://doi.org/10.1523/JNEUROSCI.0341-09.2009
25. Theis M, Jauch R, Zhuo L, Speidel D, Wallraff A, Doring B et al (2003) Accelerated hippocampal spreading depression and enhanced locomotory activity in mice with astrocyte-directed inactivation of connexin43. J Neurosci 23(3):766–776. Epub 2003/02/08
26. Spray DC (1996) Physiological properties of gap junction channels in the nervous system. In: Spary DC, Dermietzel R (eds) Gap junctions in the nervous system editors. RG Landes Company, Austin, TX, pp 35–59
27. Ransom BR (1996) Do glial gap junctions play a role in extracellular ion homeostasis? In: Spary DC, Dermietzel R (eds) Gap junctions in the nervous system editors. RG Landes Company, Austin, TX, pp 159–173. (chapter 6) Epub 1996
28. Giaume C, McCarthy KD (1996) Control of gap-junctional communication in astrocytic networks. Trends Neurosci 19(8):319–325. Epub 1996/08/01
29. Nedergaard M, Goldman S (1996;. J Neurophysiol, Chapter 19) Spreading depression—a gap junction mediated event? In: Spary DC, Dermietzel R (eds) Gap junctions in the nervous system editors. RG Landes Company, Austin, TX, pp 301–312
30. Furshpan EJ, Potter DD (1957) Mechanism of nerve-impulse transmission at a crayfish synapse. Nature 180(4581):342–343. Epub 1957/08/17
31. Kuffler SW, Potter DD (1964) Glia in the leech central nervous system: physiological properties and neuron-glia relationship. J

Neurophysiol 27:290–320.. Epub 1964/03/01. https://doi.org/10.1152/jn.1964.27.2.290

32. Muller CM. Gap-junctional communication in mammalian cortical astrocytes: development, modifiability and possible functions. In: Spray DC, Austin DR Gap junctions in the nervous system editors TX: RG Landes Company, pp 203–212. 1996
33. Ma B, Buckalew R, Du Y, Kiyoshi CM, Alford CC, Wang W et al (2016) Gap junction coupling confers isopotentiality on astrocyte syncytium. Glia 64(2):214–226. https://doi.org/10.1002/glia.22924
34. Kofuji P, Newman EA (2004) Potassium buffering in the central nervous system. Neuroscience 129(4):1045–1056. Epub 2004/11/25. https://doi.org/10.1016/j.neuroscience.2004.06.008
35. Terman D, Zhou M (2019) Modeling the role of the astroctye syncytium and K+ buffering in maintaining neuronal firing patterns. Opera Med Physiol. (in press)
36. Du Y, Ma B, Kiyoshi CM, Alford CC, Wang W, Zhou M (2015.: jn 00206 2015. Epub 2015/03/27) Freshly dissociated mature hippocampal astrocytes exhibit similar passive membrane conductance and low membrane resistance as syncytial coupled astrocytes. J Neurophysiol. https://doi.org/10.1152/jn.00206.2015
37. Du Y, Wang W, Lutton AD, Kiyoshi CM, Ma B, Taylor AT et al (2018) Dissipation of transmembrane potassium gradient is the main cause of cerebral ischemia-induced depolarization in astrocytes and neurons. Exp Neurol 303:1–11. Epub 2018/02/07. https://doi.org/10.1016/j.expneurol.2018.01.019
38. Meme W, Vandecasteele M, Giaume C, Venance L (2009) Electrical coupling between hippocampal astrocytes in rat brain slices. Neurosci Res 63(4):236–243.. Epub 2009/01/27. https://doi.org/10.1016/j.neures.2008.12.008
39. Kiyoshi CM, Du Y, Zhong S, Wang W, Taylor AT, Xiong B et al (2018) Syncytial isopotentiality: a system-wide electrical feature of astrocytic networks in the brain. Glia 66:2756. https://doi.org/10.1002/glia.23525
40. Huang M, Du Y, Kiyoshi CM, Wu X, Askwith CC, McTigue DM et al (2018) Syncytial isopotentiality: an electrical feature of spinal cord astrocyte networks. Neuroglia 1(1):271–279. Epub 2018/08/24. https://doi.org/10.3390/neuroglia1010018
41. Kuffler SW, Nicholls JG, Orkand RK (1966) Physiological properties of glial cells in the central nervous system of amphibia. J Neurophysiol 29(4):768–787. Epub 1966/07/01
42. Yang Y, Vidensky S, Jin L, Jie C, Lorenzini I, Frankl M et al (2011) Molecular comparison of GLT1+ and ALDH1L1+ astrocytes in vivo in astroglial reporter mice. Glia 59(2):200–207. Epub 2010/11/04. https://doi.org/10.1002/glia.21089
43. Zhong S, Du Y, Kiyoshi CM, Ma B, Alford CC, Wang Q et al (2016) Electrophysiological behavior of neonatal astrocytes in hippocampal stratum radiatum. Mol Brain 9(1):34. Epub 2016/03/24. https://doi.org/10.1186/s13041-016-0213-7
44. Zhou M, Schools GP, Kimelberg HK (2006) Development of GLAST(+) astrocytes and NG2(+) glia in rat hippocampus CA1: mature astrocytes are electrophysiologically passive. J Neurophysiol 95(1):134–143.. Epub 2005/08/12. https://doi.org/10.1152/jn.00570.2005
45. Glitsch HG (2001) Electrophysiology of the sodium-potassium-ATPase in cardiac cells. Physiol Rev 81(4):1791–1826.. Epub 2001/10/03. https://doi.org/10.1152/physrev.2001.81.4.1791
46. Hille B (2001) Ion channels of excitable cells. Sunderland, MA, Sinauer
47. Fiacco TA, Agulhon C, Taves SR, Petravicz J, Casper KB, Dong X et al (2007) Selective stimulation of astrocyte calcium in situ does not affect neuronal excitatory synaptic activity. Neuron 54(4):611–626. Epub 2007/05/25. https://doi.org/10.1016/j.neuron.2007.04.032
48. Olsen ML, Khakh BS, Skatchkov SN, Zhou M, Lee CJ, Rouach N (2015) New insights on astrocyte ion channels: critical for homeostasis and neuron-glia signaling. J Neurosci 35(41):13827–13835. Epub 2015/10/16. https://doi.org/10.1523/JNEUROSCI.2603-15.2015
49. Schroder W, Hinterkeuser S, Seifert G, Schramm J, Jabs R, Wilkin GP et al (2000) Functional and molecular properties of human astrocytes in acute hippocampal slices obtained from patients with temporal lobe epilepsy. Epilepsia 41(Suppl 6):S181–S184. Epub 2000/09/22
50. Zhong S, Du Y, Kiyoshi CM, Ma B, Alford CC, Wang Q et al (2016) Electrophysiological behavior of neonatal astrocytes in hippocampal stratum radiatum. Mol Brain 9:34. Epub 2016/03/24. https://doi.org/10.1186/s13041-016-0213-7
51. Schools GP, Zhou M, Kimelberg HK (2006) Development of gap junctions in hippocampal

astrocytes: evidence that whole cell electrophysiological phenotype is an intrinsic property of the individual cell. J Neurophysiol 96 (3):1383–1392.. Epub 2006/06/16. https://doi.org/10.1152/jn.00449.2006

52. Ma B, Xu G, Wang W, Enyeart JJ, Zhou M (2014) Dual patch voltage clamp study of low membrane resistance astrocytes in situ. Mol Brain 7:18. Epub 2014/03/19. https://doi.org/10.1186/1756-6606-7-18
53. Zhou M, Xu G, Xie M, Zhang X, Schools GP, Ma L et al (2009) TWIK-1 and TREK-1 are potassium channels contributing significantly to astrocyte passive conductance in rat hippocampal slices. J Neurosci. 29(26):8551–8564. Epub 2009/07/03. https://doi.org/10.1523/JNEUROSCI.5784-08.2009
54. Susaki EA, Tainaka K, Perrin D, Kishino F, Tawara T, Watanabe TM et al (2014) Whole-brain imaging with single-cell resolution using chemical cocktails and computational analysis. Cell 157(3):726–739. Epub 2014/04/22. https://doi.org/10.1016/j.cell.2014.03.042
55. Susaki EA, Tainaka K, Perrin D, Yukinaga H, Kuno A, Ueda HR (2015) Advanced CUBIC protocols for whole-brain and whole-body clearing and imaging. Nat Protoc 10 (11):1709–1727.. Epub 2015/10/09. https://doi.org/10.1038/nprot.2015.085
56. Tainaka K, Kubota SI, Suyama TQ, Susaki EA, Perrin D, Ukai-Tadenuma M et al (2014) Whole-body imaging with single-cell resolution by tissue decolorization. Cell 159 (4):911–924. Epub 2014/11/25. https://doi.org/10.1016/j.cell.2014.10.034
57. Kita Y, Kawakami K, Takahashi Y, Murakami F (2013) Development of cerebellar neurons and glias revealed by in utero electroporation: Golgi-like labeling of cerebellar neurons and glias. PLoS One 8(7):e70091. Epub 2013/07/31. https://doi.org/10.1371/journal.pone.0070091
58. Chan-Palay V, Palay SL (1972) The form of velate astrocytes in the cerebellar cortex of monkey and rat: high voltage electron microscopy of rapid Golgi preparations. Zeitschrift fur Anatomie und Entwicklungsgeschichte 138 (1):1–19. Epub 1972/01/01
59. Xu G, Wang W, Zhou M (2014) Spatial organization of NG2 glial cells and astrocytes in rat hippocampal CA1 region. Hippocampus 24 (4):383–395.. Epub 2013/12/18. https://doi.org/10.1002/hipo.22232
60. Murphy AD, Hadley RD, Kater SB (1983) Axotomy-induced parallel increases in electrical and dye coupling between identified neurons of Helisoma. J Neurosci 3(7):1422–1429. Epub 1983/07/01
61. Ransom BR, Kettenmann H (1990) Electrical coupling, without dye coupling, between mammalian astrocytes and oligodendrocytes in cell culture. Glia 3(4):258–266.. Epub 1990/01/01. https://doi.org/10.1002/glia.440030405
62. Sontheimer H, Waxman SG, Ransom BR (1991) Relationship between Na+ current expression and cell–cell coupling in astrocytes cultured from rat hippocampus. J Neurophysiol 65(4):989–1002. Epub 1991/04/01
63. Goodenough DA, Paul DL (2009) Gap junctions. Cold Spring Harb Perspect Biol 1(1): a002576. Epub 2010/01/13. https://doi.org/10.1101/cshperspect.a002576
64. Zahs KR, Newman EA (1997) Asymmetric gap junctional coupling between glial cells in the rat retina. Glia 20(1):10–22. Epub 1997/05/01
65. Harris AL (2001) Emerging issues of connexin channels: biophysics fills the gap. Q Rev Biophys 34(3):325–472. Epub 2002/02/13
66. Moreno AP, Rook MB, Fishman GI, Spray DC (1994) Gap junction channels: distinct voltage-sensitive and -insensitive conductance states. Biophys J 67(1):113–119. Epub 1994/07/01. https://doi.org/10.1016/S0006-3495(94)80460-6
67. Spray DC (1996) Physiological properties of gap junction channels in the nervous system. In: Spray DC, Dermietzel R (eds) Gap junction in the nervous system. RG Landes Company, Austin, TX, pp 39–59
68. Valiunas V, Manthey D, Vogel R, Willecke K, Weingart R (1999) Biophysical properties of mouse connexin30 gap junction channels studied in transfected human HeLa cells. J Physiol 519(Pt 3):631–644. Epub 1999/08/24
69. Muller T, Moller T, Neuhaus J, Kettenmann H (1996) Electrical coupling among Bergmann glial cells and its modulation by glutamate receptor activation. Glia 17(4):274–284.. Epub 1996/08/01. https://doi.org/10.1002/(SICI)1098-1136(199608)17:4<274::AID-GLIA2>3.0.CO;2-#
70. Houades V, Koulakoff A, Ezan P, Seif I, Giaume C (2008) Gap junction-mediated astrocytic networks in the mouse barrel cortex. J Neurosci 28(20):5207–5217.. Epub 2008/05/16. https://doi.org/10.1523/JNEUROSCI.5100-07.2008
71. Kimelberg HK, Cai Z, Schools G, Zhou M (2000) Acutely isolated astrocytes as models to probe astrocyte functions. Neurochem Int 36(4–5):359–367. Epub 2000/03/25

72. Bushong EA, Martone ME, Jones YZ, Ellisman MH (2002) Protoplasmic astrocytes in CA1 stratum radiatum occupy separate anatomical domains. J Neurosci 22(1):183–192. Epub 2002/01/05
73. Zhou M, Kimelberg HK (2000) Freshly isolated astrocytes from rat hippocampus show two distinct current patterns and different [K(+)](o) uptake capabilities. J Neurophysiol 84(6):2746–2757. Epub 2000/12/09
74. Zhou M, Kimelberg HK (2001) Freshly isolated hippocampal CA1 astrocytes comprise two populations differing in glutamate transporter and AMPA receptor expression. J Neurosci 21(20):7901–7908. Epub 2001/10/06
75. Wang W, Putra A, Schools GP, Ma B, Chen H, Kaczmarek LK et al (2013) The contribution of TWIK-1 channels to astrocyte K(+) current is limited by retention in intracellular compartments. Front Cell Neurosci 7:246. Epub 2013/12/26. https://doi.org/10.3389/fncel.2013.00246
76. Wang W, Kiyoshi CM, Du Y, Ma B, Alford CC, Chen H, Zhou M (2016) mGluR3 Activation Recruits Cytoplasmic TWIK-1 Channels to Membrane that Enhances Ammonium Uptake in Hippocampal Astrocytes. Mol Neurobiol 53 (9):6169–6182. Epub 2015/11/10. https://doi.org/10.1007/s12035-015-9496-4
77. Du Y, Kiyoshi CM, Wang Q, Wang W, Ma B, Alford CC et al (2016) Genetic deletion of TREK-1 or TWIK-1/TREK-1 potassium channels does not alter the basic electrophysiological properties of mature hippocampal astrocytes in situ. Front Cell Neurosci 10:13. Epub 2016/02/13. https://doi.org/10.3389/fncel.2016.00013

Chapter 13

Assessment of Behavioral Dysfunction Following Experimental Traumatic Brain Injury (TBI)

Shruti V. Kabadi and Kimberly R. Byrnes

Abstract

Traumatic brain injury (TBI) is a major public health concern, as it is one of the leading causes of death and disability in the USA. Over the years, several attempts have been made to reconstruct the pathophysiological events that occur during a TBI and the behavioral abnormalities that are observed in TBI survivors. Such behavioral deficits lead to long-term neurological dysfunction and often demonstrate symptoms of chronic neurodegenerative disorders known to impair the quality of life of the patients. For example, TBI patients often suffer from learning disability and memory loss similar to that observed in cases of Alzheimer's disease, motor dysfunction seen in Parkinson's disease cases, and changes in mood and emotional behavior noted in patients with anxiety and depression. Although science in the field of TBI research has shown tremendous progress and advancement, more recently, trials of therapeutic agents that have shown promise in preclinical settings have consistently failed to demonstrate improvement in behavioral outcomes after clinical TBI. Therefore, there is an urgent need to develop new models while continuing to refine existing approaches used for modeling experimental TBI and related behavioral impairment. In this chapter, we describe the concept, key features, and methodology of some of the most commonly used animal models for inducing experimental TBI and evaluating behavioral dysfunction. In addition to detailing the application of different animal models, we evaluate their relevance to clinical symptoms by examining their advantages and limitations. We acknowledge that the events that occur during and after a clinical TBI are multifactorial in nature. Therefore, while developing and utilizing such animal models, it is important to focus on the diversity of the mechanisms involved in the impact during the injury, and the complexity of the secondary injury processes that lead to long-term behavioral dysfunction.

Key words Fluid percussion, Controlled cortical impact, Diffuse axonal injury, Rotational acceleration, Blast injury, Beam walk, Morris water maze, Elevated plus maze, Open field, Novel object recognition, Statistical analysis

1 Introduction

Traumatic brain injury (TBI), or head injury, is one of the leading causes of death and disability in the USA. According to the Centers for Disease Control and Prevention (CDC), 2.5 million cases of TBI were reported in the USA in the year 2010 [1]. The total combined rates of TBI-related hospitalizations, emergency

Nicholas J. D. Wright (ed.), *Basic Neurobiology Techniques*, Neuromethods, vol. 152,
https://doi.org/10.1007/978-1-4939-9944-6_13,

department (ED) visits, and deaths drastically increased from 521.0 per 100,000 in 2001 to 823.7 per 100,000 in 2010. Falls are considered the leading cause of TBI, followed by being struck by an object and motor vehicle accidents [2]; however, motor vehicle accidents have been reported as the leading cause of fatal head injury. More recently, increased recognition of the frequency and consequences of sports-related concussive brain injury [3], and an increasing number of blast-related injuries in military personnel as well as civilians in war zones, has resulted in TBI being labeled as the "signature injury" of the wars in Iraq and Afghanistan [4].

Although TBI has been extensively studied as a major public health concern, its underlying pathogenesis is not completely understood, and it is often difficult to reconstruct the events leading to the lesions due to varying severity and regional distribution that constitute individual injuries. TBI causes cell death and neurological dysfunction through both physical disruption of the tissue (primary injury), as well as delayed molecular and cellular mechanisms that cause progressive white and grey matter damage and manifestation of behavioral deficits (secondary injury) that may potentially continue for months to years [5]. Secondary injury mechanisms, such as chronic microglial and astrocyte activation, contribute to the pathogenesis of brain injury, exacerbating neurological dysfunction [6]. Over the years, studies have demonstrated the progressive and sustained nature of neuroinflammation through microglial and astrocyte activation, particularly in the cortex, using multiple experimental models of TBI [7, 8]. Chronic inflammation following CNS trauma has provided a mechanistic link between acute and chronic neurodegeneration [7, 8]. Preclinical studies have indicated that sustained microglial and astrocyte activation after TBI may play a role in the chronic neurodegeneration and loss of neurological function [7]. Although both neuroprotective and neurotoxic microglial phenotypes have been described [9, 10], microglial activation and subsequent release of associated inflammatory factors has been proposed as an important contributing factor in chronic neurodegenerative disorders, including Alzheimer's disease [11]. Furthermore, previous studies have indicated that sustained microglial activation after TBI may play a role in neuronal cell loss following the release of neurotoxic molecules such as nitric oxide [7, 12]. Therefore, TBI must not be considered an acute or static disorder but a complex and chronic neurodegenerative condition.

Despite strong experimental data, more than 30 clinical trials of neuroprotection in patients with TBI have failed. These failures are potentially attributed to methodological differences between the clinical and animal studies, as well as inadequate preclinical evaluation using behavioral models and mechanistic tests and/or trial design problems [13]. To better understand the pathobiology of TBI and evaluate potential therapeutic approaches, efforts have been made to mimic the events that occur after clinical TBI via development and characterization of various animal models and reproduction of behavioral deficits that follow. Each model is intended to mimic certain components of clinical TBI, recognizing that it is difficult to establish consistent models leading to similar clinical behavioral dysfunction including most or all of the factors that contribute to chronic neurological abnormalities [13]. TBI represents perhaps the most heterogeneous of neurological disorders; as several factors including severity, location, invasive versus noninvasive insults, focal versus diffuse, presence or absence of intracranial bleeding, as well as differences in gender, genetic predisposition, and presence or absence of certain comorbidities may contribute to the etiology. Behavioral models used to evaluate behavioral deficits after TBI are highly versatile and can be used to assess more than one type of behavioral outcome, which sometimes makes the selection of an appropriate behavioral task for simulating TBI-induced functional deficits challenging. Therefore, in order to improve clinical translation of animal models used for experimental TBI and assessment of TBI-induced functional deficits, it is important to focus on the complexity of the behavioral dysfunction and the multifactorial nature of the events that occur during secondary injury and to develop modeling strategies directed toward mimicking these diverse consequences.

In this chapter, we describe some of the commonly used animal models of TBI that have been classified based on mechanisms involved in TBI induction. In addition, we discuss various animal behavioral models that have been designed for representing the subtle and varied neurological impairment resulting from TBI. We highlight the methodological differences between the various experimental TBI and behavioral models and provide recommendations for improving the efficiency and sensitivity of the preclinical investigation.

2 Methods

2.1 Models for Inducing Experimental TBI

Over the past 70 years or more, a number of TBI models have been developed to understand the response of the brain to trauma and the mechanism behind this response, and to evaluate the potential of therapeutic approaches. The validity of a brain injury model depends on a number of factors, including its clinical relevance, model standardization, reproducibility, and the ability of the model to reflect relevant or targeted behavior or pathology.

A recent review [14] detailed the applicability of several of the most commonly used TBI models, including the controlled cortical impact (CCI), fluid percussion (FP), penetrating injury, weight drop, and blast. Each of these models has been found to be a valid representation of some aspect of TBI. Modeling, in general, must meet at least one of 4 criteria to be considered a valid model: etiology, construct, face, and predictability [15]. This is to say that the model must represent a cause (etiology), pathogenesis (construct), pathological or behavioral outcome (face), or therapeutic response (predictability). The current review will consider the most popular and commonly used models that have met this condition, with a focus on models that have been validated at the face level.

As clinical TBI reflects a wide range of pathologies and behavioral conditions, a number of models have been created with the goal of mimicking some or all of these conditions. Currently, 80–90% of all TBI's are characterized as mild TBI or concussions [16]. Mild TBI diagnosis criteria historically has included a brief period of unconsciousness, amnesia or altered mental status. However, this has recently been revised to include mechanical injury followed by transient confusion or impaired consciousness, memory dysfunction at the time of injury, loss of consciousness of less than 30 min, acute seizure, acute irritability, lethargy or vomiting, and acute headache, dizziness, fatigue, or poor concentration [16]. Modeling of mild TBI is, therefore, expected to behaviorally model symptoms such as headache, balance problems, dizziness, visual problems, fatigue, light and noise sensitivity, confusion, memory dysfunction, disorders of executive functions (impulse control, confusion, following instructions), anxiety, depression, and sleep disorders, among other things. With more severe injuries, patients experience more severe deficits, including slurred speech, aphasia, memory and attention deficits, motivation problems, and aggression and irritability [17]. However, understandably, no one model is capable of perfectly capturing all aspects of TBI.

It is important to consider certain caveats to animal models of TBI. One important caveat is the need in several models to have a craniotomy, in order to expose the brain to the injury device. It has been shown that craniotomy itself can induce effects in the brain, which is an important consideration for interpretation of results [18]. It is also well known that TBI is often dependent on the force to which the brain is exposed and this force is dependent on the size of the human brain. Thus, scaling is a significant consideration in animal TBI modeling. Models of rotational acceleration require significantly more force to induce similar levels of injury between rodent and human brains [19]. However, altering the force can result in non-realistic loads that become inconsistent with clinical situations. Despite these issues, a number of models are well accepted in the literature, including FP, open and closed weight drop or CCI, rotational acceleration, and blast.

2.1.1 Fluid Percussion (FP)

In the literature, 1287 studies have been published through April 2017 utilizing the FP in rats, mice, dogs, cats, rabbits, or pigs of both sexes and of all age groups. Arguably one of the more reproducible models of brain injury [20], the FP model is one of the

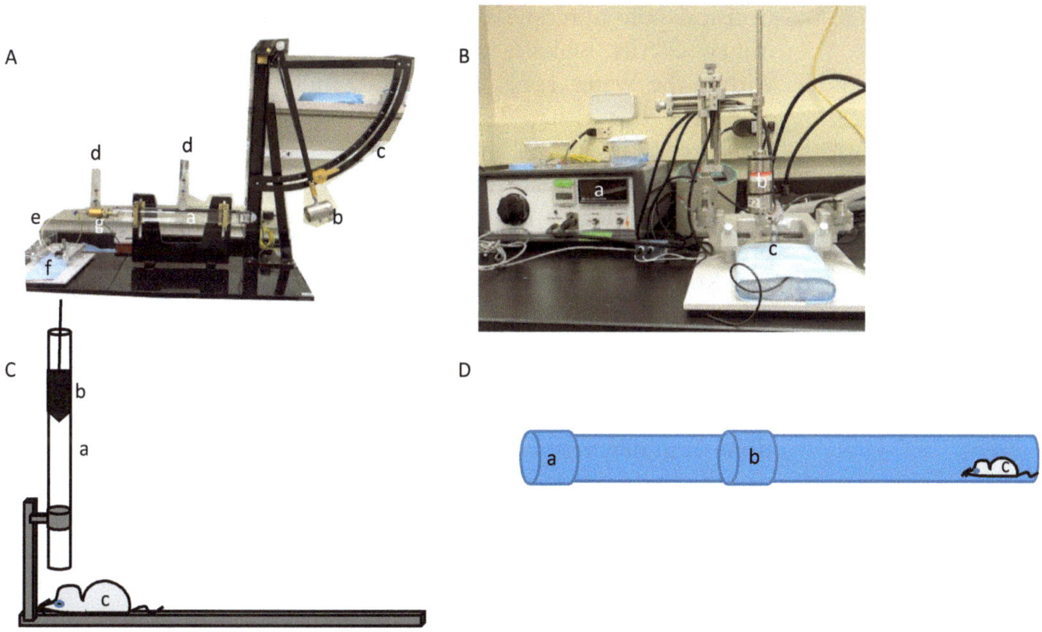

Fig. 1 Models used for inducing experimental TBI. **(A)** Fluid percussion (FP) model. (*a*) Fluid-filled tube, (*b*) pendulum, (*c*) angle set, (*d*) syringes for removal of air bubbles, (*e*) tubing for connection between a and rodent, (*f*) stereotaxic holder for rodent, (*g*) pressure gauge connector. **(B)** Controlled cortical impact (CCI) Model. (*a*) Impact one control box; (*b*) impactor piston; (*c*) temperature controlled bed for rodents, positioned on stereotaxic frame. **(C)** Closed Weight Drop Model: (*a*) Tube; (*b*) weight attached to string for retrieval; (*c*) rodent. **(D)** Blast TBI Model: (*a*) driver chamber; (*b*) membrane; (*c*) rodent

most commonly used models to replicate the diffuse aspects of TBI [21]. The FP model typically utilizes a tube filled with water or saline that is used to propagate a pressure pulse from a piston to an exposed brain (Fig. 1a); a detailed protocol for the FP injury has been previously described [22]. In this model, a craniotomy must be performed and is typically placed either in the lateral parietal cortex (lateral fluid percussion; LFP) or centrally (central fluid percussion; CFP) between lambda (intersection of the lambdoid and sagittal sutures) and bregma (intersection of the coronal and sagittal sutures). During the procedure, a plastic connector, often a female Luer connector, is cemented over the craniotomy and attached to tubing or directly to a FP device, such as the Virginia Commonwealth University (VCU) Health System Custom Design and Fabrication Model 01-B or the micro-FP device [22]. In some applications of this model, the connection of the Luer connector and the impact are completed on different days to isolate the injury from the surgical manipulations and allow for complete drying and solidifying of the connector to the skull.

While the FP model is known for its diffuse outcomes, location of the initial impact can have direct effects, and it is therefore important to consider location during experimental planning and decisions on outcome measures. Impact pressure then dictates injury severity, which is measured by a pressure transducer placed within the water tube. In general, mild injury is generated by pressures of less than 2.2 atmospheres (atm), moderate injury by pressures between 2.4 and 2.8 atm, and severe injury is generated by pressures above 2.8 atm [23]. However, tubing additions to the device, in order to facilitate surgery in a stereotaxic holder and precise location of the craniotomy, can affect pressure. For example, Dr. Kimberly Byrnes' laboratory has added 10–30 cm length of tubing (0.071 cm ID, 0.142 cm OD) between the device and the female Luer lock, which has the effect of reducing the pressure from that reported by the manufacturer. Therefore, it is important to measure pressure at the end of the tubing or to generate estimations of injury from other outcomes, such as righting reflex return or other behavioral or histological outcomes.

FP has been commonly used for modeling experimental repeated mild TBI [24]. The FP model results in a focal and diffuse injury outcome. Often at the site of TBI, FP will result in a lesion with a cavity and an inflammatory response. Diffuse from the injury site, a FP model will be marked by diffuse axonal injury (DAI), indicated by demyelination and/or beta-amyloid precursor protein (β-APP) staining and accumulation, and diffuse neuronal loss. The location of the focal injury can often dictate the functional tests utilized, but the most common behavioral outcome measures include motor assessment using rotarod or beam walk, cognitive evaluation using Morris Water Maze (MWM), and emotional and social behavior testing using forced swim, elevated plus maze and acoustic startle [14, 23]. Behavioral deficits in tests that include motor function and balance have been reported to last at least through 3 months post-injury, and memory dysfunction through 12 months post-injury [14, 23].

2.1.2 Weight Drop: Open

In the literature, 1121 studies have been published through April 2017 utilizing the CCI model in rats and mice of both sexes and of all age groups. The CCI model is an example of open skull weight drop or impact TBI [25]. The CCI model typically utilizes a weighted pole or pneumatically controlled steel piston that impacts the exposed brain directly (Fig. 1b). The speed of impact, dwell time after impact and depth of impact dictate injury severity, while the location dictates behavioral outcomes.

During injury, animals are placed in stereotaxic frames to allow for ease of craniotomy in addition to precise location of the impact. Impact is delivered using a variety of devices, including the Impact One™, Leica Microsystems; Precision Impactor 0310 device (Precision Systems & Instrumentation), the Pinpoint PCI3000 Precision Cortical Impactor (Hatteras Instruments), the Pneumatic Impact Device (Amscien Instruments) and the TBI-0310 Impactor (Precision Instruments and Instrumentation, LLC) [25, 26]. The impact tip can have a range of diameters, which also contributes to injury severity and functional effects. Research has also shown that injury outcomes depend on shape of the impactor tip [27]. Following injury, the skull flap removed during craniotomy can be replaced or the craniotomy can remain open. This difference in technique can have effects on outcome, as the skull flap can either impede swelling or press on the exposed brain [28].

CCI is known to have a focal impact, resulting in gross tissue loss and cavitation at the site of injury, as well as marked inflammation, gliosis, and neuronal damage. In addition, CCI has diffuse effects, including glucose uptake alterations at distances away from the lesion [29] and inflammation far from the lesion epicenter [30]. At the site of injury, a CCI will result in a lesion with a marked cavity and a pronounced inflammatory response. Perilesionally, a CCI can result in DAI and diffuse neuronal loss. As in FP, the location of the focal injury dictates the functional tests utilized but CCI can result in marked deficits in learning, memory, and motor function. The most common behavioral assessments utilized with the CCI model include use of sensorimotor tests, such as rotarod or beam walk, cognitive tasks, such as MWM (Fig. 2), and emotional and social behavioral tests, including forced swim, elevated plus maze, Y-maze, and acoustic startle [26].

2.1.3 Weight Drop: Closed

As of April 2017, there have been 348 journal articles published on weight drop TBI models. While currently one of the less used models, weight drop closed brain injury has been increasingly accepted as a clinically relevant model. This model, initially described by Marmarou in 1994 [25, 31], utilizes a weighted rod (approximately 450 g for rats [32]) dropped from different heights onto the closed skull of the animal (Fig. 1c). Typically, animals are placed on a foam or gel bed and the impacting rod is suspended up to 1–1.25 m above the head. The rod is then allowed to drop from a defined height onto the skin or exposed skull of the animal and prevented from impacting again (bouncing). Different applications of the model involve impact onto non-incised skin or incision of the skin and location of a target based on bregma coordinates. In addition, studies have investigated the use of a metal 'helmet' affixed to the skin or the skull to abbreviate the impact force and protect against skull fracture. Additional modifications to the model have included replacement of the weighted rod with the CCI impactor described above [33].

The resulting injury can be graded from mild to moderate, depending on the height from which the rod is dropped; however weight drop TBI is known to have higher variability in outcome than the open skull models previously described [14, 20]. The weight drop closed model is known for its diffuse outcomes, although location of the initial impact can have direct effects, particularly in the event of a skull fracture making it important to

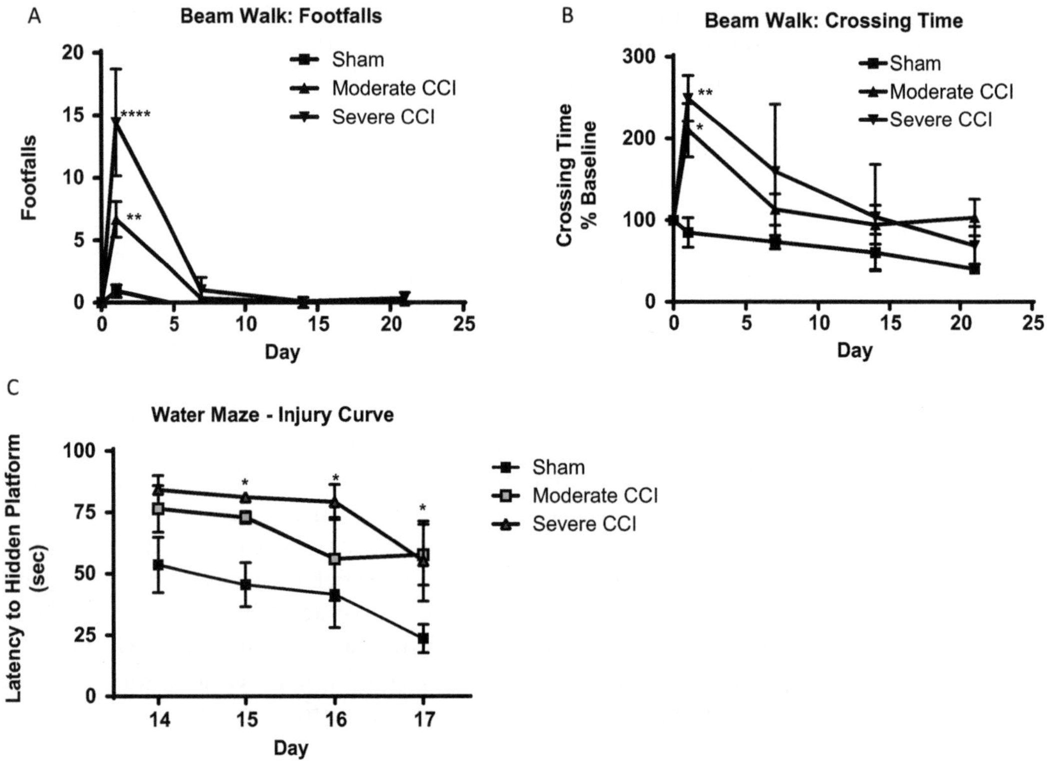

Fig. 2 Effect of moderate (5 m/s impact speed, 2 mm impact depth) and severe (6 m/s impact speed, 3 mm impact depth) CCI on beam walk and Morris water maze behavioral outcomes. Testing occurred at days 1, 7, 14, and 21 post-injury (**A**, **B**, Beam Walk) or from days 14 to 17 (**C**, Morris water maze) after injury. Two-way repeated measures ANOVA demonstrated significant effects of injury on number of footfalls ($****p < 0.0001$ sham v severe; $**p < 0.01$ sham v moderate), crossing time ($**p < 0.01$ sham v severe; $*p < 0.05$ sham v moderate) and latency to platform ($*p < 0.05$ sham v moderate and severe) in both the moderate and severe injured groups. $N = 4–5$/group. Points represent mean ± SEM

consider location during experimental planning and decisions on outcome measures [34].

An important benefit to the noninvasive nature of the weight drop model is its utility for repeated injury. The weight drop model has been used to evaluate the effect of variables, such as time between injuries [35, 36], on outcomes. Repeating a weight drop injury has been shown to increase the pathophysiology and functional impairments, such as increased inflammation, increased axonal damage, and worsened functional outcome (either further impaired or lasting for longer times) [32].

The weight drop model primarily results in a diffuse injury outcome with little to no gross alterations or changes noted by magnetic resonance imaging or gross histology [34, 37]. The most common outcome is DAI, including demyelination and/or β-APP staining and accumulation, oxidative stress, mild to moderate neuroinflammation and astrogliosis, and diffuse neuronal loss, depending on injury severity [34, 35]. After weight drop injury, a progressive worsening of outcome has been noted, particularly in anxiety and balance [34, 37]. The most common behavioral models used for evaluating functional outcomes after weight drop injury include rotarod, MWM, elevated plus maze [37], active place avoidance [33], Barnes maze, and Y-maze [32].

2.1.4 Rotational Acceleration

Clinical TBI is often the result of rapid rotational acceleration and deceleration. To more closely model the etiology of TBI, rotational acceleration TBI models have become more popular over the past decade. As of April 2017, there have been 93 journal articles published on rotational acceleration injury models. Currently, there is not a universally accepted injury device for rotational acceleration; most models are laboratory derived and thus very difficult to replicate between laboratories. For example, the Medical College of Wisconsin (MCW) has developed a rotational accelerator, which consists of a helmet that can be impacted by an impactor, leading to rotation around the cervical spine [38]. This model can produce an injury from differing speed and duration, thus resulting in different "severities" of injury. In this model, behavioral deficits in emotionality (measured by elevated plus maze) have been observed, but memory function impairments have not [38]. While the MCW model has been developed for rats, very few rotational acceleration models are available for rodents due to the size of the brain and the physics of the force needed to produce a comparable injury in the small brain compared to the larger human brain without causing irreparable damage to the body or becoming an unmanageable amount of force [39].

However, one model has become available for testing by multiple labs, the Closed-Head Impact Model of Engineered Rotational Acceleration (CHIMERA). This model is known for inducing DAI and diffuse microglial and astrocyte activation [40]. Like the weight drop, the CHIMERA model has been utilized in repeated injury [41]. Repeating a rotational acceleration injury has been shown to result in transient motor deficits and acute and chronic memory impairment [41]. The most commonly used behavioral models used for assessment of functional impairment after rotational acceleration injury include beam walk, MWM [41], and neurological severity scoring [40].

2.1.5 Blast

The most common cause of TBI in the military is impact through blast [42]. Experimental blast TBI models are becoming more common in the laboratory, in order to investigate mechanism, diagnostic approaches, and therapeutic options for service members. As of April 2017, there have been 195 journal articles published on blast brain injury models. These models are wide-ranging, encompassing everything from live explosive exposure to blast tube to blast wave generation. A number of different species have been utilized in these models, including both gyrencephalic (pigs, ferrets) [43, 44] and lissencephalic (rodents) [45, 46]. With this model, the subject is placed near a blast source (Fig. 1d). Depending on model, the subject's body is either protected, such as by a Kevlar body wrap, or exposed to the blast [44]. Orientation of the blast to the head is of continued discussion—as placement lateral to the head, above the head, or facing the nose can alter the outcome of the study [47].

Debate is ongoing as to the exact mechanism of injury resulting from blast exposure, and models have been developed to explore different mechanisms. For example, there are conflicting reports on whether the blast overpressure wave is directly responsible for the injury. As a result, models that induce an overpressure wave but no other features of blast [48] and those that eliminate only the overpressure wave [49] have been developed. The elimination model utilized exposure of rats to an explosive force through a pair of 2.5–5 cm thick aluminum platforms that mimics the propelling force of a blast under an armored car [49]. Blast in this model results in diffuse microglial activation, loss of hippocampal neurons, and reduced tight junctions and synaptic proteins. Such pathophysiological changes are associated with transient memory deficits and chronic anxiety examined using Y-maze and elevated plus maze, respectively. Shock or blast tubes can be also used to somewhat isolate the blast overpressure (BOP) exposure, although there is debate about the influence of flying debris from the model's torn membrane.

Overall, blast injury models are known for their diffuse nature, although the model is plagued by problems with reproducibility. Widespread blood–brain barrier dysfunction is noted often [50], as well as changes in intracranial pressure [50], cortical neuronal damage [51], and axonal damage [51]. Like the weight drop, the blast injury models have also been utilized in repeated injury [44]. Repeating a blast injury has been shown to increase neuronal loss, particularly in the hippocampus [44].

Blast TBI models have the benefit of a direct clinical correlation—humans are often exposed to very similar situations and blast pressure exposures. As a result, humans demonstrate emotional and cognitive impairments such as impaired executive function, sleep disturbances, learning and memory dysfunction and impaired social interactions [52]. Therefore, the most common behavioral tests to reflect and model these alterations include Barnes maze, elevated plus maze, and MWM [53].

In summary, in an effort to replicate a number of different aspects of brain injury, including mechanism, pathology, and functional results, several experimental TBI models have been developed. Each model has a number of variables, such as location of the impact, protections applied during the procedure, and severity of injury, which will influence behavioral outcomes. Despite these variables, various behavioral models have been found that consistently detect differences between injured and non-injured animals, and reflect changes in function that are observed in the clinic. These behavioral models will now be discussed.

2.2 Models for Assessing Experimental TBI-Induced Behavioral Dysfunction

Over the years, several behavioral models have been developed to mimic functional aspects of neurological disorders and evaluate therapeutic potential of different pharmacological agents. Behavioral models rely on evaluation of particular function(s) based on quantitative measures, such as velocity, distance, and time, as well as qualitative parameters, such as strategies used and preference for a particular object, location, or environment. The validity of an animal model for behavioral assessments depends on the following factors:

- The ability to demonstrate intensity- or severity-dependent behavioral changes.
- The sensitivity to distinguish between behavioral outcomes observed in treated versus control groups.
- The efficiency to determine pharmacological and/or toxicological behavioral consequences of test compounds.

We describe the applications, advantages, and limitations of some behavioral tests widely used to assess functional impairments induced by experimental TBI and to evaluate the therapeutic potential of new pharmacological treatments. Furthermore, we provide a representation of the behavioral tests with information on time frame of testing and species generally selected for these tests (Table 1).

2.2.1 Motor Function

Several behavioral tests have been developed to model and assess TBI-induced impairments in motor function and to determine the effectiveness of treatment strategies against such functional deficits. Prior to induction of TBI, the animals are trained on most of these tasks for learning the task, familiarizing them with the apparatus, and detecting any motor impairment in naive animals that may interfere with performance in the motor tasks. The analysis of results from behavioral tests involves statistical comparisons of data from treated groups versus injured, sham-operated (subjected to craniotomy but not injured) and naïve (uninjured) animals. The sensitivity of the test is evaluated by statistically significant differences between injured versus sham-operated and naïve groups. Some of the commonly used motor tests are summarized below.

(a) Beam walk test: The beam walk test is used in rodents to model and detect sensorimotor and locomotor dysfunction. The goal of this test is to evaluate the ability of a rat or a mouse to stay upright and walk across an elevated narrow beam (Fig. 3a) to a safe platform [54–60]. The beam walk test has

Table 1
Models for assessing experimental TBI-induced behavioral dysfunction

Behavioral model/task	Function	Timeframe of testing (duration of impairment)	Species
Motor behavior			
Beam walk	Sensorimotor and locomotor	Acute and chronic (days to months)	Rodents
Rotarod	Sensorimotor	Acute (days to a month)	Rodents
Inclined plane or angle board	Sensorimotor and vestibulomotor	Acute and chronic (days to weeks)	Rodents
Limb placing	Sensorimotor	Acute (days to a month)	Rats (only)
Adhesive removal (sticky tape)	Hand–eye coordination (fine motor ability)	Acute and chronic (days to months)	Rodents
Object retrieval (reach-and-grasp)	Hand–eye coordination (fine motor ability)	Acute and chronic (days to months)	Rodents and nonhuman primates (monkeys, gorillas, chimpanzees, baboons)
Open field	Exploration-based locomotor behavior	Acute and chronic (days to months)	Rodents
Bungalow test	Exploration-based locomotor behavior	Acute and chronic (days to months)	Nonhuman primates (monkeys, gorillas, chimpanzees, baboons)
Combined neuroscore (battery of tests)	All types of motor behavior	Acute and chronic (days to months)	Rodents

Cognition (learning and memory)			
Morris water maze	Working memory, spatial learning and memory, retention memory, associative learning, reversal learning	Acute and chronic (days to months)	Rodents
Reward-based: T-maze, Y-maze, radial maze	Reference and working memory	Acute and chronic (days to months)	Rodents
Novel object recognition	Retention memory	Acute and chronic (days to months)	Rodents and nonhuman primates (monkeys, gorillas, chimpanzees, baboons, pigs)
Fear conditioning (active and passive avoidance)	Retention and emotional memory	Acute and chronic (days to months)	Rodents and nonhuman primates (monkeys, gorillas, chimpanzees, baboons, pigs)
Emotional and social behavior (anxiety and depression)			
Elevated plus maze	Anxiety	Acute and chronic (days to months)	Rodents
Tail suspension	Depression	Acute and chronic (days to months)	Mice (only)
Porsolt's forced swim	Depression	Acute and chronic (days to months)	Rodents
Social interaction	Sociability, communication and self-awareness (psychiatric	Acute and chronic (days to months)	Rodents and nonhuman primates (monkeys, gorillas, chimpanzees, baboons, pigs)

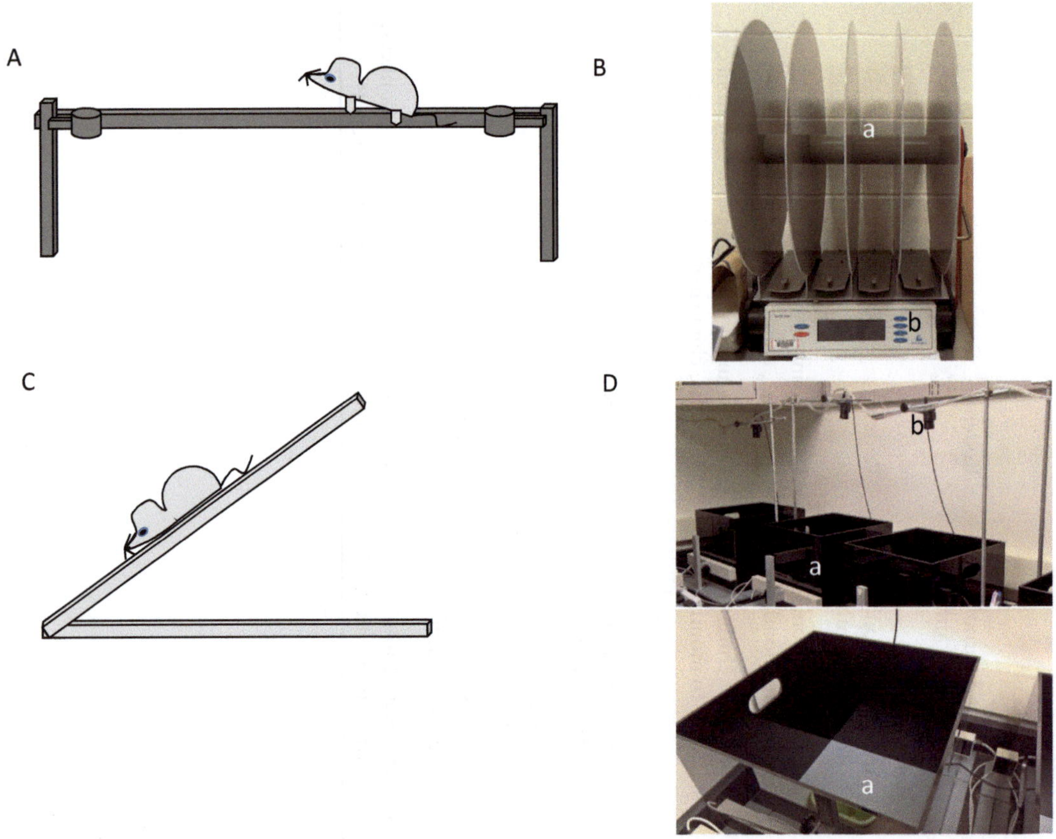

Fig. 3 Behavioral tests used for assessing motor function. (**A**) Beam walk, demonstrating rodent traversing a 1–4 mm beam. Example of forepaw step with hind paw slip. (**B**) Rotarod (*a*) rod for rodent placement; (*b*) control panel for speed and start/stop instructions. Partitions allow for testing of up to four rodents at a time. (**C**) Angle Board or Inclined Plane. (**D**) Open field. (*a*) Open field box; (*b*) camera

been used widely to model the degree and rate of motor dysfunction and impaired motor coordination after TBI [55–59, 61]. The motor impairment is recorded and analyzed using parameters such as number of foot-slips or falls or faults the animal shows while walking across the beam and time taken for an animal to walk across the beam. Prior to induction of TBI, animals are trained on the beam walk to identify any motor impairment in animals selected for the study and to help them learn the task. After induction of TBI, animals show increased foot faults and take longer to complete the task than sham-operated and naïve animals. Data can then be reported as raw number or as a percentage of baseline (pre-injury) values (Fig. 1). Repeated assessment of motor performance in this test provides a measure for quantification of rate and degree of motor recovery. Over time, some degree of recovery is observed in TBI-induced groups as the injured animals begin to show improved performance in the beam walk task at later time points, which is evaluated by comparing their behavior with sham-operated and naïve animals. More recently, modified forms of the beam walk task have been developed that involve introduction of hurdles in the path of beam traversal in the forms of pegs [62] or reduced thickness of the beam [63]. These tests represent more rigorous forms of the beam walk task developed to detect highly sensitive and subtle impairments in fine motor coordination associated particularly with mild or repeated mild TBI.

Notes: The animals should be trained on the beam walk prior to the induction of experimental TBI to facilitate learning of the task and to identify any motor impairment in the animals selected for the study. During training, observers can facilitate the function by gently redirecting animals that turn around or delay in crossing; however, it is important to avoid physical interaction with the animals during the testing phase to avoid inadvertent manipulation of the data. It is important to remember that the injured animals tend to recover over time and show improved performance in the beam walk test. The degree of impairment or improvement in motor performance in the beam walk test is determined by the statistically significant differences in the outcomes between groups.

(b) Rotarod test: The rotarod test is used to evaluate motor coordination as an indicator of cerebellar dysfunction in rodents [60, 64–68]. The goal of this test is to evaluate the ability of an animal to balance itself on a rod, beam, or pole (Fig. 3b) that is rotating at a constant speed or accelerating with time. The rotarod test is commonly used to model and demonstrate impaired motor coordination and to evaluate neuroprotective strategies in experimental TBI models [69, 70]. Fine motor coordination using the rotarod apparatus is assessed in terms of the overall time for which the animal is able to balance itself or latency to fall off the rotating beam. In addition to being used for evaluation of motor gait performance, the rotarod test is designed to determine the degree of motor skill learning which improves over repeated assessment in uninjured animals. Similar to the beam walk test, the rotarod test requires the animals to be trained prior to the beginning of the study and evaluated at designated time points after the induction of TBI to evaluate the rate and degree of recovery. In general, the recovery of injured animals is faster in rotarod test than other motor tests, such as beam walk, indicating that this is a simpler task for the animals to learn, and therefore, may not be an appropriate model for assessing the subtle long-term motor deficits observed after mild and repeated mild TBI. Rotarod is however a highly reliable task for detecting acute deficits in motor coordination observed during the first 7 to 10 days after induction of TBI.

Notes: The animals should be trained on the rotarod prior to the induction of experimental TBI to facilitate learning of the task and to identify any motor impairment in the animals selected for the study. The animals show improved performance in the rotarod test at earlier time points due to the simplicity of the task. Therefore, rotarod should only be used for evaluating experimental TBI-induced short-term motor impairment.

(c) Inclined plane or angle board test: Inclined plane uses an angle board (Fig. 3c) to assess to the ability of rodents to maintain their position on a stationary inclined plane or board [60, 71–74]. The inclined plane test is commonly used to estimate measures of sensorimotor and vestibulomotor function in rodents, evaluated in terms of first fall angle, threshold angle, total falls in the inclined plane after repeated assessment, and mean latencies to fall. Inclined plane test is used for modeling neurological impairments following TBI, including diffuse and repeated TBI, and evaluating the therapeutic potential of treatment approaches in such experimental TBI models [72, 74].

Notes: No training of the animals prior to the induction of TBI is required; however, performance is evaluated once before starting a study for identifying any inherent motor deficits which could interfere with performance in the task. During testing, it is important that the room remains quiet to avoid distraction of the animal.

(d) Limb placing task: The limb placing task provides a sensitive and reliable measure of sensorimotor response to visual, tactile and proprioceptive placing and is commonly used in rats [60, 67, 75]. The paw or foot placement on a tabletop upon visual, tactile, and proprioceptive stimulation is graded in terms of reaction time. A modified form of this test involves eliciting a sensorimotor reflex or limb placing in response to stimulation through the vibrissae as a measure of cross-midline motor function [76]. The limb placing test has been used in determining acute neurological impairment and evaluating treatments for TBI [75, 76]; however the efficiency of the test may be reduced by investigator variability and the tightness of the hold around the animals while the test is being executed.

Notes: Training prior to TBI induction is required to familiarize the animals with the task and for detecting any motor impairment. The performance of animals in this test is very sensitive to the tightness of hold around the animals. Therefore, it is very important for the same investigator to perform this test throughout a study as investigator variability can affect the outcome of the test.

(e) Hand–eye coordination models: Hand–eye coordination is the ability of the vision to coordinate the information through the eyes which is followed by guiding, controlling, and executing motor responses required for accomplishment of a particular task [60]. An example of hand–eye coordination-based tasks is the adhesive removal test, also known as the sticky tape test. This test provides a sensitive method for assessing sensorimotor deficits in rodents [60, 67, 77, 78]. The test protocol involves applying adhesive tape on each forepaw of the animal and determining the time elapsed between contact and removal. The adhesive removal test evaluates motor ability and sensitivity as well as dexterity. Furthermore, it can be used to measure the animal's sensory asymmetries by varying the size of adhesive tapes on each limb [67]. The adhesive removal test has been used for evaluating therapeutic potential of treatment strategies for TBI [78]. Another behavioral test based on hand–eye coordination used in rodents [67] as well as nonhuman primates [79–82] is the object retrieval test, also known as reach-and-grasp. In this test, a reward such as a type of food is suspended from rods of different complexity and fine motor ability is determined by parameters such as latency to retrieve the reward and frequency of left or right paw or hand use [79, 81].

Notes: No training for this type of test is required prior to TBI induction. Inter-investigator variability may be observed in adhesive removal test due to the variability in tightness of adhesive application. Therefore, it is very important for the same investigator to perform this test throughout a study.

(f) Exploration-based locomotor models: Exploration-based models are used to assess locomotor activity-related to spontaneous exploratory behavior [60]. These tasks can also be used to measure fatigue, anxiety, and attention. An example of exploration-based tasks is the open field test [60, 83, 84]. In this test, animals are allowed to explore an open field (Fig. 3d) for a particular period of time and their motor function is assessed in terms of distance traveled, velocity, and time of immobility versus activity. With the introduction of automated recording by using systems such as AnyMaze (Stoelting Co; Wood Dale, IL) and Ethovision (Noldus; Wageningen, The Netherlands) for data collection, the field can be divided into different imaginary zones and more advanced parameters can be recorded for evaluating subtle motor deficits such as time of mobility versus immobility. More recently, studies have been published using a modified version of this test in which the apparatus is divided into different sections, such as central and peripheral compartments, or light and dark chambers

[85]. Behavioral performance in each section along with the number of lines crossings between sections is measured to further determine place preference and locomotor function. The open field test has been used to evaluate neuroprotective potential of treatments after experimental CNS injuries, such as TBI by observing and evaluating differences in degree of exploratory behavior between treated injured and untreated injured animals [61, 69]. A similar exploration-based task using a bigger arena has been developed for nonhuman primates. This test is often referred to as the Bungalow test [81, 86] in which the animals are allowed to explore an arena divided into different compartments with partitions or doors.

Notes: No training is required for this task prior to TBI induction. This test is very sensitive to test-environmental factors such as temperature, lighting, and odor. as the behavioral outcomes of the test are dependent on the activity of the animals. Therefore, it is important for the investigator to ensure that there are no changes in the testing environment throughout the duration of a study and the room used for testing is soundproof.

(g) Combined neuroscore: A single behavioral test is oftentimes not sufficiently sensitive to detect subtle and complex motor impairments. Using a combination of behavioral tests in a study could provide a more accurate evaluation of motor deficits in experimental TBI models [22, 87]. Therefore, investigators often resort to a battery of tests which comprises of a series of motor tasks to better understand the rate and degree of motor dysfunction. One approach of using a composite scoring system includes a combination of certain individually scored sensitive tests, each scored using an ordinal scale ranging from zero (severe impairment) to five (normal function) [22, 87]. The total composite neurologic score (0–35) is obtained by combining the scores of the following tests: (1) lateral pulsion (left and right), (2) forelimb flexion (left and right), and (3) inclined plane (two vertical and two horizontal positions for assessing maximum angle at which the animal can stand for 5 seconds; scoring: $>50° = 5$, $45–50° = 4$, 40 to $45° = 3$, $35–40° = 2$, $30–35° = 1$, and $<29° = 0$. Developing and implementing a reliable and reproducible scoring system, such as the composite neuroscore, provides an accurate and efficient integrated measure of assessing motor impairments and evaluating the potential of neuroprotective therapies as well as chronic effects of neurotoxic agents.

Notes: Each motor test, described above, evaluates a different aspect of motor dysfunction observed after experimental TBI. Therefore, using an appropriate battery of motor tests improves the efficiency and sensitivity of the behavioral assessment. However, it is important for the investigator to not choose a large number of motor tests as excessive testing may tire the animals and yield "*false positive*" or "*false negative*" results. Furthermore, tests that evaluate the same motor function must not be included together under a testing system as combining scores from similar tests would imply repetition of the testing strategy instead of integrating outcomes from varied motor tests. For example, adhesive removal test should not be used with object retrieval test in rodents after experimental TBI as both tests evaluate hand-to-eye coordination and selecting these tasks under the same battery of testing would not improve the accuracy of the results.

2.2.2 Cognitive Function

Behavioral models are commonly used to simulate impairments in learning and memory observed after TBI and to screen drug candidates for their therapeutic potential.

(a) Morris water maze (MWM) task: The MWM in the most commonly used cognitive test, which was developed by Richard G. Morris at the University of St. Andrews in 1981 to study spatial learning and distinguish it from other forms of associative learning. This highly reliable and reproducible behavioral model serves as the "gold standard" for assessing learning and memory function in rodents and primarily simulates impairments in spatial and reference memory dependent on escape motivation after single and repeated experimental TBI [8, 22, 24, 57–59, 67, 88, 89]. The test uses a circular pool (Fig. 4a) which is divided into four imaginary quadrants using a video tracking software, such as AnyMaze (Stoelting Co) and Ethovision (Noldus), with a hidden escape platform in one of the quadrants. The apparatus is set up in a room with distinct visual cues displayed on the walls surrounding the pool. The animals are trained to find the hidden platform when introduced from different points of the pool using a certain number of trials every day for a particular number of days, generally 4 trials per day for 4–5 days with well-defined trial duration of 60 or 90 s. This training phase is defined as the "*acquisition*" stage of the MWM task. Animals with cognitive impairments fail to learn the task and therefore, are described to be deficient in spatial learning and memory function. The cognitive function is assessed in terms of navigational parameters, such as latency to find the platform, distance traveled, velocity, and time spent in different quadrants. The inter-trial learning on a day is often referred to as "short-term" memory. On the final day, the platform is

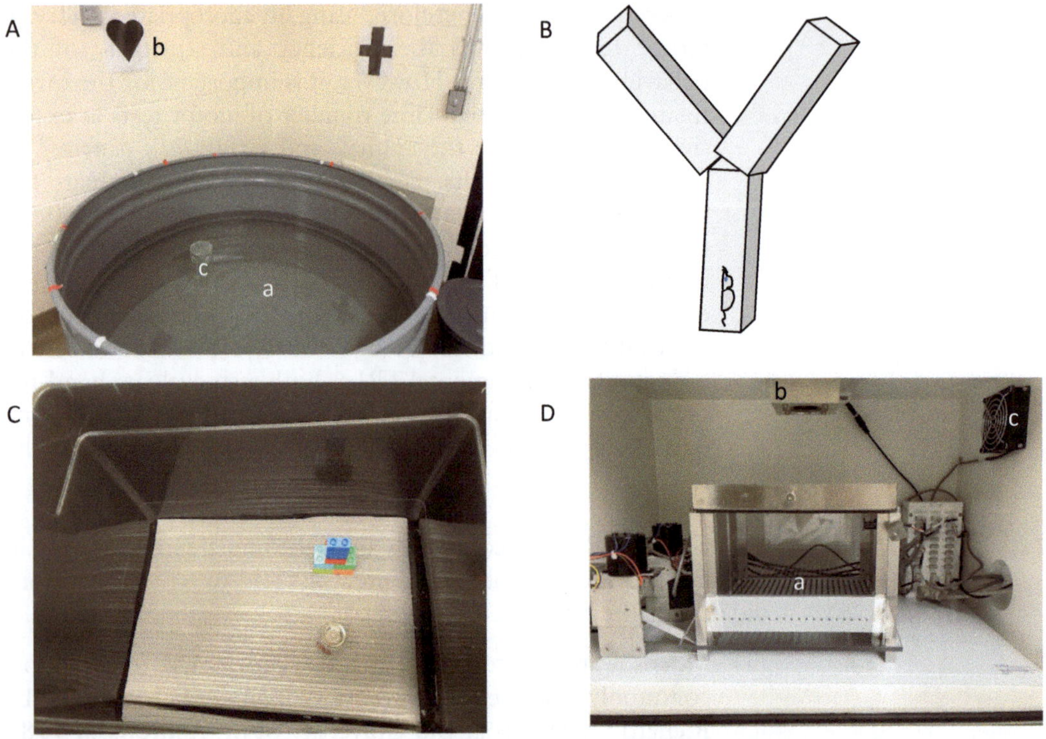

Fig. 4 Behavioral tests used for assessing cognitive function. (**A**) MWM: (*a*) tank; (*b*) images or cues for navigation; (*c*) hidden platform (in this case clear in clear water). (**B**) Y-Maze, with rodent placed in start position. (**C**) Novel object: 2 objects (Legos and a shot glass). (**D**) Fear conditioning setup, including chamber with grid floor (*a*), light source (*b*), and sound source (*c*). Not shown: alternative location for cued behavior testing

removed and the time spent in the target quadrant where the platform was hidden for the acquisition stage is recorded. This phase, called the "*probe*" trial, is used to evaluate reference or "long-term" learning and memory. Since the test design is highly dependent on visual acuity of the animals, a flag is placed on the platform which is returned to the pool following the probe trial and the latency to find the flagged platform is determined. This is called "*visual probe*" or "*cue*" test. Several modifications of this test have been developed over the years. Switching positions of the platform between trials during an acquisition phase provides a measure of "working" memory [88, 90], whereas moving the platform to the quadrant opposite to the one where it was initially hidden after completion of full acquisition phase evaluates "reversal" learning [91]. More recently, "*search strategy*" analysis was developed to provide a qualitative assessment of the strategies used by the animals to locate the hidden platform. The search strategy analysis provides a more thorough approach for evaluating spatial learning and memory, as animals with compromised cognitive function loop around (looping behavior) or spend time in irrelevant quadrants (systematic behavior), whereas animals with normal cognitive function use the cues on the walls to navigate their way straight to the platform (spatial behavior) [58, 92].

Notes: The testing protocol of the MWM test can be modified to evaluate different aspects of cognition. This task is generally performed by conducting 4 trials per day for 4–5 days with trial duration of 60 or 90 seconds. Subjecting the animals to this task for more number of days or more trials per day would not improve the sensitivity of the behavioral outcome and excessive testing may in fact tire the animals and yield "*false positive*" or "*false negative*" results. Some laboratories perform the "*visual probe*" or "*cue*" test on the first day of testing to evaluate the visual acuity of the animals, which is acceptable as long as the same protocol is followed throughout a study. The search strategy analysis is a novel method of assessment of cognitive function under the MWM test. A second investigator blinded to groups should be involved to confirm the type of search strategy assigned per trial due to the qualitative nature of the assessment. The MWM task is very sensitive to test-environmental factors such as temperature, lighting, and odor. Therefore, it is important for the investigator to ensure that there are no changes in the testing environment throughout the duration of a study and the room used for testing is soundproof.

(b) Reward-based cognitive models: Several cognitive behavioral models that rely on appetitive motivation have been developed for evaluating experimental TBI-induced cognitive dysfunction in rodents [60]. These models use an apparatus of a well-defined shape and dimensions. For example, the T-maze or the Y-maze (Fig. 4b) includes a simple three-arm maze which is shaped as a "T" or "Y," respectively, whereas the radial maze has a central platform with 8 arms radiating from the center to create a circular apparatus [67, 90, 93, 94]. The animals are generally fasted overnight, and a reward, such as food or sucrose water, is hidden in one of the arms. The latency to find the reward over certain number of trials on a day or different days provides a measure of learning and memory. Similar to the MWM task, the location of the reward can be switched in between trials to determine preference for arms and forced alternation to evaluate "*reference*" and "*working*" memory functions [90, 94].

Notes: The reward-based cognitive models are primarily dependent on appetitive motivation. Therefore, it is important for the investigator to ensure that the fasting conditions are consistent across groups throughout the duration of a study. These tests are very sensitive to test-environmental factors such as temperature, lighting, and odor. Thus, the testing conditions should be consistent across trials throughout the duration of a study and the room used for testing should be soundproof.

(c) Exploration-based cognitive models: Exploration-based cognitive models provide a measure of "*retention*" or "*intact*" memory function in rodents [8, 58, 95–97] as well as nonhuman primates [98, 99] after experimental TBI. The most widely used exploration-based model uses mechanisms of spatial novelty and delayed non-matching and involves exploration of a novel object [8, 24, 58, 60, 95–97]. Therefore, this task has been named the novel object recognition or exploration test. An open field assay is used to habituate the animals prior to testing. Training involves exposure of the animals to two identical objects for a defined period of time for the purpose of familiarization, followed by testing in which one of the objects is replaced by a novel object (Fig. 4c) after a well-defined trial interval. This test assesses the "*retention*" memory function in terms of time spent exploring the novel object versus old or familiar object. Furthermore, it also evaluates the ability of the animals to distinguish between the two objects, calculated as the "*discrimination index*" [8, 24, 58, 97]. Modifications of the standard novel object recognition protocol have been developed that involve changing the location or position of the object to enable assessment of "*spatial*" memory [100] or increasing the level of difficulty by performing more trials or introducing more number of novel objects [101].

Notes: The exploration-based models are very sensitive to test-environmental factors such as temperature, lighting, and odor. Thus, the testing conditions should be consistent across trials throughout the duration of a study and the room used for testing should be soundproof. Familiarization with the apparatus improves the efficiency of these tasks. Therefore, an open field test is often performed prior to beginning the novel object trials to habituate the animals to the testing environment and the set-up. Furthermore, the use of bright-colored and large-sized objects should be avoided in such models as they may intimidate the animals and cause "*avoidance*" instead of "*exploration*" of the objects.

(d) Fear-based cognitive models: Fear-based models are based on the principle of fear conditioning [60, 102, 103]. Fear conditioning is a behavioral paradigm which enables avoidance learning in response to an aversive stimulus to rodents [104, 105] as well as nonhuman primates [106, 107]. In other words, the animals learn that a certain response will avoid an aversive stimulus. Fear conditioning tasks are of two types: active avoidance, when an animal is trained to jump over a barrier when a signal prior an aversive stimulus is generated; and passive avoidance, when an animal is trained not to elicit a certain response to avoid an aversive stimulus. In a typical fear conditioning paradigm (Fig. 4d), one stimulus or signal is paired with aversive stimulus when presented in isolation and is not paired with an aversive stimulus when presented together with a second stimulus. Thus, after training one stimulus or signal elicits a fear response, whereas two stimuli elicit less of a fear response. Therefore, reduction of fear to both stimuli together is a result of inhibitory fear learning to the aversive stimulus that counteracts the fear response that would otherwise results if the first stimulus occurred by itself [103]. A fear conditioning model is used to assess "*retention*" and "*emotional*" memory based on fear. The memory function is evaluated using certain parameters such as time taken to elicit a certain response following the stimulus, and latency to and time of freezing. An auditory method of fear conditioning has been commonly used to modulate prepulse inhibitory behavior in different models of neurological disorders [108–111]. Prepulse inhibition (PPI) is the attenuation of a startle response by applying a preceding non-startling sensory stimulus prior to aversive stimulus caused by contextual fear conditioning. PPI provides an acoustic startle paradigm that reflects a protective function of reducing disruptive influences to the processing of prepulse signals and is considered as a model of sensorimotor gating [108, 110].

Notes: The aversive stimulus should be tested on naïve animals prior to beginning the study to ensure that it does not cause any long-term harm or pain to the animals. Unlike the MWM and the exploration-based tasks, multiple trials are not required for evaluating the cognitive function in the fear-based models as excessive testing may tire or intimidate the animals yielding "*false positive*" or "*false negative*" results. The fear-based models are very sensitive to test-environmental factors such as temperature, lighting, and odor. Therefore, it is important for the investigator to ensure that there are no changes in the testing environment throughout the duration of a study and the room used for testing is soundproof.

2.2.3 Emotional and Social Behavior

Several behavioral paradigms have been developed to model impairments in emotional and social behavior associated with experimental TBI.

(a) Anxiety-based models: The anxiety-based models have been used to screen the anti-anxiety effects of pharmacological agents and steroid hormones following experimental TBI and to better understand the mechanisms and brain regions involved in anxiety-related behavior. The most commonly used anxiety-based assay is an elevated plus maze (EPM). The EPM task is widely used in rodents and involves an apparatus with four arms (two open and two enclosed) [60, 112]. This assay is based on approach-avoidance mechanism: the animal's preference for dark and enclosed spaces or approach and an unconditioned fear of heights/open space (avoidance).

Notes: The anxiety-based models are very sensitive to test-environmental factors such as temperature, lighting, and odor. Therefore, it is important for the investigator to ensure that there are no changes in the testing environment throughout the duration of a study and the room used for testing is soundproof.

(b) Depression-based models: The depression-based behavioral models are used in rodents to simulate depressive-like behavior based on level of activity and are often defined as tests of hopelessness and despair. The classical test for depression used in mice is the tail suspension test [60, 113, 114]. The protocol involves suspending a mouse by the tail from a lever and recording the movement of the animal. Depressive-like symptoms are assessed by period of agitations versus time of immobility. This test is commonly used to screen antidepressants and when coupled with measurement of locomotor activity it can be used to distinguish between the "locomotor stimulant" dose from the "antidepressant dose." The classical test for depression for rats, now also used in mice, is the Porsolt's swim, or forced swim test [60, 113]. The protocol involves scoring the activity (swimming and climbing or struggle) versus passive (immobility or floating) when rodents are forced to swim in a cylinder from which there is no escape for extended period of time, such as 5 or 10 minutes. The reduction in immobility in both tail suspension and forced swim tests is interpreted as an antidepressant-like effect.

Notes: The depression-based models are very sensitive to test-environmental factors such as temperature, lighting, and odor. Therefore, it is important for the investigator to ensure that there are no changes in the testing environment throughout the duration of a study and the room used for testing is soundproof. The assessment of depressive behavior is to some extent qualitative in nature as the scoring is dependent on subtle changes in behavior, such as mobility versus immobility. Therefore, a second investigator blinded to groups should be involved to confirm the type and duration of behavior assigned per trial. More recently, computerized systems have been used to detect changes in movement or mobility for evaluating depressive-like symptoms in such models. These systems are very sensitive to any movement in the testing arena.

Therefore, the investigator should avoid any movement in the room when a trial is being conducted and perform manual scoring to confirm the results recorded by the computer software.

(c) Social interaction-based models: Impaired sociability and communicative behavior, and lack of self-awareness are a hallmark of TBI and many neuropsychiatric disorders that increase the risk of depression. Social communication in rodents is generally assessed by olfactory interaction or time spent sniffing olfactory stimuli from novel animals and identification of novel versus familiar animals through olfactory cues in an open field setting [115–117]. It can also be assessed in rodents by vocal communication via ultrasonic vocalization recordings emitted during social interactions [118], particularly observed as an early communicative behavior of the pup-mother dyad. Several social interaction-based models have been developed using nonhuman primates to assess communicative behavior with others as well as self [119]. These tests rely on vocal communication and social interactions in the form of behavioral parameters such as grooming, mutual tolerance at feeding sites, formation of aggressive alliances or social bonding, reaction to stressors or threats due to cage dominance, and voice recognition [119]. Furthermore, social self-awareness is studied in monkeys by using a mirror test that is based on recognition of an animal's reflection in a mirror [119]. The social interaction-based models exhibit great potential for understanding the underlying pathologic mechanisms as well as evaluating the pharmacological or toxicological effects of compounds on sociability [120, 121].

Notes: The social interaction-based models rely on subtle changes in behavioral interaction or communication. Therefore, it is important for the investigator to ensure that there are no changes in the testing environment throughout the duration of a study and the room used for testing is soundproof. The assessment of these tasks is qualitative in nature. Therefore, a second investigator blinded to groups should be involved to confirm the type and duration of behavioral observation.

Behavioral models are highly versatile in their applicability as every model developed for assessment of a particular function can be used to simulate different behavioral traits. For example, the MWM test of learning and memory consists of a locomotor functional component and the open field test consists of locomotor and exploratory behavior. Both are dependent on the emotional state of the animal. Therefore, using an integrated system comprising of a series of behavioral models (such as using the MWM and the open field test) is recommended for a more accurate and reliable assessment of function and associated abnormalities. Furthermore, data analysis of behavioral tasks should be performed using parameters directed not only toward the primary behavioral function under investigation but also other functions that can be studied using the same test.

2.2.4 Statistical Analysis of Behavioral Tests

Statistical analysis plays a very important role in not only the experimental design but also the determination of the degree of behavioral impairment induced by an experimental TBI model. Additionally, the effectiveness of a potential therapeutic candidate in improving behavioral outcomes relies heavily on sensitivity and accuracy of the statistical analysis and comparisons.

Statistical power analysis, which forms an important aspect of experimental study design, needs to be performed to determine an appropriate sample size (number of animals per group) prior to beginning a study. The estimation of statistical power enables the investigator to determine the sample size required to detect an effect of a given size with a given degree of confidence. Once the data is collected, the quantitative raw data should be expressed as means with standard deviation (±SD) or standard error of means (±SEM). The choice of an appropriate statistical test depends on the parameters assessed, conditions of testing, and distribution of the collected data [122]. The method of statistical analysis chosen for evaluating behavioral data should reflect the purpose of the study, the design of the experiment, and the nature of the collected data [122]. More importantly, the selected statistical test should define the hypothesis to be tested.

Quantitative data can be analyzed using parametric or nonparametric tests. Parametric statistics are based on the assumptions that the residuals have a normal distribution, the variances are approximately same in each group, and the observations are independent of each other [123]. Such tests are generally recommended for data representing a normal distribution, homogeneity of variance, and independent sampling. In contrast, when the assumptions of approximately equal variation in each group and approximate normality of the residuals are not valid, and no scale transformation is available to correct the heterogeneity of variance and/or non-normality, a nonparametric test is recommended to compare the equality of population means and medians [123]. Therefore, nonparametric statistics do not rely on the distribution of data, that is, the methods do not make any assumptions that the data are drawn from a given probability distribution.

A determination of the degree of behavioral dysfunction induced by an experimental TBI model is achieved by demonstrating a statistical significance in impairment of functional outcomes in the injured group versus sham-operated and naïve animals. The significance level is defined to identify the chance of obtaining false positive results because of sampling error, which is usually set to 5%.

Furthermore, the effectiveness of a treatment in a TBI model in terms of behavioral improvement is demonstrated by statistical significance in recovery of functional outcomes in the treated injured groups versus injured untreated groups. Generally, for analysis of quantitative behavioral data, such as number of foot faults in the beam walk, latency to the find the platform (in seconds) in the MWM, and discrimination index (percentage) in the novel object recognition task, from a study with more than two groups, analysis of variance (ANOVA) methods are used to initially test the overall hypothesis by analyzing the differences among group means and their associated procedures, that is, variation within and between groups. ANOVA methods provide multiple comparison/contrast test, called post hoc tests that allow the investigator to identify patterns of statistical significance between different combinations of groups. When a test statistic falls under normal distribution in case of comparison between two groups of data sets, Student's t-test is commonly applied to determine statistical difference between the two groups. These tests could be used to evaluate an alternative hypothesis by selecting a two-tailed test which indicates that two means or proportions differ or by using a one-tailed test by specifying the direction of the difference. For example, a two-tailed test will determine if there is a statistically significant difference between treatment and control, whereas a one-tailed test will determine whether the treatment causes statistically significant improvement in the outcomes being evaluated versus control. Two-tailed tests require larger sample size.

In contrast, the analysis of qualitative data is performed by a categorical measurement that is not expressed in terms of numbers but rather by means of language description. These data can be expressed as "categorical" data that may have a structure or order and therefore could be analyzed as ordinal variables such as size-related (small, medium or large) or specific responses to questions (strongly disagree, disagree, neutral, agree or strongly agree). For example, while analyzing data from search strategy analysis performed in the MWM model, the qualitative behavioral data are expressed as looping, systematic, and spatial to define the degree of spatial learning and memory dysfunction after experimental TBI. Such qualitative data are analyzed by using a chi-squared (χ^2) test, also called Pearson's χ^2 or goodness of fit test. This statistical test assesses the goodness of fit between observed values and those expected theoretically. It is applied to sets of categorical data to evaluate how likely it is that any observed difference between the sets occurred by chance.

The changes observed in behavioral tests after induction of experimental TBI are often subtle, particularly if the severity of the injury is mild or repeated mild. Therefore, the selection of an appropriate statistical test determines the robustness of the results and therefore, validates the experimental TBI and behavioral models used.

3 Conclusion

This chapter summarizes commonly used experimental TBI and behavioral assessment models developed over the years to mimic the events that occur during and after the induction of a TBI. It is challenging to reconstruct the events that are involved in the induction of a clinical TBI and manifestation of behavioral dysfunction. Furthermore, no one model of experimental TBI or behavioral assessment can represent all aspects of a clinical TBI and related behavioral impairment. This has resulted in the failure of potential neuroprotective agents in clinical trials conducted on patients with TBI, despite strong supporting preclinical data. Therefore, there is a need for new models and refining of existing models to closely simulate pathophysiological and behavioral conditions observed in clinical TBI cases as much as possible. In this chapter, we discussed the application of different animal models for induction of experimental TBI and determining the degree of resulting behavioral impairment while evaluating their relevance to clinical symptoms and examining their advantages and limitations. We discuss the current status of TBI research conducted by utilizing animal models of experimental TBI and behavioral assessment, and provide recommendations for improving the sensitivity of these models by focusing on specific characteristics of experimental design, data collection, and statistical analysis.

In conclusion, this chapter has been written to provide guidance to investigators in the field of TBI research on the selection, development and application of experimental TBI and behavioral assessment models. While developing and utilizing an animal model for induction of experimental TBI and/or for assessing related behavioral dysfunction, it is important to understand the multifactorial nature of events involved at the time of the brain injury as well as the complexity of secondary injury mechanisms that lead to chronic behavioral impairment. The selection of an appropriate experimental TBI and behavioral assessment model may depend on the feature(s) of clinical TBI and behavioral dysfunction that the investigator is interested in simulating under a preclinical setting.

Acknowledgments

This work was supported by the Uniformed Services University Rat Behavioral Core and the Center for Neuroscience and Regenerative Medicine.

Disclaimer (Shruti V. Kabadi): The interpretations expressed in this book chapter represent that of the author and not necessarily that of the US Food and Drug Administration (US FDA).

References

1. CDC (2010) National Hospital Discharge Survey (NHDS); National Hospital Ambulatory Medical Care Survey (NHAMCS); National Vital Statistics System (NVSS)
2. Frieden TRH, D.; Baldwin, G. (2013) Traumatic brain injury in the United States: epidemiology and rehabilitation. Report to Congress
3. Ellemberg D, Henry LC, Macciocchi SN, Guskiewicz KM, Broglio SP (2009) Advances in sport concussion assessment: from behavioral to brain imaging measures. J Neurotrauma 26(12):2365–2382. https://doi.org/10.1089/neu.2009.0906
4. Hoge CW, McGurk D, Thomas JL, Cox AL, Engel CC, Castro CA (2008) Mild traumatic brain injury in U.S. soldiers returning from Iraq. N Engl J Med 358(5):453–463. https://doi.org/10.1056/NEJMoa072972
5. Bramlett H, Dietrich W (2007) Progressive damage after brain and spinal cord injury: pathomechanisms and treatment strategies. Prog Brain Res 161:125–141
6. Holmin S, Soderlund J, Biberfeld P, Mathiesen T (1998) Intracerebral inflammation after human brain contusion. Neurosurgery 42 (2):291–298. discussion 298-299
7. Byrnes KR, Loane DJ, Stoica BA, Zhang J, Faden AI (2012) Delayed mGluR5 activation limits neuroinflammation and neurodegeneration after traumatic brain injury. J Neuroinflammation 9:43. https://doi.org/10.1186/1742-2094-9-43
8. Kabadi SV, Stoica BA, Loane DJ, Luo T, Faden AI (2014) CR8, a novel inhibitor of CDK, limits microglial activation, astrocytosis, neuronal loss, and neurologic dysfunction after experimental traumatic brain injury. J Cereb Blood Flow Metab 34(3):502–513. https://doi.org/10.1038/jcbfm.2013.228
9. Loane DJ, Byrnes KR (2010) Role of microglia in neurotrauma. Neurotherapeutics 7 (4):366–377. https://doi.org/10.1016/j.nurt.2010.07.002
10. Soltys Z, Ziaja M, Pawlinski R, Setkowicz Z, Janeczko K (2001) Morphology of reactive microglia in the injured cerebral cortex. Fractal analysis and complementary quantitative methods. J Neurosci Res 63(1):90–97. https://doi.org/10.1002/1097-4547(20010101)63:1<90::AID-JNR11>3.0.CO;2-9
11. Eikelenboom P, Bate C, Van Gool WA, Hoozemans JJ, Rozemuller JM, Veerhuis R, Williams A (2002) Neuroinflammation in Alzheimer's disease and prion disease. Glia 40(2):232–239. https://doi.org/10.1002/glia.10146
12. Byrnes KR, Faden AI (2007) Role of cell cycle proteins in CNS injury. Neurochem Res 32 (10):1799–1807. https://doi.org/10.1007/s11064-007-9312-2
13. Kabadi SV, Faden AI (2014) Neuroprotective strategies for traumatic brain injury: improving clinical translation. Int J Mol Sci 15 (1):1216–1236. https://doi.org/10.3390/ijms15011216
14. Berkner J, Mannix R, Qiu J (2016) Clinical traumatic brain injury in the preclinical setting. Methods Mol Biol 1462:11–28. https://doi.org/10.1007/978-1-4939-3816-2_2
15. Anisman H, Matheson K (2005) Stress, depression, and anhedonia: caveats concerning animal models. Neurosci Biobehav Rev 29(4–5):525–546. https://doi.org/10.1016/j.neubiorev.2005.03.007
16. McCrea MA (2007) Mild traumatic brain injury and postconcussion syndrome- the new evidence base for diagnosis and treatment. Oxford University Press, New York
17. Dixon KJ (2017) Pathophysiology of traumatic brain injury. Phys Med Rehabil Clin N Am 28(2):215–225. https://doi.org/10.1016/j.pmr.2016.12.001
18. Cole JT, Yarnell A, Kean WS, Gold E, Lewis B, Ren M, McMullen DC, Jacobowitz DM, Pollard HB, O'Neill JT, Grunberg NE, Dalgard CL, Frank JA, Watson WD (2011) Craniotomy: true sham for traumatic brain injury, or a sham of a sham? J Neurotrauma 28(3):359–369. https://doi.org/10.1089/neu.2010.1427
19. Panzer MB, Wood GW, Bass CR (2014) Scaling in neurotrauma: how do we apply animal experiments to people? Exp Neurol 261:120–126. https://doi.org/10.1016/j.expneurol.2014.07.002
20. Rostami E (2016) Traumatic brain injury models in animals. Methods Mol Biol 1462:47–59. https://doi.org/10.1007/978-1-4939-3816-2_4
21. Van KC, Lyeth BG (2016) Lateral (parasagittal) fluid percussion model of traumatic brain injury. Methods Mol Biol 1462:231–251. https://doi.org/10.1007/978-1-4939-3816-2_14
22. Kabadi SV, Hilton GD, Stoica BA, Zapple DN, Faden AI (2010) Fluid-percussion-induced traumatic brain injury model in rats.

Nat Protoc 5(9):1552–1563. https://doi.org/10.1038/nprot.2010.112

23. Sell SL, Johnson K, DeWitt DS, Prough DS (2017) Persistent behavioral deficits in rats after parasagittal fluid percussion injury. J Neurotrauma 34(5):1086–1096. https://doi.org/10.1089/neu.2016.4616
24. Aungst SL, Kabadi SV, Thompson SM, Stoica BA, Faden AI (2014) Repeated mild traumatic brain injury causes chronic neuroinflammation, changes in hippocampal synaptic plasticity, and associated cognitive deficits. J Cereb Blood Flow Metab 34(7):1223–1232. https://doi.org/10.1038/jcbfm.2014.75
25. Chauhan NB, Gatto R (2010) Synergistic benefits of erythropoietin and simvastatin after traumatic brain injury. Brain Res 1360:177–192. https://doi.org/10.1016/j.brainres.2010.09.010
26. Osier ND, Bales JW, Pugh B, Shin S, Wyrobek J, Puccio AM, Okonkwo DO, Ren D, Alexander S, Conley YP, Dixon CE (2017) Variation in PPP3CC genotype is associated with Long-term recovery after severe brain injury. J Neurotrauma 34 (1):86–96. https://doi.org/10.1089/neu.2015.4343
27. Pleasant JM, Carlson SW, Mao H, Scheff SW, Yang KH, Saatman KE (2011) Rate of neurodegeneration in the mouse controlled cortical impact model is influenced by impactor tip shape: implications for mechanistic and therapeutic studies. J Neurotrauma 28 (11):2245–2262. https://doi.org/10.1089/neu.2010.1499
28. Glover LE, Tajiri N, Lau T, Kaneko Y, van Loveren H, Borlongan CV (2012) Immediate, but not delayed, microsurgical skull reconstruction exacerbates brain damage in experimental traumatic brain injury model. PLoS One 7(3):e33646. https://doi.org/10.1371/journal.pone.0033646
29. Brabazon F, Wilson CM, Shukla DK, Mathur S, Jaiswal S, Bermudez S, Byrnes KR, Selwyn R (2017) [18F]FDG-PET combined with MRI elucidates the pathophysiology of traumatic brain injury in rats. J Neurotrauma 34(5):1074–1085. https://doi.org/10.1089/neu.2016.4540
30. Jacobowitz DM, Cole JT, McDaniel DP, Pollard HB, Watson WD (2012) Microglia activation along the corticospinal tract following traumatic brain injury in the rat: a neuroanatomical study. Brain Res 1465:80–89. https://doi.org/10.1016/j.brainres.2012.05.008
31. Marmarou A, Foda MA, van den Brink W, Campbell J, Kita H, Demetriadou K (1994) A new model of diffuse brain injury in rats. Part I: Pathophysiology and biomechanics. J Neurosurg 80(2):291–300. https://doi.org/10.3171/jns.1994.80.2.0291
32. McAteer KM, Corrigan F, Thornton E, Turner RJ, Vink R (2016) Short and long term behavioral and pathological changes in a novel rodent model of repetitive mild traumatic brain injury. PLoS One 11(8):e0160220. https://doi.org/10.1371/journal.pone.0160220
33. Grin'kina NM, Li Y, Haber M, Sangobowale M, Nikulina E, Le'Pre C, El Sehamy AM, Dugue R, Ho JS, Bergold PJ (2016) Righting reflex predicts Long-term histological and behavioral outcomes in a closed head model of traumatic brain injury. PLoS One 11(9):e0161053. https://doi.org/10.1371/journal.pone.0161053
34. Yates NJ, Lydiard S, Fehily B, Weir G, Chin A, Bartlett CA, Alderson J, Fitzgerald M (2017) Repeated mild traumatic brain injury in female rats increases lipid peroxidation in neurons. Exp Brain Res 235:2133. https://doi.org/10.1007/s00221-017-4958-8
35. Acabchuk R, Briggs DI, Angoa-Perez M, Powers M, Wolferz R Jr, Soloway M, Stern M, Talbot LR, Kuhn DM, Conover JC (2016) Repeated mild traumatic brain injury causes focal response in lateral septum and hippocampus. Concussion 1(3):CNC13. https://doi.org/10.2217/cnc-2015-0001
36. Meehan WP 3rd, Zhang J, Mannix R, Whalen MJ (2012) Increasing recovery time between injuries improves cognitive outcome after repetitive mild concussive brain injuries in mice. Neurosurgery 71(4):885–891. https://doi.org/10.1227/NEU.0b013e318265a439
37. Hou J, Nelson R, Wilkie Z, Mustafa G, Tsuda S, Thompson FJ, Bose PK (2017) Mild and mild to moderate traumatic brain injury (TBI)-induced significant progressive and enduring multiple comorbidities. J Neurotrauma 34:2456. https://doi.org/10.1089/neu.2016.4851
38. Stemper BD, Shah AS, Budde MD, Olsen CM, Glavaski-Joksimovic A, Kurpad SN, McCrea M, Pintar FA (2016) Behavioral outcomes differ between rotational acceleration and blast mechanisms of mild traumatic brain injury. Front Neurol 7:31. https://doi.org/10.3389/fneur.2016.00031
39. Prange AJ Jr (1999) Thyroid axis sustaining hypothesis of posttraumatic stress disorder. Psychosom Med 61(2):139–140
40. Namjoshi DR, Cheng WH, Bashir A, Wilkinson A, Stukas S, Martens KM,

Whyte T, Abebe ZA, McInnes KA, Cripton PA, Wellington CL (2017) Defining the biomechanical and biological threshold of murine mild traumatic brain injury using CHIMERA (closed head impact model of engineered rotational acceleration). Exp Neurol 292:80–91. https://doi.org/10.1016/j.expneurol.2017.03.003

41. Chen H, Desai A, Kim HY (2017) Repetitive closed-head impact model of engineered rotational acceleration induces Long-term cognitive impairments with persistent astrogliosis and microgliosis in mice. J Neurotrauma 34:2291. https://doi.org/10.1089/neu.2016.4870
42. Armistead-Jehle P, Soble JR, Cooper DB, Belanger HG (2017) Unique aspects of traumatic brain injury in military and veteran populations. Phys Med Rehabil Clin N Am 28(2):323–337. https://doi.org/10.1016/j.pmr.2016.12.008
43. Bauman RA, Ling G, Tong L, Januszkiewicz A, Agoston D, Delanerolle N, Kim Y, Ritzel D, Bell R, Ecklund J, Armonda R, Bandak F, Parks S (2009) An introductory characterization of a combat-casualty-care relevant swine model of closed head injury resulting from exposure to explosive blast. J Neurotrauma 26(6):841–860. https://doi.org/10.1089/neu.2009-0898
44. Goodrich JA, Kim JH, Situ R, Taylor W, Westmoreland T, Du F, Parks S, Ling G, Hwang JY, Rapuano A, Bandak FA, de Lanerolle NC (2016) Neuronal and glial changes in the brain resulting from explosive blast in an experimental model. Acta Neuropathol Commun 4(1):124. https://doi.org/10.1186/s40478-016-0395-3
45. Kamnaksh A, Kwon SK, Kovesdi E, Ahmed F, Barry ES, Grunberg NE, Long J, Agoston D (2012) Neurobehavioral, cellular, and molecular consequences of single and multiple mild blast exposure. Electrophoresis 33 (24):3680–3692. https://doi.org/10.1002/elps.201200319
46. Ahmed FA, Kamnaksh A, Kovesdi E, Long JB, Agoston DV (2013) Long-term consequences of single and multiple mild blast exposure on select physiological parameters and blood-based biomarkers. Electrophoresis 34(15):2229–2233. https://doi.org/10.1002/elps.201300077
47. Dal Cengio Leonardi A, Keane NJ, Bir CA, Ryan AG, Xu L, Vandevord PJ (2012) Head orientation affects the intracranial pressure response resulting from shock wave loading in the rat. J Biomech 45(15):2595–2602. https://doi.org/10.1016/j.jbiomech.2012.08.024
48. McCabe JT, Moratz C, Liu Y, Burton E, Morgan A, Budinich C, Lowe D, Rosenberger J, Chen H, Liu J, Myers M (2014) Application of high-intensity focused ultrasound to the study of mild traumatic brain injury. Ultrasound Med Biol 40 (5):965–978. https://doi.org/10.1016/j.ultrasmedbio.2013.11.023
49. Tchantchou F, Fourney WL, Leiste UH, Vaughan J, Rangghran P, Puche A, Fiskum G (2017) Neuropathology and neurobehavioral alterations in a rat model of traumatic brain injury to occupants of vehicles targeted by underbody blasts. Exp Neurol 289:9–20. https://doi.org/10.1016/j.expneurol.2016.12.001
50. Kawoos U, Gu M, Lankasky J, McCarron RM, Chavko M (2016) Effects of exposure to blast overpressure on intracranial pressure and blood-brain barrier permeability in a rat model. PLoS One 11(12):e0167510. https://doi.org/10.1371/journal.pone.0167510
51. Lucke-Wold BP, Logsdon AF, Turner RC, Huber JD, Rosen CL (2017) Endoplasmic reticulum stress modulation as a target for ameliorating effects of blast induced traumatic brain injury. J Neurotrauma 34:S-62. https://doi.org/10.1089/neu.2016.4680
52. Agoston DVK, A. (2015) Modeling the neurobehavioral consequences of blast-induced traumatic brain injury Spectrum disorder and identifying related biomarkers. In: Kobeissy FH (ed) Brain Neurotrauma: molecular, neuropsychological, and rehabilitation aspects. CRC Press/Taylor & Francis, Boca Raton, FL
53. Kovesdi E, Kamnaksh A, Wingo D, Ahmed F, Grunberg NE, Long JB, Kasper CE, Agoston DV (2012) Acute minocycline treatment mitigates the symptoms of mild blast-induced traumatic brain injury. Front Neurol 3:111. https://doi.org/10.3389/fneur.2012.00111
54. Fleming SM, Salcedo J, Fernagut PO, Rockenstein E, Masliah E, Levine MS, Chesselet MF (2004) Early and progressive sensorimotor anomalies in mice overexpressing wild-type human alpha-synuclein. J Neurosci 24(42):9434–9440. https://doi.org/10.1523/JNEUROSCI.3080-04.2004
55. Goldstein LB (2003) Model of recovery of locomotor ability after sensorimotor cortex injury in rats. ILAR J 44(2):125–129
56. Goldstein LB, Davis JN (1990) Beam-walking in rats: studies towards developing

an animal model of functional recovery after brain injury. J Neurosci Methods 31 (2):101–107

57. Kabadi SV, Stoica BA, Byrnes KR, Hanscom M, Loane DJ, Faden AI (2012) Selective CDK inhibitor limits neuroinflammation and progressive neurodegeneration after brain trauma. J Cereb Blood Flow Metab 32(1):137–149. https://doi.org/10.1038/jcbfm.2011.117
58. Kabadi SV, Stoica BA, Hanscom M, Loane DJ, Kharebava G, Murray Ii MG, Cabatbat RM, Faden AI (2012) CR8, a selective and potent CDK inhibitor, provides neuroprotection in experimental traumatic brain injury. Neurotherapeutics 9(2):405–421. https://doi.org/10.1007/s13311-011-0095-4
59. Kabadi SV, Stoica BA, Loane DJ, Byrnes KR, Hanscom M, Cabatbat RM, Tan MT, Faden AI (2012) Cyclin D1 gene ablation confers neuroprotection in traumatic brain injury. J Neurotrauma 29(5):813–827. https://doi.org/10.1089/neu.2011.1980
60. Wu Y-C, Kabadi SV, Neal-Kluever A (2017) Developmental neurotoxicity considerations for food additive safety. In: Sachan A, Hendrich S (eds) Food toxicology current advances and future challenges. Apple Academic Press, p 495
61. Merkler D, Metz GA, Raineteau O, Dietz V, Schwab ME, Fouad K (2001) Locomotor recovery in spinal cord-injured rats treated with an antibody neutralizing the myelin-associated neurite growth inhibitor Nogo-A. J Neurosci 21(10):3665–3673
62. Singleton RH, Yan HQ, Fellows-Mayle W, Dixon CE (2010) Resveratrol attenuates behavioral impairments and reduces cortical and hippocampal loss in a rat controlled cortical impact model of traumatic brain injury. J Neurotrauma 27(6):1091–1099. https://doi.org/10.1089/neu.2010.1291
63. Selwyn R, Hockenbury N, Jaiswal S, Mathur S, Armstrong RC, Byrnes KR (2013) Mild traumatic brain injury results in depressed cerebral glucose uptake: an (18) FDG PET study. J Neurotrauma 30 (23):1943–1953. https://doi.org/10.1089/neu.2013.2928
64. Carter RJ, Morton J, Dunnett SB (2001) Motor coordination and balance in rodents. Curr Protoc Neurosci. Chapter 8:unit 8 12. https://doi.org/10.1002/0471142301.ns0812s15
65. Caston J, Jones N, Stelz T (1995) Role of preoperative and postoperative sensorimotor training on restoration of the equilibrium behavior in adult mice following cerebellectomy. Neurobiol Learn Mem 64 (3):195–202. https://doi.org/10.1006/nlme.1995.0002
66. Lalonde R, Bensoula AN, Filali M (1995) Rotorod sensorimotor learning in cerebellar mutant mice. Neurosci Res 22(4):423–426
67. Schaar KL, Brenneman MM, Savitz SI (2010) Functional assessments in the rodent stroke model. Exp Transl Stroke Med 2(1):13. https://doi.org/10.1186/2040-7378-2-13
68. Shiotsuki H, Yoshimi K, Shimo Y, Funayama M, Takamatsu Y, Ikeda K, Takahashi R, Kitazawa S, Hattori N (2010) A rotarod test for evaluation of motor skill learning. J Neurosci Methods 189 (2):180–185. https://doi.org/10.1016/j.jneumeth.2010.03.026
69. Desai A, Kevala K, Kim HY (2014) Depletion of brain docosahexaenoic acid impairs recovery from traumatic brain injury. PLoS One 9 (1):e86472. https://doi.org/10.1371/journal.pone.0086472
70. Lindner MD, Plone MA, Cain CK, Frydel B, Francis JM, Emerich DF, Sutton RL (1998) Dissociable long-term cognitive deficits after frontal versus sensorimotor cortical contusions. J Neurotrauma 15(3):199–216. https://doi.org/10.1089/neu.1998.15.199
71. Doeppner TR, Kaltwasser B, Bahr M, Hermann DM (2014) Effects of neural progenitor cells on post-stroke neurological impairment-a detailed and comprehensive analysis of behavioral tests. Front Cell Neurosci 8:338. https://doi.org/10.3389/fncel.2014.00338
72. Lyeth BG, Dixon CE, Hamm RJ, Jenkins LW, Young HF, Stonnington HH, Hayes RL (1988) Effects of anticholinergic treatment on transient behavioral suppression and physiological responses following concussive brain injury to the rat. Brain Res 448(1):88–97
73. Murphy MP, Rick JT, Milgram NW, Ivy GO (1995) A simple and rapid test of sensorimotor function in the aged rat. Neurobiol Learn Mem 64(2):181–186. https://doi.org/10.1006/nlme.1995.1057
74. Yonemori F, Yamaguchi T, Nakayama H, Narita K, Hojo S, Tamura A (2000) Effect of JTP-2942, a novel thyrotropin-releasing hormone analog, on motor deficits after chronic focal cerebral ischemia in rats. J Cereb Blood Flow Metab 20(1):74–81. https://doi.org/10.1097/00004647-200001000-00011
75. Jeong SW, Chu K, Jung KH, Kim SU, Kim M, Roh JK (2003) Human neural stem cell transplantation promotes functional recovery in rats with experimental intracerebral

hemorrhage. Stroke 34(9):2258–2263. https://doi.org/10.1161/01.STR.0000083698.20199.1F
76. Woodlee MT, Asseo-Garcia AM, Zhao X, Liu SJ, Jones TA, Schallert T (2005) Testing forelimb placing "across the midline" reveals distinct, lesion-dependent patterns of recovery in rats. Exp Neurol 191(2):310–317. https://doi.org/10.1016/j.expneurol.2004.09.005
77. Bouet V, Boulouard M, Toutain J, Divoux D, Bernaudin M, Schumann-Bard P, Freret T (2009) The adhesive removal test: a sensitive method to assess sensorimotor deficits in mice. Nat Protoc 4(10):1560–1564. https://doi.org/10.1038/nprot.2009.125
78. Freret T, Bouet V, Leconte C, Roussel S, Chazalviel L, Divoux D, Schumann-Bard P, Boulouard M (2009) Behavioral deficits after distal focal cerebral ischemia in mice: usefulness of adhesive removal test. Behav Neurosci 123(1):224–230. https://doi.org/10.1037/a0014157
79. Lacreuse A, Diehl MM, Goh MY, Hall MJ, Volk AM, Chhabra RK, Herndon JG (2005) Sex differences in age-related motor slowing in the rhesus monkey: behavioral and neuroimaging data. Neurobiol Aging 26(4):543–551. https://doi.org/10.1016/j.neurobiolaging.2004.05.007
80. Pessiglione M, Guehl D, Agid Y, Hirsch EC, Feger J, Tremblay L (2003) Impairment of context-adapted movement selection in a primate model of presymptomatic Parkinson's disease. Brain 126(Pt 6):1392–1408
81. Philippens IH, Melchers BP, Roeling TA, Bruijnzeel PL (2000) Behavioral test systems in marmoset monkeys. Behav Res Methods Instrum Comput 32(1):173–179
82. Sutcliffe JS, Beaumont V, Watson JM, Chew CS, Beconi M, Hutcheson DM, Dominguez C, Munoz-Sanjuan I (2014) Efficacy of selective PDE4D negative allosteric modulators in the object retrieval task in female cynomolgus monkeys (Macaca fascicularis). PLoS One 9(7):e102449. https://doi.org/10.1371/journal.pone.0102449
83. Hutter-Saunders JA, Gendelman HE, Mosley RL (2012) Murine motor and behavior functional evaluations for acute 1-methyl-4-phenyl-1,2,3,6-tetrahydropyridine (MPTP) intoxication. J Neuroimmune Pharmacol 7(1):279–288. https://doi.org/10.1007/s11481-011-9269-4
84. Ijomone OM, Olaibi OK, Biose IJ, Mba C, Umoren KE, Nwoha PU (2014) Performance of motor associated behavioural tests following chronic nicotine administration. Ann Neurosci 21(2):42–46. https://doi.org/10.5214/ans.0972.7531.210203
85. Klejbor I, Ludkiewicz B, Turlejski K (2013) Effect of light-dark changes on the locomotor activity in open field in adult rats and opossums. Folia Morphol (Warsz) 72(4):300–305
86. Ferguson SA, Bowman RE (1990) Effects of postnatal lead exposure on open field behavior in monkeys. Neurotoxicol Teratol 12(2):91–97
87. Kabadi SV, Faden AI (2014) Selective CDK inhibitors: promising candidates for future clinical traumatic brain injury trials. Neural Regen Res 9(17):1578–1580. https://doi.org/10.4103/1673-5374.141779
88. Vorhees CV, Williams MT (2006) Morris water maze: procedures for assessing spatial and related forms of learning and memory. Nat Protoc 1(2):848–858. https://doi.org/10.1038/nprot.2006.116
89. Vorhees CV, Williams MT (2014) Value of water mazes for assessing spatial and egocentric learning and memory in rodent basic research and regulatory studies. Neurotoxicol Teratol 45:75–90. https://doi.org/10.1016/j.ntt.2014.07.003
90. Dudchenko PA (2004) An overview of the tasks used to test working memory in rodents. Neurosci Biobehav Rev 28(7):699–709. https://doi.org/10.1016/j.neubiorev.2004.09.002
91. Zhao Z, Loane DJ, Murray MG 2nd, Stoica BA, Faden AI (2012) Comparing the predictive value of multiple cognitive, affective, and motor tasks after rodent traumatic brain injury. J Neurotrauma 29(15):2475–2489. https://doi.org/10.1089/neu.2012.2511
92. Brody DL, Holtzman DM (2006) Morris water maze search strategy analysis in PDAPP mice before and after experimental traumatic brain injury. Exp Neurol 197(2):330–340. https://doi.org/10.1016/j.expneurol.2005.10.020
93. O'Keefe J, Dostrovsky J (1971) The hippocampus as a spatial map. Preliminary evidence from unit activity in the freely-moving rat. Brain Res 34(1):171–175
94. Shoji H, Hagihara H, Takao K, Hattori S, Miyakawa T (2012) T-maze forced alternation and left-right discrimination tasks for assessing working and reference memory in mice. J Vis Exp 60. https://doi.org/10.3791/3300
95. Antunes M, Biala G (2012) The novel object recognition memory: neurobiology, test procedure, and its modifications. Cogn Process

13(2):93–110. https://doi.org/10.1007/s10339-011-0430-z
96. Bevins RA, Besheer J (2005) Novelty reward as a measure of anhedonia. Neurosci Biobehav Rev 29(4–5):707–714. https://doi.org/10.1016/j.neubiorev.2005.03.013
97. Bevins RA, Besheer J (2006) Object recognition in rats and mice: a one-trial non-matching-to-sample learning task to study 'recognition memory'. Nat Protoc 1 (3):1306–1311. https://doi.org/10.1038/nprot.2006.205
98. Buckmaster CA, Eichenbaum H, Amaral DG, Suzuki WA, Rapp PR (2004) Entorhinal cortex lesions disrupt the relational organization of memory in monkeys. J Neurosci 24 (44):9811–9825. https://doi.org/10.1523/JNEUROSCI.1532-04.2004
99. Peissig JJ, Singer J, Kawasaki K, Sheinberg DL (2007) Effects of long-term object familiarity on event-related potentials in the monkey. Cereb Cortex 17(6):1323–1334. https://doi.org/10.1093/cercor/bhl043
100. Dix SL, Aggleton JP (1999) Extending the spontaneous preference test of recognition: evidence of object-location and object-context recognition. Behav Brain Res 99 (2):191–200
101. Kinnavane L, Albasser MM, Aggleton JP (2015) Advances in the behavioural testing and network imaging of rodent recognition memory. Behav Brain Res 285:67–78. https://doi.org/10.1016/j.bbr.2014.07.049
102. Maren S (2001) Neurobiology of Pavlovian fear conditioning. Annu Rev Neurosci 24:897–931. https://doi.org/10.1146/annurev.neuro.24.1.897
103. Rescorla RA (1969) Conditioned inhibition of fear resulting from negative CS-US contingencies. J Comp Physiol Psychol 67 (4):504–509
104. Myers K, Goulet M, Rusche J, Boismenu R, Davis M (2004) Inhibition of fear potentiated startle in rats following peripheral administration of secretin. Psychopharmacology 172 (1):94–99. https://doi.org/10.1007/s00213-003-1633-5
105. Myers KM, Davis M (2004) AX+, BX- discrimination learning in the fear-potentiated startle paradigm: possible relevance to inhibitory fear learning in extinction. Learn Mem 11(4):464–475. https://doi.org/10.1101/lm.74704
106. Kazama AM, Heuer E, Davis M, Bachevalier J (2012) Effects of neonatal amygdala lesions on fear learning, conditioned inhibition, and extinction in adult macaques. Behav Neurosci 126(3):392–403. https://doi.org/10.1037/a0028241
107. Winslow JT, Parr LA, Davis M (2002) Acoustic startle, prepulse inhibition, and fear-potentiated startle measured in rhesus monkeys. Biol Psychiatry 51(11):859–866
108. Du Y, Wu X, Li L (2010) Emotional learning enhances stimulus-specific top-down modulation of sensorimotor gating in socially reared rats but not isolation-reared rats. Behav Brain Res 206(2):192–201. https://doi.org/10.1016/j.bbr.2009.09.012
109. Ishii D, Matsuzawa D, Fujita Y, Sutoh C, Ohtsuka H, Matsuda S, Kanahara N, Hashimoto K, Iyo M, Shimizu E (2010) Enhancement of acoustic prepulse inhibition by contextual fear conditioning in mice is maintained even after contextual fear extinction. Prog Neuro-Psychopharmacol Biol Psychiatry 34(1):183–188. https://doi.org/10.1016/j.pnpbp.2009.10.023
110. Li L, Du Y, Li N, Wu X, Wu Y (2009) Top-down modulation of prepulse inhibition of the startle reflex in humans and rats. Neurosci Biobehav Rev 33(8):1157–1167. https://doi.org/10.1016/j.neubiorev.2009.02.001
111. Li N, Ping J, Wu R, Wang C, Wu X, Li L (2008) Auditory fear conditioning modulates prepulse inhibition in socially reared rats and isolation-reared rats. Behav Neurosci 122 (1):107–118. https://doi.org/10.1037/0735-7044.122.1.107
112. Walf AA, Frye CA (2007) The use of the elevated plus maze as an assay of anxiety-related behavior in rodents. Nat Protoc 2 (2):322–328. https://doi.org/10.1038/nprot.2007.44
113. Slattery DA, Cryan JF (2012) Using the rat forced swim test to assess antidepressant-like activity in rodents. Nat Protoc 7 (6):1009–1014. https://doi.org/10.1038/nprot.2012.044
114. Steru L, Chermat R, Thierry B, Simon P (1985) The tail suspension test: a new method for screening antidepressants in mice. Psychopharmacology 85(3):367–370
115. Bakker J, Honda S, Harada N, Balthazart J (2002) Sexual partner preference requires a functional aromatase (cyp19) gene in male mice. Horm Behav 42(2):158–171
116. McGraw LA, Young LJ (2010) The prairie vole: an emerging model organism for understanding the social brain. Trends Neurosci 33 (2):103–109. https://doi.org/10.1016/j.tins.2009.11.006

117. Moretti P, Bouwknecht JA, Teague R, Paylor R, Zoghbi HY (2005) Abnormalities of social interactions and home-cage behavior in a mouse model of Rett syndrome. Hum Mol Genet 14(2):205–220. https://doi.org/10.1093/hmg/ddi016
118. Malkova NV, Yu CZ, Hsiao EY, Moore MJ, Patterson PH (2012) Maternal immune activation yields offspring displaying mouse versions of the three core symptoms of autism. Brain Behav Immun 26(4):607–616. https://doi.org/10.1016/j.bbi.2012.01.011
119. Rendall D, Cheney DL, Seyfarth RM (2000) Proximate factors mediating "contact" calls in adult female baboons (Papio cynocephalus ursinus) and their infants. J Comp Psychol 114(1):36–46
120. Satoh Y, Endo S, Nakata T, Kobayashi Y, Yamada K, Ikeda T, Takeuchi A, Hiramoto T, Watanabe Y, Kazama T (2011) ERK2 contributes to the control of social behaviors in mice. J Neurosci 31(33):11953–11967. https://doi.org/10.1523/JNEUROSCI.2349-11.2011
121. Hsiao EY, Patterson PH (2011) Activation of the maternal immune system induces endocrine changes in the placenta via IL-6. Brain Behav Immun 25(4):604–615. https://doi.org/10.1016/j.bbi.2010.12.017
122. Chiarotti F (2005) Statistical analysis of behavioral data. Curr Protoc Toxicol. chapter 13:Unit13 18. https://doi.org/10.1002/0471140856.tx1308s25
123. Festing MF, Altman DG (2002) Guidelines for the design and statistical analysis of experiments using laboratory animals. ILAR J 43(4):244–258

Index

A

Adeno-associated virus (AAV)....... 19–41, 135, 145, 148
Adenosine triphosphate (ATP).................... 55, 151, 240, 242, 244, 246, 255, 262, 270
Aging 22, 30, 113, 136, 147, 189, 198, 202, 205, 209, 216, 219, 289, 318, 319
Amyloid-beta (Aβ) 69, 72, 87–92
Animal behavior 21, 202, 207, 317
Anticonvulsant................... 186, 189, 201, 203, 204, 209
Anxiety.............................213–215, 221, 222, 224, 225, 229, 230, 318, 323, 324, 331, 334
Astrocytes 2, 6, 7, 99–101, 104, 289, 293, 297–299, 301, 303, 306–308, 316, 322
Astrocyte syncytium 285–309
ATP, *see* Adenosine triphosphate (ATP)
At-risk AD model learning and memory.................74, 84

B

Bacteria 240, 242, 246, 253–255
BBM, *see* Blood-brain barrier (BBB)
Beam walk.................................. 320–322, 324, 329, 341
Blood–brain barrier (BBB).................99, 100, 102, 104, 106–112, 114, 116–122, 124, 237
Brain..............................3, 19, 45, 70, 99, 131, 186, 230, 235, 280, 285, 315
  injury.....................................91, 197–201, 204, 209, 316–318, 321, 322, 324, 342
  slices 7, 11, 30, 34, 36, 39, 136, 139, 235–237, 240, 245, 248–253, 272, 280, 287, 290–301
Braincubator............. 237, 238, 245, 249, 250, 252–254

C

Calcium..................... 2, 46, 84, 131, 164, 240, 261, 286
Calcium response 58, 62, 132, 143, 145, 146, 151
CCI, *see* Controlled cortical impact (CCI)
Central nervous system (CNS)............... 19, 88, 99–124, 214, 316, 332
Chemical screens 102, 106–110, 115–116
Chemogenetics.. 19–41, 286
CNS, *see* Central nervous system (CNS)
Computational modeling 287, 296, 297
Confocal microscope 263–265, 270, 273
Controlled cortical impact (CCI) 317–319, 321
Convulsions 196, 198, 208
Cryoanesthesia 22, 25, 26, 28, 40

D

DAI, *see* Diffuse axonal injury (DAI)
Dendritic spines................. 261, 263, 272, 273, 275, 280
Depression 40, 81, 87, 134, 213–215, 221, 222, 226, 227, 230, 318, 338, 339
Diffuse axonal injury (DAI) 320–322

E

EEG, *see* Electroencephalography (EEG)
Electroencephalography (EEG) 186, 187, 189–194, 196–198, 200–203, 205–209
Electrophysiology............................... 3, 7–13, 21, 29–34, 156, 161, 165, 168, 170, 172, 174, 187, 192, 254, 291, 293
Elevated plus maze (EPM) 221, 222, 226, 227, 229, 320–324, 338
Epilepsy 205
EPM, *see* Elevated plus maze (EPM)

F

Fluid percussion (FP)..................................169, 317–321
Fluorescence 1, 30, 46, 115, 132, 158, 263, 292
Fluorescent proteins.......................... 1, 3, 7, 26, 47, 106, 158, 164, 165, 167–169, 172, 173, 175, 277
FP, *see* Fluid percussion (FP)

G

Gap junctions 286, 287, 289, 294, 297, 302, 303
GCaMP...47, 62, 263
GCaMP6s 49, 54, 61, 63, 135, 147–151
Genetics 2, 99–104, 106–110, 113–115, 121, 133, 145, 171, 175, 286, 289, 317
Genetic screens 100, 102, 107, 109, 113–115, 117
Goldman-Hodgkin-Katz (GHK) equation 287, 289, 296

I

Incubation 5, 15, 30, 82, 121, 147, 163, 164, 169, 238, 240, 242, 245, 246, 248–251, 253–255, 281, 292

Nicholas J. D. Wright (ed.), *Basic Neurobiology Techniques*, Neuromethods, vol. 152, https://doi.org/10.1007/978-1-4939-9944-6,

In vivo imaging 21, 45, 47, 60, 65
Ion channels 19, 99, 155–179
Ischemia 91, 185, 186, 188, 195–199, 201, 204, 209, 236, 240, 246, 255

K

K^+ channels 170

L

Live imaging 110, 111, 173

M

Membrane biophysics 156
Membrane potential................... 2, 3, 10–16, 33, 36, 37, 170, 178, 236, 286, 287, 289, 292, 296, 303, 306
Microscopy 1, 3, 30, 37, 108, 132, 139, 145, 157–161, 280, 281
Morris water maze (MWM) 71, 320–324, 328, 335, 337, 339, 341
Mouse auditory cortex 51, 53
Mouse models 186, 194, 197, 203, 205, 206
MWM, *see* Morris water maze (MWM)

N

Nanoparticle 2, 3
Neonatal mouse 19–41
Neurons 2, 19, 45, 69, 99, 131, 170, 189, 235, 261, 286, 316
Novel object recognition 336, 341

O

OF, *see* Open field (OF)
Open field (OF) 221, 225, 226, 229, 331, 332, 336, 339
Optogenetics 2, 19–41, 151, 286

P

Patch-clamp 155–179, 287, 291
Patch-clamp fluorometry (PCF) 155–179
Patch electrode 131–151, 291, 294, 301
PCF, *see* Patch-clamp fluorometry (PCF)
Photometry 131–151
Photomultiplier tube (PMT) 133, 137, 140, 143–145, 148, 175
Posttraumatic stress disorder (PTSD) 213–215, 217, 223, 224, 230
PTSD, *see* Posttraumatic stress disorder (PTSD)

Q

QD, *see* Quantum dot (QD)
Quantum dot (QD) 3, 281

R

Rat subclinical AD model 90
Rodents....................................20, 45, 72, 99, 102, 104, 189, 195, 206, 214, 217, 221, 223, 225, 229, 230, 248, 292, 318, 322, 324, 328–331, 333, 335, 336, 338, 339
Rotational acceleration 318, 322

S

Seizures 91, 198, 318
Signaling molecules...................77, 84, 85, 87, 170, 287
Spectrometer 132, 133, 137, 140–144, 147, 148
Spectroscopy..1, 141, 158
Statistical analysis 82, 338–342
Stress 70, 109, 213, 240, 322
Stroke..185, 186, 189, 204, 307
Surgery..21, 25, 26, 40, 48, 51, 162, 187, 320
Synaptic plasticity 70, 76–77, 80, 84, 85, 87, 88, 91, 131
Syncytial isopotentiality 286, 287, 294–300, 307

T

Tonotopic organization 54, 55, 60, 64
Trauma.. 216, 219, 220, 240, 248, 315–342

U

Uncaging 262–264, 267–269, 272, 280

V

Voltage 3, 5, 7, 10, 13, 16, 33, 34, 37, 45, 47, 144, 156–158, 165, 166, 170–175, 178, 203, 270, 287, 291, 293, 294, 297, 301–303, 305, 307

Z

Zebrafish 99–124

MIX
Papier aus verantwortungsvollen Quellen
Paper from responsible sources
FSC® C105338

If you have any concerns about our products,
you can contact us on
ProductSafety@springernature.com

In case Publisher is established outside the EU,
the EU authorized representative is:
Springer Nature Customer Service Center GmbH
Europaplatz 3, 69115 Heidelberg, Germany

Printed by Libri Plureos GmbH
in Hamburg, Germany